PSYCHOPHYSIOLOGY AND EXPERIMENTAL PSYCHOPATHOLOGY

A TRIBUTE TO SAMUEL SUTTON

ANNALS OF THE NEW YORK ACADEMY OF SCIENCES
Volume 658

PSYCHOPHYSIOLOGY AND EXPERIMENTAL PSYCHOPATHOLOGY

A TRIBUTE TO SAMUEL SUTTON

Edited by David Friedman and Gerard Bruder

The New York Academy of Sciences
New York, New York
1992

Cover: *The P300 component of the event-related potential (ERP) (see page 206).*

Library of Congress Cataloging-in-Publication Data

Psychophysiology and experimental psychopathology a tribute to Samuel Sutton / edited by David Friedman and Gerard Bruder.
p. cm.—(Annals of the New York Academy of Sciences, ISSN 00778923 : v. 658)
Includes bibliographical references and index.
ISBN 0-89766-647-X (cloth : alk. paper).—ISBN 0-89766-648-8 (pbk. : alk. paper)
1. Mental illness—Physiological aspects—Congresses. 2. Evoked potentials (Electrophysiology)—Congresses. 3. Psychophysiology—Congresses.
4. Schizophrenia—Physiological aspects—Congresses.
5. Brain—Psychophysiology—Congresses. I. Sutton, Samuel, 1921–1986.
II. Friedman, David, 1944 Feb. 13- III. Bruder, Gerard. IV. Series.
[DNLM: 1. Mental Disorders—physiopathology—congresses. 2. Psychology, Experimental—congresses. 3. Psychopathology—congresses.
4. Psychophysiology—congresses. W1 AN626YL v. 658 / WN 100 P989275]
Q11.N5 vol. 658
[RC455.4.B5]
616.89—dc20
DNLM/DLC
for Library of Congress 92-18837
CIP

Bi-Comp/PCP
Printed in the United States of America
ISBN 0-89766-647-X (cloth)
ISBN 0-89766-648-8 (paper)
ISSN 0077-8923

ANNALS OF THE NEW YORK ACADEMY OF SCIENCES

Volume 658
July 1, 1992

PSYCHOPHYSIOLOGY AND EXPERIMENTAL PSYCHOPATHOLOGY

A TRIBUTE TO SAMUEL SUTTON[a]

Editors

DAVID FRIEDMAN AND GERARD BRUDER

Conference Chair

JOSEPH ZUBIN

Conference Organizers

MITCHELL KIETZMAN, DAVID FRIEDMAN, AND JOSEPH ZUBIN

Advisory Board

F. FRANK LEFEVER AND HELEN RICHTER

CONTENTS

[a] The papers in this volume were presented at a conference entitled the Samuel Sutton Memorial Conference, which was held at the New York Academy of Sciences in New York City on March 18, 1987.

Financial assistance was received from:

- THE SAMUEL SUTTON RESEARCH AWARD FUND

Samuel Sutton 1921–1986

A Tribute to Samuel Sutton

Someone has suggested that a criterion for determining whether one has made a contribution can be whether one's concepts are referred to without mentioning one's name. This certainly is true of Samuel Sutton. His strength was in methodology more than in theory. That is why he did not leave a "school" behind him. However, the authors of this volume are in a sense the scholars of his school, following the paths he laid out.

Although it has been six years since he left us, his influence still permeates us (as Bonnie Spring points out) as we respond to a missing or absent stimulus, a paradigm which he discovered.

As I advanced in years with Sam still by my side, I often anticipated the day when I would organize a celestial host to greet him at the Pearly Gates. Unfortunately, he preceded me. But what would have happened if the reverse had been true? I can see Sam modestly waiting in the antechamber where according to Plato the souls are triaged. Finally, the admitting angel reaches Sam's name and calls him up. Sam moves forward haltingly, embarrassed by the sudden attention focused on him. The angel lifts up the record and leafs through it and says, "Well, you seem to have contributed to a variety of fields, inner ear function in physiology, reaction time studies in schizophrenia, prognosis, event-related potentials, methodological problems—which of these entitle you to admission to the celestial academy?" "I am not asking for admission to the academy, I just want to be permitted to continue my studies in some corner," answers Sam. Just then, Berger, Penfield, Jaspers, and Heinrich Klüver stand up, as one, and announce, "He has come closer than any of us to showing the connection between the mind and the brain through the discovery of P300." At this, the admitting angel closes the record and escorts Sam to his proper seat in the academy.

While the papers in this memorial volume attest to Samuel Sutton's ingenious contribution to the field, perhaps, the one thing he will be long be remembered for is P300, a phenomenon which, though still baffling as to its origin and significance, has become established as one of the cornerstones of brain physiology.

Joseph Zubin

A Personal Note

I am deeply honored to have been invited to write about Samuel Sutton at the beginning of this book, dedicated to him as a thoughtful and meticulous scientist, and as a friend and colleague of its contributors.

Sam was my close friend and my closest colleague for thirty years. Even this says something about his breadth, since my own work involving social networks is remote from his. He learned enough and thought enough about this area to be my most insightful critic.

Sam was an intellectual in the old-fashioned sense. While he was committed to his own very specialized field, toward which he contributed notably, he saw science—or perhaps I should say knowledge and understanding—as a truly grand guest. He was dedicated to truth, to discovering it and speaking it. While somewhat embarrassed to say so—scientists should be too sophisticated to speak of "truth"—truth, broadly, was what Sam sought.

With his discovery twenty-five years ago of a basic mode of operation of P300, he inspired a new arena of work that has continued to produce significant insights into brain/mind function. And he had an abiding interest in the relationship of brain and mind, seeking always to comprehend those concepts not as a duality but in concert.

I need not mention to those who knew him his wit, his humor, his kindness, his respect for and enjoyment of both women and men. Working with him almost daily over many years was not only rewarding but in itself a joy.

While I and so many others will continue to miss him, I cannot help feeling that he died fulfilled. He worked intensely but with pleasure, producing work he himself respected and loved to do.

Muriel Hammer

Preface

This volume is the result of a conference held in March of 1987 (a year after Samuel Sutton's death) at the New York Academy of Sciences, which was organized to honor Sam's contributions to psychophysiology and psychopathology. Most of us remember Sam for his discovery of the P300 component of the event-related brain potential (ERP), but are less familiar with the extent to which he contributed to and influenced the field of experimental psychopathology. The speakers at the conference, a majority of whom had worked with Sam in one capacity or another, were chosen to illustrate the breadth of Sam's influence in basic behavioral and psychophysiological research, as well as experimental psychopathology. As can be seen in the Table of Contents, the wide range of topics nicely define Sam's eclectic approach.

The impetus for Sam's original use of the ERP was to aid in understanding what went on in the schizophrenic's brain that led to an increase in reaction time when stimuli (lights and tones) were shifted from one sensory modality to the other. This method of integrating behavioral and psychophysiological techniques in the study of psychopathology has become a standard approach in the field. This is documented not only in the paper by Cohen and Rist, who measured ERPs of schizophrenic patients during the modality shift paradigm, but also in the paper by Ford and colleagues, who review the status of P3 research in schizophrenia, and in that by Bruder dealing with P3 in affective and anxiety disorders. In the studies exemplified by this methodology, ERPs and behavioral measures are obtained while psychiatric patients are actively engaged in different cognitive tasks. An alternative approach is illustrated in the paper by John and Prichep, in which ERPs are recorded under passive, "no task" conditions.

Sam's enthusiasm for psychophysiological recording was not limited to the ERP, but also extended to pupillary motility. The integration of these central and autonomic measures and its application to psychopathology is detailed in the paper by Steinhauer and Hakerem. One of the early contributors to Sam's work with P300, Tueting, along with her colleagues, considers the use of a psychopharmacologic approach as a means for understanding the contributions of both biological and cognitive mechanisms to the ERP.

Following the discovery of P300, Sam and his colleagues proceeded to perform a number of studies aimed at better understanding the psychological variables that modulate P300. The findings of these early studies remain today as a firm foundation for the psychological basis of the P300 component. As the paper by Ritter and Ruchkin makes quite clear, these early studies also led to the discovery of several additional ERP components that are affected by cognitive variables. The discovery of P300 was followed, some years later, by studies applying this technique not only to psychopathology but also to fields such as cognitive development and aging, as described in the paper by Friedman. One of Sam's earlier collaborators in the area of auditory psychophysics, Babkoff, and his colleagues present the results of studies of sleep deprivation on cognitive function.

Some of the early interest in the possible use of cognitive measures as markers of vulnerability to schizophrenia grew out of Zubin and Sutton's work on reaction time in schizophrenia. Subsequently, researchers have used a variety of behavioral measures as potential markers, as elegantly described in the paper by Spring. One of the pioneers in the psychophysiological study of psychopathology, Venables, also did one of the early investigations using a putative psychophysiological marker. He here reviews a variety of evidence supporting the hypothesis of hippocampal involvement in the etiology of schizophrenia. Pribram, another pioneer in examining the relationship between brain and behavior, musters his own evidence, as well as that of others, in support of a model of attentional processing.

Finally, Salzinger reminds us that we shall be at a loss in understanding brain mechanisms without first adequately dealing with basic behavioral mechanisms. This point was lucidly examined by Sam in his landmark paper published in 1969, entitled "The Specification of Psychological Variables in an Average Evoked Potential Experiment." In short, Sam's ideas and body of scientific work have had a profound influence, not only on those of us who worked with him, but also on the fields of psychophysiology and psychopathology. This volume should give the reader a sense of Samuel Sutton's legacy.

David Friedman
Gerard Bruder

A Review of Event-Related Potential Components Discovered in the Context of Studying P3[a]

WALTER RITTER[b,c,e] AND DANIEL S. RUCHKIN[d]

[b]*Department of Neuroscience*
Albert Einstein College of Medicine
Bronx, New York 10461

and

[c]*Department of Psychology*
Lehman College
City University of New York
Bronx, New York 10468

[d]*Department of Physiology*
School of Medicine
University of Maryland
Baltimore, Maryland 21201

In a seminal paper published in 1965, Sutton, Braren, Zubin and John first reported the P3 component recorded from the scalp of human subjects. The method used was to have subjects guess prior to each trial whether a randomly presented tone or flash of light would occur, and to have the subject discover whether the guess was correct or not when the stimulus for that trial was delivered. In another condition, the same stimulus was presented on all trials and the subject so informed. The result was that P3, a positive going wave with a peak latency of about 300 msec, was elicited when the subject was uncertain as to which stimulus would occur. Since the stimuli provided the subject with information concerning the correctness of the guess, Sutton *et al.* suggested that P3 was associated with information delivery.

In a second study, Sutton, Tueting, Zubin and John (1967) presented one or two clicks, separated by 580 msec, on a random trial-by-trial basis. In addition, the clicks of a given trial were randomly loud or soft. In one condition, the subject guessed prior to each trial whether one or two clicks would be delivered. In another condition, the subject guessed whether the clicks would be loud or soft. Thus, in the first condition, the point in time when the subject learned the correctness of the guess was when the second click was, or was not, presented. It was found that the P3 was elicited by the second click and its absence. When guessing loudness, however, the subject learned the correctness of the guess when the first click was presented (the second click, if present, always had the same intensity as the first). In this condition, the presence or absence of the second click was not associated with P3. Hence, P3 was elicited by a stimulus (or the absence of a stimulus) that delivered relevant information.

[a] This work was supported in part by United States Public Health Service Grants NS30029 to the first author and NS11199 to the second author.

[e] Address for correspondence: 750 Kappock St., Bronx, NY 10463.

In their 1967 study, Sutton *et al.* had a condition where one or two clicks were presented on each trial, as just described, except that the subject was told prior to each trial what would occur. Again, predictable stimuli did not elicit P3. Sutton and colleagues therefore concluded that P3 was associated with resolution of uncertainty.

The elicitation of P3 by the presence or absence of a stimulus led Sutton *et al.* (1967) to the view that P3 is endogenous in nature. Since P3 occurred whether the stimulus was auditory or visual, and its latency was comparable when adjustment was made for arrival time at the cortex by lining up the event-related potential (ERP) waveforms at N1 (Sutton et al., 1965), it was concluded that P3 was not specific to a particular stimulus modality, in contrast to earlier latency exogenous components. Sutton *et al.* (1965) also found that P3 was probability sensitive, since if either the tone or the flash was made less probable than the other, that stimulus elicited the larger P3.

These findings of Sutton and colleagues constitute the most fundamental properties of P3, and set off an extensive research effort. In addition, investigations of P3 led to the discovery of other cognitive ERP components that accompany P3, some preceding and some following it in latency.

In a relatively short time, P3 was associated with such activities as target selection, decision making, sensory discrimination and match-mismatch processes (for a review, see Tueting, 1978). However, Ritter, Simson and Vaughan (1972) reported that, whereas the latency of P3 and reaction time (RT) are correlated, an earlier component (N2) was probably associated with these processes and hence may reflect brain events causal to RT. Indeed, P3 is often too late to be associated with discriminative and decision-related processes on the trials in which it is elicited (Ritter, 1978). This invited efforts toward examining ERP components that were elicited in the same experimental situations as P3 but were earlier in latency. The main components were N2 and the mismatch negativity (MMN), each described in more detail below. The MMN can onset as early as 50 msec after a stimulus and is thought to reflect the outcome of a memory operation that determines whether a stimulus is different than immediately preceding stimuli. Thus, the effort to delineate and characterize these early components focused on processes that could be completed in very short periods of time.

In 1975, N. Squires, K. Squires and Hillyard reported that P3 is also accompanied by slow wave activity that overlaps with and continues past P3. The initial studies of slow wave activity continued to use experimental tasks where the subject's performance on a given trial was determined by processes which often occurred earlier in time than P3. Later, investigators turned to tasks where the performance on a given trial was based on processes that were very long in duration and persisted well beyond P3. For example, a subject might be shown a set of numbers and required to perform some operation (such as arithmetic calculations) on the numbers. The numbers would deliver critical information in performing the task, and hence would elicit P3. But the slow waves that followed P3 appeared to be associated with the subsequent, long duration processes that lead to the subject's response.

The present paper reviews these two offshoots of P3 research. The first author (WR), who played a role in the work that examined components shorter in latency than P3, reviews this line of investigation. The second author (DSR), who collaborated with Sutton in pioneering work on slow waves, reviews that line of investigation. In each case, the review begins by sketching how the two lines of research began, and then concentrates on more recent investigations.

Early Negative Endogenous Components

Probably the single most common stimulus arrangement used for studying P3 has been the oddball paradigm, in which one stimulus occurs frequently (termed the standard) and the other stimulus occurs infrequently (termed the deviant). One reason for the popularity of the oddball paradigm is that it provides a low probability event, which is known to increase the size of P3. The amplitude of P3 is inversely related to the probability of the eliciting stimulus, as was initially shown by Sutton *et al.* (1965) using stimuli of two different probabilities, and later demonstrated in parametric fashion by Tueting and Sutton (1970) and Duncan-Johnson and Donchin (1977). It is by no means necessary to use the oddball paradigm for obtaining P3. Sutton *et al.* (1965; 1967) and many subsequent studies have obtained P3 using equiprobable stimuli. However, since one of the criteria used for identifying P3 is that its amplitude is probability sensitive, the oddball paradigm is useful because it can satisfy that criterion by comparing the ERPs elicited by the standard and deviant stimuli. Moreover, use of the oddball paradigm led to the discovery of other cognitive ERP components that are also inversely related in amplitude to stimulus probability (which, in turn, became one of the defining characteristics for those components).

As with P3, the other components studied with the use of the oddball paradigm were found through serendipity. In 1968, Klinke, Fruhstorfer and Finkenzeller delivered vibratory stimuli at a constant rate of one every 0.86 sec and, for the oddball event, infrequently omitted the stimulus. As had been shown by Sutton *et al.* (1967), the unpredictable omission of the stimulus elicited P3. However, the rapid pace and steady rhythm of the standards, which induced more accurate anticipation by the subjects of the point in time when each standard was to be presented, yielded more effective time-locking of endogenous components associated with the oddball (in this case, omitted) event. The P3 was found to be preceded by a negative component, subsequently termed N2. In a seminal paper, Squires *et al.* (1975) showed that N2 is probability sensitive and that the oddball stimulus also elicits at least two other components, P3a and Slow Wave, which are also inversely related in amplitude to the probability of the eliciting stimulus. The latter components were both maximally positive at Pz, but could be differentiated from P3 in latency and topography. The P3a was earlier in latency and more frontal than P3 (subsequently termed P3a and P3b, respectively), whereas the Slow Wave activity peaked later than P3b and, while largest in the parietal region as was P3b, shifted to a negative polarity frontally.

In 1978, again serendipitously, Näätänen, Gaillard and Mäntysallo found that the oddball stimulus elicits a negative component that is even earlier in latency than N2, termed mismatch negativity (MMN). Like P3, probably none of these components requires an oddball paradigm for its elicitation and identification. Sams, Alho and Näätänen (1983), for example, have shown that when local sequences of equiprobable stimuli are examined (*i.e.*, the sequences of the stimuli that immediately precede the eliciting stimulus), both the MMN and N2 can be identified. Nevertheless, the oddball paradigm provides a convenient method for enhancing both.

An important modification of the oddball paradigm is the use of two intermixed oddball sequences in the study of selective attention (Hillyard, Hink, Schwent & Picton, 1973). In this arrangement, the subject's task is to attend to one of the two concurrently presented oddball sequences and respond differentially, either by counting or making a motor response, to the deviant stimulus in one of the sequences, and to ignore the stimuli of the other sequence. The two sequences usually differ by a simple physical feature (for example, ear of delivery).

A selective review of some of the components found through the use of a single or double oddball paradigm follows. When a double oddball paradigm is used, as just mentioned, subjects are usually instructed to attend to one sequence and ignore the other. A superficially analogous arrangement for a single oddball paradigm is to have subjects attend to the stimuli and differentially respond to the standard and deviant stimuli in one condition, and to ignore the stimuli in another condition (usually while reading a book or performing an unrelated task). One focus of this review is the circumstances under which the components are and are not elicited. In the main, this will entail comparison of which components are obtained when subjects do and do not attend to the stimuli. Another focus concerns the differences that are found between single and double oddball arrangements when attended and ignored sequences are compared.

The Mismatch Negativity

Using auditory stimuli in an oddball paradigm, the MMN is elicited by tones that deviate from the standards in various ways, including changes in pitch, intensity, location and duration (see Näätänen, 1990, for a detailed review). The latency of the MMN is inversely related, and its amplitude positively related, to the magnitude of the difference between the standard and the deviant stimuli (Näätänen & Gaillard, 1983). Its amplitude grows as the number of standards that precede a deviant stimulus increases (Sams *et al.*, 1983), and diminishes as the interstimulus interval (ISI) is lengthened (Mäntysalo & Näätänen, 1987). The MMN is largest over the frontocentral region and inverts in polarity along a coronal line from the frontal midline to the mastoid, when the nose is used as the reference (Alho, Paavilainen, Reinikainen, Sams & Näätänen, 1986). On the basis of magnetic recordings (Hari, Hämäläinen, Ilmoniemi, Kaukoranta, Reinikainen, Salminen, Alho, Näätänen & Sams, 1984), dipole analysis of electrical recordings in humans (Scherg, Vajsar & Picton, 1989), and intracranial recordings in cats (Csepe, Karmos & Molnar, 1987) and primates (Javitt, Schroeder, Arezzo & Vaughan, in press), the generator of the MMN has been localized as being within or in the immediate vicinity of primary auditory cortex.

The MMN has been reported to be different from the other components discussed in the present paper in two important ways: it can be elicited by predictable stimuli and by either attended or unattended stimuli. Each of these situations is examined in turn.

Scherg *et al.* (1989) presented tones in two ways while subjects were reading and ignored the stimuli. In one arrangement, tones of two different pitches were presented randomly, with one of the tones occurring on 80% of the trials and the other tone occurring on 20% of the trials. In the other arrangement, the probability of the two tones was the same, but they were presented in a regular manner, the deviant tone always following after four standard tones. No difference was found in the MMN obtained in the two ways of presenting the stimuli. This contrasts with N2 and P3 which are not elicited (or are very small) for completely predictable stimuli. Sutton *et al.* (1965; 1967), it will be recalled, had found that when stimuli were presented in a predictable manner, P3 was small or absent. Indeed, they concluded that P3 was associated with reduction in (stimulus) uncertainty. In general, endogenous components are smaller or absent when stimuli are predictable than when unpredictable. The MMN, however, appears not to be affected by the predictability of the deviant stimulus.

The study by Scherg *et al.* (1989) also illustrates the other way in which the MMN is different from other endogenous components: subjects need not attend to the stimuli for the MMN to be elicited. In general, the MMN is elicited by infrequent changes of auditory stimuli both when subjects are attending as well as not attending the stimuli (Näätänen, 1990). The latter is true whether a single oddball paradigm is used (and subjects attend to the stimuli in one condition and ignore the stimuli in another condition) or a double oddball paradigm is used (and

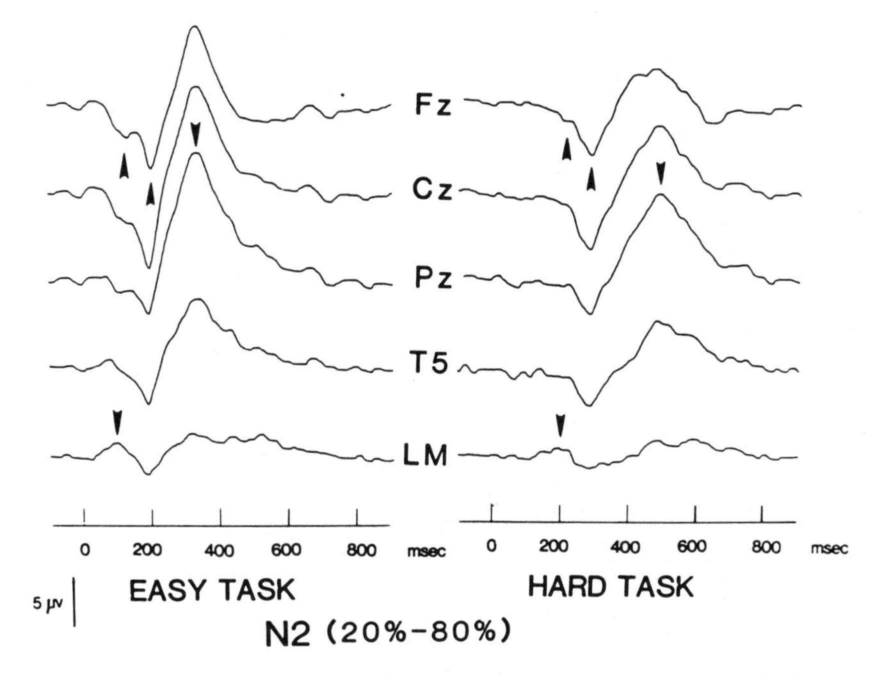

FIGURE 1. MMN and N2 difference wave forms in a choice reaction time task: ERPs to target minus ERPs to standard tones. For both tasks, *arrowheads* indicate the MMN and N2 at Fz, P3 at Pz, and the MMN at LM. Positive is *up*. (From Novak *et al.*, 1990. Reprinted by permission from *Electroencephalography and Clinical Neurophysiology*.)

subjects attend to one of the sequences while concurrently ignoring the other sequence). There is a potential problem in studying the MMN in the attend conditions, however, because the deviant stimuli will elicit N2 that overlaps the MMN. This issue is tractable, however, because N2 does not invert in polarity at the mastoid as does the MMN. FIGURE 1 illustrates a situation where both the MMN and N2 were elicited by the deviant stimuli in a single oddball, discrimination task (Novak, Ritter, Vaughan & Wiznitzer, 1990). In this experiment, there was an easy and a hard pitch discrimination between the standard and deviant stimuli. The nose was employed as the reference in the recordings. The figure shows the grand mean difference waveforms obtained by subtracting the ERPs associated with the standard tones from the ERPs associated with the deviant tones. The left and right panels present the results for the easy and hard tasks, respectively. At Fz, two upward pointing arrows designate two negative-going deflections, separated by a small positive peak or shoulder, and a downward pointing arrow at Pz designates P3. The earlier latency negative deflection is the MMN and the later is N2. As can be seen, the MMN and N2 overlap in time. However, at the left mastoid (LM), the MMN (indicated by the downward pointing arrow) inverts in polarity, whereas the N2 does not.

Note in FIGURE 1 that at the left mastoid the MMN is about 100 msec longer in latency for the hard than the easy discrimination task, and that similar increments in latency are seen at the midline recordings for N2 and P3 (RT also increased about the same amount). Since both the onset and peak of the MMN at the left mastoid increased by about the same amount from the easy to the hard task (and therefore the duration of the MMN was about the same in the two tasks), Novak *et al.* (1990) suggested that the neural activity underlying the MMN reflects the outcome of a mismatch operation rather than the mismatch process itself.

The circumstance that the MMN can be elicited with or without the subject attending to the stimuli led Näätänen (1982) to propose that it is associated with an automatic process. Näätänen (1990) reported results which tentatively suggested that in certain conditions the MMN might be reduced in amplitude, but not eliminated, in the absence of attention. Subsequently, Woldorff, Hackley and Hillyard (1991) reported that the MMN could be eliminated if the experimental arrangement promoted a strongly focused attentional state with regard to the relevant stimuli. Tone pips of 13 msec duration were presented at very rapid rates (ISIs ranging from 65 to 205 msec, with a rectangular distribution). The tones were delivered dichotically on an equiprobable basis to the two ears, the tones delivered to one ear being 2000 Hz and the tones to the other ear being 6000 Hz. Infrequent deviant tones were presented to both ears which were fainter (the standards were 90 dB SL and the deviants averaged 20 dB less across subjects) and adjusted so that the subjects detected the intensity change with 75% accuracy. The task was to attend to the stimuli of one ear and to press a button when a deviant tone was presented. This arrangement induced a strong attentional focus on the relevant stimuli for several reasons: the fast rate of stimulation made it costly for the subject to attend to the irrelevant ear; the intensity discrimination required to detect targets was difficult; the dichotic presentation and the large difference in pitch of the tones delivered to the two ears made it easy to distinguish the tones of the relevant from the irrelevant ear. Woldorff *et al.* (in press) concluded that the MMN is not independent of attention. Using ISIs of 70–200 msec in a selective attention task, Näätänen, Paavilainen, Tiitinen, Jiang and Alho (submitted) also found that strongly focused attention on the relevant stimuli was associated with elimination of the MMN for a tone embedded within the irrelevant stimuli that deviated in intensity. It may be important in understanding these results that the data used by Woldorff *et al.* (in press) are from the same experiment by Woldorff, Hansen and Hillyard (1987), which found a selective attention effect that began at a latency of about 20 msec. Such an early effect suggests gating of the irrelevant tones. Accordingly, a more detailed conclusion that could be drawn is that when gating occurs the input to the comparator mechanism upon which the MMN depends is insufficient, and in this sense the MMN is not independent of attention. However, in those circumstances where gating does not occur (for example, when a strong attentional focus is not required), the MMN is independent of attention. The reports that the MMN occurs in sleep in cats (Csepe *et al.*, 1987), newborn infants (Alho, Sainio, Sajaniemi, Reinikainen & Näätänen, 1990) and adult humans (Campbell, Bell & Basstien, in press) are consistent with this interpretation.

Further support for the notion that the MMN can be independent of attention derives from studies of selective attention. It has been shown that when a double oddball paradigm is used, the difficulty of discriminating the between sequence cue has no effect on the MMN elicited by within sequence deviant stimuli. In an intramodal selective attention task, Novak, Ritter and Vaughan (in press b) found that neither the amplitude nor the latency of the MMN was affected by the difficulty of the discrimination between the relevant and irrelevant tones. Using intermodal selective attention tasks where auditory and visual stimuli were intermixed, Alho, Woods, Algazi and Näätänen (in press) found that when the visual stimuli were attended, neither the amplitude nor the latency of the MMN elicited by deviant tones was affected by the difficulty of the discrimination between standard and deviant visual stimuli, or by performing a simple (visual) RT task in which the standards were removed and only the deviants presented. In both studies, allocating additional attentional resources (in order to discriminate between relevant and irrelevant tones in Novak *et al.*, or in order to discriminate between standard and deviant visual stimuli in Alho *et al.*) had no effect on the MMN elicited by deviant auditory stimuli.

At present, the MMN has not been demonstrated for any sensory modality other than auditory. Nyman, Alho, Laurinen, Paavilainen, Radil, Reinikainen, Sams and Näätänen (1990) attempted to obtain a MMN for visual stimuli, but were unsuccessful. A number of investigators are currently attempting to resolve the question of whether the MMN is unique to the auditory system.

Given the characteristics of the MMN just described, and its early latency (its onset can be in the vicinity of 50 msec—see the left mastoid (LM) recording in the left panel of FIG. 1), the MMN appears to be based on a memory that is accessed very early in the processing of auditory stimuli. Näätänen (1990) has proposed that the memory in question is sensory memory. Support for that view was obtained by Winkler, Reinikainen and Näätänen (submitted), who found that the MMN could be eliminated by employing masking. Both standard and deviant tones (which differed in pitch) were preceded or followed by a masker tone. Increasing the amount of time between the masker and the test tones gradually diminished the masking effect on the MMN. Moreover, the amplitude of the MMN across conditions correlated with the capacity of the subjects to identify deviant stimuli.

Whether or not subsequent research confirms that the memory upon which the MMN depends is sensory memory, it seems clear that it is based on the operation of a memory that is accessed early in stimulus processing. An especially interesting prospect is that examination of the circumstances that modulate the amplitude and latency of the MMN will allow investigation of this memory and the operation of the mismatch detector with which it is associated. For example, by gradually increasing the ISI and determining at what point the MMN is no longer elicited, Mäntysalo and Näätänen (1987) and Näätänen, Paavilainen, Alho, Reinikainen and Sams (1987) have inferred that the memory only lasts a few seconds. In a different vein, Winkler, Paavilainen and Näätänen (submitted) concluded on the basis of the behavior of the MMN that the memory is capable of storing at least two different traces simultaneously.

Finally, in view of the ability to obtain the MMN without requiring cooperation on the part of the subject (for example, while reading or even sleeping) and the localization of the generator of the MMN to primary or immediately adjacent auditory cortex, the prospect for clinical use of the MMN appears to be considerable.

The MMN and N2

The N2, often associated with the MMN, is also inversely related in amplitude to the probability of the eliciting stimulus (Squires *et al.*, 1975) and varies in latency as a function of the magnitude of the difference between the standard and deviant stimuli of an oddball paradigm (Ritter, Simson, Vaughan & Friedman, 1979; Renault & Leserve, 1979). In contrast to the MMN, N2 has been identified for both the auditory and visual modalities, being modality specific in its scalp distribution (Simson, Vaughan and Ritter, 1976; 1977). In 1978, Näätänen *et al.* separated the auditory N2 deflection into two subcomponents, N2a and N2b. Subsequently, the term MMN largely replaced the term N2a (hence, in the present paper, the term N2 will be retained for N2b).

The major difference between N2 and the MNN (aside from the earlier latency of the MMN, when both components are elicited by the same deviant stimulus, and the difference in their topography) is that the MMN exhibits automatic characteristics whereas N2 does not. The distinction between N2 and the MMN was initially drawn by Näätänen *et al.* (1978) in the context of a selective attention experiment. In this situation, the MMN is found for both the attended and the unattended oddball sequence, whereas the N2 is generally only found for the attended sequence (as is also true of P3). Analogously, when a single oddball sequence is used, the MMN is obtained in conditions where subjects both attend and ignore the stimuli, but the N2 is generally only found when subjects attend to the stimuli. An exception is if deviant stimuli in an unattended sequence of stimuli are so salient as to trigger an orienting response, in which case N2 and P3 would be elicited (Näätänen, Simpson and Loveless, 1982). More generally, N2 is only found when subjects attend to the stimuli, whether via active or passive attention. However, N2 is not always obtained when subjects attend to the stimuli (see below).

Although the MMN and N2 are usually found to occur together, the N2 can occur alone and thus is not dependent on the prior occurence of the MMN. The reason why the N2 can occur alone is because of the difference in the psychological processes with which it is associated. The MMN is elicited by deviant stimuli that infrequently differ physically from standard stimuli: so is the N2. Since oddball paradigms are commonly used in ERP research, the two components are often found together. However, as Breton, Ritter, Simson and Vaughan (1988) pointed out, physical deviance and stimulus probability are confounded in the usual oddball paradigm, since the deviant stimuli are also the infrequent stimuli. Thus, the question arises whether the N2 is associated with physical deviance, *per se*, as is the MMN, or with expectation, as is P3 (cf. Friedman, Ritter and Simson, 1978).

The finding of Scherg *et al.* (1989), described above, that the MMN is not affected by whether stimulus sequences are predictable or unpredictable, indicates that the MMN is not associated with expectation. Ritter, Paavilainen, Lavikainen, Reinikainen and Näätänen (submitted) presented tones of five different pitches on a random basis, and, also randomly, on 20% of the trials any of the five tones could repeat. In one condition, the stimuli were delivered while the subjects ignored the tones and read material of their own choosing. In another condition, the subject's task was to count the number of times that the pitch of the tones repeated. The

MMN was not obtained for repetitions in either the ignore or the count condition. N2 and P3, however, were obtained in the count condition. Thus, N2 as well as P3 is associated with unpredictable stimuli, whether they match or deviate from preceding stimuli. And both components can be obtained in the absence of the MMN.

N2 and N400

Kutas and Hillyard (1980) found that words that ended sentences in an analomous manner were associated with a negative component that peaked around 400 msec (N400). The original intent of this experiment was to examine P3 elicited by unexpected words contained in sentences (Hillyard, personal communication). "The finding that words out of context elicited an N400 wave was unexpected in light of the evidence that stimuli which disconfirm a subject's expectancies generally elicit an enhanced late positive component (P300)..." (Kutas & Hillyard, 1980, p. 204). Thus, as with the other negative components reviewed above, N400 was encountered serendipitously while studying P3. Subsequent work has shown that N400 is not associated specifically with anomalous words presented in the context of sentences, but rather with unexpected or semantically unrelated words, as, for example, in an otherwise unrelated series of words in a lexical decision task (*e.g.*, Bentin, McCarthy & Wood, 1985).

There have been several experiments that appear to show that N400 is automatically elicited (Fischler, Bloom, Childers, Roucos & Perry, 1983; Bentin, McCarthy & Wood, 1984; Kutas & Hillyard, 1989). In each of these studies, the semantic relatedness of words was not relevant to the task performed by the subjects, yet semantically unrelated words elicited N400. In Bentin *et al.* (1984), for example, each trial consisted of a memory set of five words, presented one at a time visually, followed by a probe word that either was or was not contained in the memory set. The task was to respond differentially on the basis of whether the probe word was contained in the memory set. A small percentage of adjacent words in the memory sets were semantically related, and the subjects were not informed of this. It was found that adjacent words in the memory sets that were semantically unrelated elicited an enhanced N400 compared to adjacent words that were semantically related. Upon questioning after the experiment, subjects reported that they either did not notice that there were words in the memory sets that were semantically related or, if they did, vastly underestimated the number of semantically related words that were presented (Bentin, personal communication). Thus, one interpretation of these results is that N400 was automatically elicited since subjects were not attending to the semantic relatedness of the words in the memory sets and usually did not notice them.

In a somewhat similar vein, Ritter, Simson and Vaughan (1983) reported that N2 may be associated with automatic processing under some circumstances. They delivered angles or brackets in a simple reaction time (RT) task, where on 20% of the trials an angle or bracket reversed (*e.g.*, from < to >). The infrequent trials were associated with N2 (without a subsequent P3). Since noticing that the angles or brackets had changed was not relevant to the task, it appeared that N2 could be related to automatic processing. (Since N2 had the same peak latency in the simple RT task as when differential responses were required between the stimuli, this argues against the N2 actually being a MMN). The result bears a resemblence to obtaining P3 to task-irrelevant changes in the size of words (Kutas & Hillyard, 1980) and the sex of the person speaking words (McCallum, Farmer & Pocock,

1984), when subjects are processing sentences for a subsequent test of memory for those sentences.

In contrast to these results, Deacon, Breton, Ritter and Vaughan (1991) found that neither N2 nor N400 appear to be related to automatic processing. Words were presented one at a time and infrequently the size or the semantic category of the words changed in the same run. In one condition, subjects responded whenever the size of the words changed. In another condition, subjects responded whenever the semantic category of the words changed. When size was the relevant stimulus dimension, infrequent changes in the size of the words elicited N2 (which peaked at a latency of about 320 msec) followed by P3, whereas infrequent changes in the semantic category of the words did not elicit N400 or P3. When semantic category was the relevant stimulus dimension, infrequent changes in the semantic category of words elicited N400 (which peaked at a latency of about 400 msec) followed by P3, whereas infrequent changes in the size of the words did not elicit N2 or P3. Consequently, this study found results for N2, N400 and P3 that appear to conflict with the results found for these components described in the preceding two paragraphs. (The circumstance that Deacon *et al.* addressed the controversial question as to whether N2 and N400 may be the same component at different latencies is immaterial to the present discussion. What matters is that when the size task was performed there was no enhanced negativity at around 400 msec when irrelevant changes in the semantic category of the words occurred, and when the semantic category task was performed there was no enhanced negativity at around 320 msec when irrelevant changes in the size of the words occurred.

Conflicting N2 Results. With regard to N2, Deacon *et al.* suggested that the presence of N2 in the simple RT task of Ritter *et al.* (1983) may have been because the task was so easy that subjects paid attention to the infrequent size changes. In contrast, the category task of Deacon *et al.*, being more difficult, may have precluded attending to the infrequent size changes.

Conflicting N400 Results. Concerning N400, in the studies by Fischler *et al.* (1983), Bentin *et al.* (1984) and Kutas and Hillyard (1989), it was either necessary or much more efficient for the subjects to process the meaning of the words in order to perform their tasks. In the experiment of Bentin *et al.*, for example, the most efficient way for subjects to perform the task was to process the five words in each memory set for their meaning and to store that information in memory for comparison with the probe word (versus storing in memory the individual letters of the words in the memory set, or some other strategy that did not require processing the meaning of the words). In the study by Deacon *et al.*, no benefit would be gained by processing the words for their meaning when performing the size discrimination task. Moreover, the absence of N400 when the subjects responded to the size of the words is consistent with behavioral studies (Smith, 1979; Henik, Friedrich & Kellogg, 1983) showing that semantic priming effects are not obtained when words are not processed for their meaning. Consequently, it appears that what was essential to obtaining the semantic priming effects on N400 amplitude in Fischler *et al.* (1983), Bentin *et al.* (1984) and Kutas and Hillyard (1989) was whether the words were processed for their meaning, not whether the semantic category of the words was task relevant. This interpretation is in line with the findings that the ERP repetition effect found with words is much smaller (Rugg, Furda and Lorist, 1988) or unobservable (Hamberger & Friedman, 1990) when words are processed for the case of their letters compared to processing the words for their semantic category.

A recent study, however, has found that when sentences are spoken to subjects while performing a phonological task (counting the number of words that begin with the letter P), words that end the sentences in an anomalous manner elicit N400 (Woodward, Ford & Hammett, in preparation). Thus, even though processing the meaning of the words would not provide any benefit in performing the task, the anomalous words elicited N400. The subjects, therefore, must have processed the words for their meaning. Even more striking, in order for N400 to be elicited by the words that ended sentences anomalously, the subjects not only had to process the individual words for their meaning but also the relationship between the meaning of the words in order to derive the meaning of the sentences.

The results of the study by Woodward *et al.* appear to contradict the results of Deacon *et al.* as well as the behavioral studies cited above. In Smith (1979), when subjects conducted a letter search for the prime and primed words, no priming effect was obtained on RT. In Henik *et al.* (1983), RT priming effects were obtained when subjects performed a word-naming task but not when they performed a letter-search task. The study by Woodward *et al.*, on the other hand, also used a letter-search task. There were two critical elements that appear to distinguish the study of Woodward *et al.* from those of Deacon *et al.*, Smith and Henik *et al.*: 1) the former presented words in the auditory modality whereas the latter three all presented words in the visual modality; 2) the former presented words that formed sentences whereas that was not true for any of the latter three. Concerning the sentences used by Woodward *et al.*, a related factor may have been that the anomalous sentences were often intrinsically interesting because of an element of humor (*e.g.*, consider the sentence "He spread the warm bread with socks."). Once a subject noticed one or two of the anomalous sentences, there could be an intent to pay attention to the sentences.

Conflicting P3 Results. It is not clear how to interpret the different results obtained by Deacon *et al.* versus Kutas and Hillyard (1980) and McCallum *et al.* (1984) for P3 with regard to task-irrelevant physical changes of the stimuli. Irrelevant changes in the size of words in Deacon *et al.* did not elicit P3. In contrast, irrelevant changes in the font of words in Kutas and Hillyard elicited P3, and irrelevant changes in the voice quality of the words in McCallum *et al.* elicited P3. Since Deacon *et al.* and Kutas and Hillyard used visual stimuli, while McCallum *et al.* used spoken words, the sensory modality employed cannot provide the answer. The element that may distinguish these studies is that in Kutas and Hillyard and McCallum *et al.* the physical changes of the stimuli only occurred for the last (seventh) words of sentences, which also were the only words that rendered some sentences anomalous. Hence, subjects may have paid more attention to these words than others in the sentences, and may thereby have become attuned to noting both the unexpected semantic (*i.e.*, anomalous) meaning and physical changes of the seventh words. In Deacon *et al.*, the irrelevant size changes occurred randomly throughout a run. Thus, N2, P3 and N400 appear to be elicited when subjects pay attention to relevant stimulus dimensions and otherwise are not automatically elicited.

The issue of what paying attention to a relevant stimulus dimension constitutes requires further elaboration. When two oddball sequences are concurrently presented during selective attention, deviant stimuli in the unattended sequence do not usually elicit N2 or P3, even though they may deviate in the identical way as the deviant stimuli in the attended sequence. On the other hand, Roth, Ford and Kopell (1978) found that if a single oddball sequence contains two, equiprobable (10%) deviant stimuli that differ from the standards along the same dimension (for example, one deviant may be higher in pitch, and the other deviant equivalently lower in pitch, than the standards), and one deviant is designated a target and the other deviant designated a nontarget, both deviants will elicit P3 (cf. Courchesne,

Hillyard & Courchesne, 1977). Using an essentially similar stimulus arrangement, Näätänen, Simpson and Loveless (1982) showed that N2 as well as P3 is elicited by both the target and the nontarget deviants. In the double oddball arrangement, it seems that the absence of N2 and P3 for the irrelevant sequence is related to the minimal amount of attention given these stimuli, and that the presence of these components for the relevant sequence is associated with attending to these stimuli. For the study by Roth *et al.*, it could be that the subjects used a strategy of noting any pitch change (since half of the time a pitch change was a target), or it could be that any rare pitch in an attended sequence of stimuli will elicit P3 (*i.e.*, the subject cannot help noticing the change in pitch). However, if the relevant sequence contains deviant stimuli that differ from the standards along two different physical dimensions (for example, one of the deviant stimuli differs in pitch and the other deviant stimulus differs in intensity from the standards) and one of the deviants is a target and the other is not, both deviants will also elicit N2 and P3 (Näätänen *et al.*, submitted). A similar ambiguity in interpreting these results applies as in the study by Roth *et al.*: it could be that the subjects used a strategy of noting any stimulus change because any given stimulus change had a 50% chance of being a target, or it could be that any rare change in an attended sequence will elicit P3 (*i.e.*, the subjects cannot help noticing it). Since the two ways in which deviance occurred were along different stimulus parameters, the strategy interpretation seems somewhat less likely (when a pitch change is a target, a change in intensity is never a target). In any case, in these studies P3 was elicited by rare changes in an attended sequence whether or not the changes were targets. In contrast, it will be recalled, in Deacon *et al.* (1991) a single oddball sequence elicited either N2 and P3 or N400 and P3 depending on whether subjects were processing the stimuli for their size or their semantic category. In the latter study, subjects had to pay attention to all of the stimuli in both conditions in order to perform the tasks. Thus, attending to the stimuli appears to be a necessary but not a sufficient basis for eliciting the components under discussion. The data reviewed so far suggest that when subjects attend to physical changes of stimuli, any physical change of the attended stimuli will elicit N2 and P3. On the other hand, when attending to physical changes, changes in the semantic category of the stimuli do not elicit N400 and P3, when the stimuli are presented visually and/or do not form sentences. When attending to changes in semantic category, physical changes *per se* do not elicit N2 and P3 (*per se* is used because Kutas & Hillyard (1980) and McCallum *et al.* (1984) obtained P3 for physical changes when subjects process sentences for their meaning). Thus, attending to stimuli is usually a necessary prerequisite for eliciting N2, N400 and P3; however, which stimuli will elicit these components depends on which aspect of the stimuli is attended (physical or semantic) and possibly the sensory modality and the manner with which the stimuli are presented, *i.e.*, whether they form sentences or not (and, perhaps especially, interesting sentences).

An intriguing and unexpected result is contained in the study by Alho *et al.* (in press) of intermodal selective attention described earlier. Small and large deviants were included among both the auditory and visual stimuli. For the auditory modality, the standards were 1000 Hz and rare deviants were 1050 Hz and 1500 Hz. In the visual modality, rare deviants were either slightly shorter or much shorter than the standards in height. Subjects attended to one or the other modality and responded to one of the deviants in the attended modality in a given condition (making for a total of four conditions, one for each deviant as the target). For the large deviants in both modalities, it is not surprising that they elicited P3 whether they were targets or not, so long as subjects were attending the modality in which they were presented (cf. Roth *et al.*, 1978 and Näätänen *et al.*, 1982). However, for both modalities, the small deviants did not elicit P3 when the large deviants of the same modality were targets. Specifically, when the 1500 Hz deviant was the target, the 1050 Hz deviant did not elicit N2 or P3 (FIG. 2); similar results were obtained for the visual stimuli. Nontarget deviant stimuli that were in the same modality as the target deviant impaired performance significantly if they occurred on the trial immediately preceding the target (the hit rate went down from over 90% to about 50% and RT increased). The result may not be surprising when a

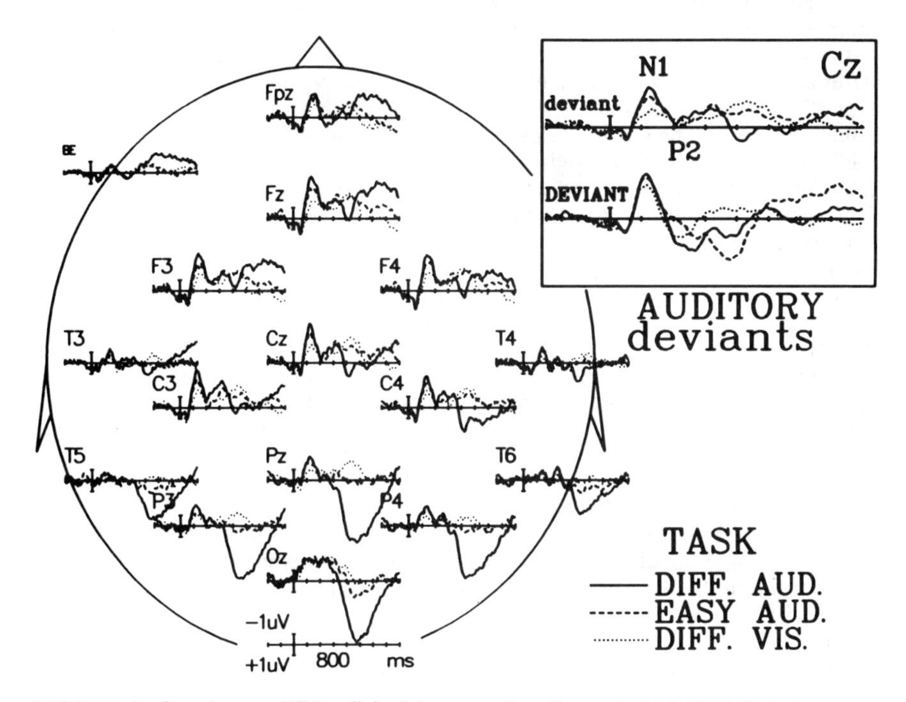

FIGURE 2. Grand mean ERPs elicited by a small auditory deviant (1050 Hz) during an intermodal selective attention task. *Solid lines*: ERPs elicited when the small auditory deviant was the target. *Dashed lines*: ERPs elicited when the large auditory deviant (1500 Hz) was the target. Notice the absence of P3 at Pz in the latter condition. Negative *up*. (Unpublished figure reprinted by permission from David Woods.)

small deviant was the target and it was preceded by a large deviant in the same modality, especially since the latter elicited N2 and P3 (FIG. 2). However, performance was equivalently impaired when the large deviant was the target and it was preceded by a small deviant in the same modality which elicited neither N2 nor P3. It is striking that a deviant stimulus that is capable of eliciting P3 when it is a target (the 1050 Hz deviant in FIG. 2), does not do so when it is not a target, even though the stimuli are attended (the standards, small and large deviants all were associated with ERP indicies of selective attention, discussed in the next section, regardless of which deviant was the target). In this study it would appear that when a large, nontarget deviant was in the attended modality it elicited P3 because the subjects could not help noticing it, whereas that was not true of the small deviant. Hence, the latter results favor the hypothesis that P3 was elicited by nontarget deviants in Roth *et al.* (1978) and Näätänen *et al.* (1982) because the subjects could not help noticing them.

Processing Negativity

Beginning in the early 1960s, investigators sought ERP correlates of selective attention (*e.g.*, Haider, Spong & Lindsley, 1964). This effort began before the first published report of P3, and therefore was not done in the context of studying P3. After a number of unsuccessful efforts (reviewed by Näätänen, 1975), Hillyard *et al.* (1973) finally devised an experimental paradigm that yielded reliable selective attention effects. The key to their success was the use of two intermixed, auditory oddball sequences that were independent of one another, with subjects alternately attending each of the sequences. The findings, in other words, were not serendipitous, as with the other components discussed so far, but due to a concerted effort from several laboratories to find ERP indices of selective attention. Subtracting the ERPs associated with the standard tones of the irrelevant stimulus sequence from the ERPs associated with the standard tones of the attended sequence delineates a selective attention effect, termed Nd by Hansen and Hillyard (1980). Näätänen *et al.* (1978), in the same paper that first reported the MMN, found that the principle effect of selective attention on ERPs was a long-lasting negtivity that spanned the N1 and P2 components and beyond, termed Processing Negativity. According to the latter view, the timing and amplitude of the Nd reflects the difference between the Processing Negativity associated with the relevant versus the irrelevant stimulus sequence.

Hansen and Hillyard (1980) found that as the separation between two stimulus sequences was reduced (in this study, pitch) the peak latency of Nd became longer. However, across conditions there was no significant difference in RT to the target deviant stimuli (the deviants were longer than the standards, the standards being 51 msec, and the deviants 102 msec, in duration). The peak latency of P3 also remained constant. The constant RT suggested that the processing of the cue that separated the stimulus sequences (pitch) occurred in parallel with the cue that separated the standard and deviant stimuli within a sequence (stimulus duration). Since the latter discrimination was handicapped relative to the former (a minimum of 51 msec had to elapse before a deviant could be discriminated from a standard), this suggests it set a rate-limiting step, such that increments in the time required to discriminate the pitch of the tones did not result in increases in RT. Support for this interpretation was obtained by Hansen and Hillyard (1983) in which four intermixed oddball sequences were used. Two of the sequences were presented to one ear (one high and one low in pitch) and another two sequences were presented to the other ear (again, one high and one low in pitch). Embedded within each sequence were deviant stimuli which were longer in duration, just as in the previous study. The subject's task was to attend to tones of a given pitch and location and press a button whenever a longer duration (target) tone occurred. For one group of subjects the pitch and location discriminations were easy; for a second group the pitch discrimination was easy but the location discrimination was hard; for a third group the location discrimination was easy but the pitch discrimination was hard.

When the four stimulus sequence tasks were delivered and one between sequence cue was easy to discriminate (*e.g.*, pitch), and the other between sequence cue was hard to discriminate (*e.g.*, location), there was a later Nd associated with the hard discrimination that persisted beyond the earlier Nd associated with the easy discrimination. But this was only on the trials when the easy discrimination indicated that the tone fulfilled one of the criteria of being a target. When the easy discrimination indicated that the tone failed one of the criteria (*i.e.*, that it came from an irrelevant location or was of an irrelevant pitch), and therefore could not be a target, then the later Nd was terminated. Hansen and Hillyard suggested that when the easy discrimination signaled that the stimulus had one of the target criteria, the subject completed analysis of the hard discrimination. But when the easy discrimination signaled that the stimulus did not fulfill one of the criteria of a target (and hence could not be a target), then analysis of the hard discrimination was terminated. Moreover, whereas the deviant (longer duration) tones in the relevant sequence elicited P3, the deviant tones in the three irrelevant sequences elicited little or no P3. Hansen and Hillyard concluded that there was an hierarchical organization of these brain events in that the activity of each subsequent event was contingent on the outcome of the preceding event.

As in the earlier (Hansen & Hillyard, 1980) study, there was no significant difference in RT across conditions. It was suggested that all the relevant features of the tones were processed in parallel (as opposed to first processing the between-sequence cues and then the within sequence cues), and that the latency of P3 is associated with the terminal decision that a stimulus fulfills all the criteria of being a target. It was again thought that the discrimination of stimulus duration set the rate-limiting step, or what Hansen and Hillyard called "the bottleneck in target identification" (1983, p. 16). This issue is discussed further in the next section.

Single Compared to Double Oddball Sequences

In the main, similar components are elicited in single and double oddball sequences with similar effects associated with attention. The MMN is elicited by deviant stimuli whether attended or not, and N2 and P3 usually only occur for attended sequences. However, there are a number of ways in which different results are obtained.

Processing Negativity. When the ERPs of the standards from a single oddball sequence are compared across attend and ignore conditions, Nd is generally not observed (*e.g.*, Sams, Paavilainen, Alho & Näätänen, 1985; Ritter *et al.*, submitted). N1 and P2 are often larger for the attend than ignore condition, possibly due to increased arousal (Picton, Campbell, Baribeau-Braun & Proulx, 1978). However, this makes clear that Processing Negativity is not associated with attention *per se*, since subjects must attend to the stimuli when actively discriminating between the standard and deviant stimuli in a single oddball sequence. Another reason why this result is significant is because Processing Negativity is a candidate brain activity which could help account for why the MMN is followed by N2 and P3 when subjects attend to the stimuli and not when subjects do not attend to the stimuli. In the experiment by Hansen and Hillyard (1983) which used four intermixed oddball sequences, it was concluded that an hierarchical relationship existed between Nd for an easy between-sequence discrimination, a later Nd for a hard between-sequence discrimination, and P3. If the first did not satisfy a target criterion, neither of the other two occurred. If the first satisfied a target criterion, then the second occurred, but if it did not satisfy a target criterion then P3 did not occur. P3 only occurred, in other words, when the first two brain activities were associated with identification of stimulus features that fulfilled target criteria. Hence, the brain activity underlying the earliest Processing Negativity (associated with the easy between sequence cue) would seem to be capable of determining whether the longer lasting Processing Negativity (associated with the hard between-sequence discrimination) would occur, and the brain activity underlying both capable of determining whether P3 would occur. "Thus, the early selection indexed by Nd appears to block or attenuate further processing of material in the irrelevant channel" (Hillyard & Picton, 1987, p. 538). Consequently, it could be that the Processing Negativity plays a role in determining whether or not the MMN, which is an automatic response, is followed by N2 and P3, which are usually dependent on attention for their elicitation. However, this hypothesis is weakened by the absence of Processing Negativity for a single oddball sequence during a discrimination condition. In other words, the brain activity underlying the Processing Negativity need not be present for the MMN to be followed by N2 and P3. On the other hand, it could be that whereas the brain activity underlying the Processing Negativity need not be present for the elicitation of N2 and P3, it may play a role in inhibiting their elicitation.

The N2 Component. Alho, Sams, Paavilainen, Reinikainen and Näätänen (1989) recently found no observable N2 in a double oddball sequence during selective attention. The deviant stimuli for both the relevant and irrelevant sequences elicited the MMN, but no additional negativity intervened between it and the subsequent P3 for the relevant sequence. They point out that similar results can be found in the figures of Näätänen *et al.* (1978 and 1982) and state that N2 "appears to be much more readily elicited by deviant stimuli in a one-channel oddball task than in a two- or multi-channel, selective attention task" (p. 526). Just why this is so requires clarification, especially why N2 is present for some double oddball sequences used in selective attention tasks and not others. The possibility that a step in information processing that is present for a single oddball sequence would not be present for a comparable discrimination in a double oddball sequence is intriguing.

The Latencies of P3 and Earlier Negative Components. It will be recalled that Hansen and Hillyard (1983) proposed that for their selective attention task, which employed a double oddball sequence, P3 was associated with the terminal decision that a stimulus is a target. This entails the corollary that P3 should be longer in latency for a double than a single oddball sequence, when the within-sequence difference between the standard and the deviant is identical. The reason for this is that when a discrimination is performed on a single sequence, detection of a deviant provides all the information necessary for target identification. In contrast, a selective attention task using a double oddball sequence requires combining the between-sequence information with the within-sequence information in order to identify a target. In Hansen and Hillyard (1980), this included one between-sequence cue and one within-sequence cue, whereas in Hansen and Hillyard (1983) this included two between-sequence cues and one within-sequence cue.

TABLE 1 presents data from two different experiments. In one study (Novak, Ritter & Vaughan, in press a), there was a discrimination task which employed a single oddball sequence, where 80% of the tones were 100 msec in duration and 20% of the tones were randomly 170 msec in duration. The subjects were required to respond whenever a tone of the longer duration occurred. In the other study (Novak, *et al.*, in press b), the same stimulus duration discrimination was used for the within-stimulus sequence discrimination in a selective attention task which employed a double oddball sequence. The between-stimulus sequence cue was pitch. In an easy condition, the tones were equiprobably 1000 and 2000 Hz, and in a hard condition the tones were equiprobably 1000 and 1030 Hz. The overall percentages of the various tones are given in the table.

The MMN, N2 and P3 had about the same latency in all of the conditions of the two experiments. The almost identical mean latency values of the MMN, N2 and P3 suggests that the discrimination of stimulus duration took about the same time in all three conditions. These data suggest that the processing of stimulus duration occurred in parallel with the processing of pitch in the double oddball sequence, in support of Hansen and Hillyard (1983). However, the large increase in RT from the single oddball sequence to the double oddball sequences, combined with the lack of change in the latency of P3, indicates that the latter does not reflect the terminal decision that the stimulus is a target. A similar conclusion was drawn by Wijers, Otten, Feenstra, G. Mulder and L. Mulder (1989) in a selective attention task, who remarked that "P3b latency reflects the durations of only certain subsystems of stimulus evaluation" (page 465).

The increase in RT from the single to the double sequence tasks presumably is

TABLE 1. Mean Latency (in msec) of MMN, N2, P3 and RT in Single and Double Oddball Sequences

Hz	Percent	Duration	MMN	N2	Nd	P3	RT
Single oddball sequence							
1000	80	100					
1000	20	170	197	274		432	390
Easy double oddball sequence							
1000	40	100					
1000	10	170	177	262	267	400	512
2000	40	100					
2000	10	170					
Hard double oddball sequence							
1000	40	100					
1000	10	170	186	265	352	421	514
1030	40	100					
1030	10	170					

related to the time required to put the two relevant pieces of information together necessary to define a target. Triesman and Gelade (1980) have proposed that an attention-dependent "conjunction operation" follows independent feature analyses when the features are in separable dimensions. The criterion for separable, as opposed to integral, dimensions is that processing of one is not affected by processing of the other. In the studies under discussion, the between-stimulus sequence feature (pitch) and the within-sequence feature (stimulus duration) appear to be separable dimensions since alterations in one (presence or absence, as well as ease or difficulty, of the pitch discrimination) had no effect on the latency of the MMN, N2 or P3. In contrast, the difficulty of the pitch discrimination affected the peak latency of Nd. The lack of a change in RT from the easy to the hard double sequence task is similar to the RT results of Hansen and Hillyard (1980, 1983). However, the data do not support the interpretation that the discrimination of stimulus duration set the rate-limiting step of target identification. In the hard double oddball sequence condition, both the peaks of the MMN and N2 preceded the peak of Nd by more than 100 msec. It is possible that the lack of change in RT between the easy and hard double oddball task was due to subjects emphasizing speed at the cost of accuracy, since the percentage of hits was lower, and the number of false alarms higher, in the hard than the easy condition.

Conclusion

The search for components which precede P3 in latency has proved to be fruitful and has expanded understanding of the role of various brain activities critical to information processing. The present review has concentrated on the use of the oddball paradigm in this endeavor. In this regard, it is important to point out that McCallum, Barrett and Pocock (1989) have cautioned against exclusive reliance on the oddball paradigm. They used a double stimulus sequence in a selective attention task, except that equiprobable tones of four different pitches were presented to each ear in a dichotic manner. The subjects were required to respond to tones of a given pitch and ear. It was found that P3 was elicited by all of the eight tones. The amplitude of P3 was actually significantly larger for the tones delivered to the unattended than the attended ear, but this may have been due to overlap with the processing negativity which was larger for the relevant stimuli. Nevertheless, robust P3 components were elicited by the unattended stimuli despite the presence of Nd. Fortunately, as was mentioned above, none of the components reviewed above is dependent upon the oddball paradigm for its elicitation. Nevertheless, it is clear that the strategies subjects use to perform different tasks will alter the pattern of ERP results obtained.

Slow Waves

Samuel Sutton was deeply involved in slow wave research, entering it via a P3 path. In the early collaboration between Sam and the author primarily responsible for this portion of the present paper (DSR), from 1972–1979, we focussed upon the emitted P3 elicited by stimulus absence. In many of our experiments, P3 was followed by a long duration positivity which sometimes was nearly as large as the P3. At first we had no explanation for the slow wave. It appeared to be undesirable variance to me, which I wished would go away. Sam had more *sang froid*. He was quite confident that it was there for a reason and we would get a handle on it, especially after the pioneering paper on slow wave by Nancy Squires and her colleagues (N. Squires, K. Squires, and Hillyard, 1975). Our break-through came in the spring of 1979. We were analyzing the results of two P3 experiments involving signal detection. A late slow positivity was cluttering up the data. To my despair, in some conditions the slow wave almost completely overshadowed the P3. And then it happened—we suddenly realized that the ugly overshadowing was systematic—it occurred under conditions of high task demand. Although we continued to be interested in P3, the direction of our research began to change, as we became more and more intrigued with what study of slow waves could produce. Sam's influence upon slow wave research has continued beyond his passing away in 1986. Many of our experiments conducted since then were first conceptualized in interactions with Sam.

Initial Findings

Slow wave activity was first reported by Squires *et al.* (1975) in an experiment in which subjects counted target stimuli in a random series of standard and target auditory stimuli. The focus of their paper was primarily upon the first evidence that P3 was not a unitary component. Squires *et al.* were able to separate P3 into two components (P3a and P3b) with overlapping, but not identical, topographies and functional behaviors. They also found that the P3 complex was followed by a slow wave which was maximally positive over posterior scalp, small at vertex and negative over frontal scalp. Although the slow wave topography differed from that of P3b, its functional behavior with respect to stimulus probability and whether the stimulus was attended was similar to that of P3b (larger for low probability targets and negligible for unattended stimuli). The presence of slow wave activity was noted in subsequent experiments with designs similar to those of Squires *et al.* However, since slow wave and P3 were closely correlated, slow wave provided no new information. Thus relatively little attention was given to it (Donchin, Ritter and McCallum, 1978).

Evidence of dissociation of slow wave activity and P3 was first reported by Roth, Ford and Kopell (1978) in a go/no-go reaction time (RT) paradigm. P3 and a following positive slow wave were affected in the same way by probability but had an opposite relationship to RT. Smaller P3s and larger slow waves were associated with longer RTs. The findings of Roth *et al.* suggested that slow wave activity might be directly related to task demand, since the longer RTs may have been an indication that the subjects were experiencing more difficulty in performing the task (also see Friedman, 1984).

Perceptual Demand and Positive Slow Waves

The relationship between slow wave and level and type of task demand became more evident in signal detection and recognition experiments (Ruchkin, Sutton and Stega, 1980a; Ruchkin, Sutton, Kietzman and Silver 1980b; Kok and de Jong, 1980; Parasuraman, Richer and Beatty, 1982). The slow wave findings in the two experiments of Ruchkin *et al.* were serendipitous. Sam Sutton and I designed these experiments to study apparent differences between P3 in signal detection and prediction tasks and to further explore emitted P3s in correct rejection and false alarm trials. In the first study, we found that the apparent difference between P3 in detection and prediction was actually due to overlap with the onset of a posterior positive slow wave whose topography varied as a function of task, while P3 did not vary significantly between these tasks (Ruchkin *et al.*, 1980a). In the second study, where the task was detection of near threshold clicks, as detection became more difficult, P3 amplitude decreased while later centro-parietal slow wave activity became more positive (Ruchkin *et al.*, 1980b). This increase in post-P3 slow wave activity as difficulty level of task demand increased led Sam and me to postulate that slow wave was associated with further processing related to the increased task demand, beyond that reflected by P3.

The studies by Kok and de Jong (1980) and Parasuraman *et al.* (1982) further supported and extended the construct of post-P3 slow wave activity reflecting further processing in response to perceptual task demand. Kok and de Jong used a visual discrimination task in which the stimuli were letters that were either intact (easy) or degraded (difficult). Positive slow wave activity was larger over centro-parietal scalp for the difficult stimuli. Besides showing that increased slow wave activity also occurred with increased task demand when the task involved visual pattern recognition, the study by Kok and De Jong ruled out differential preparation as an explanation for the slow wave effects, since level of difficulty was varied randomly from trial-to-trial. This was important, since in our experiments (Ruchkin *et al.*, 1980a, 1980b) blocked designs were employed. Thus it was possible that differences in pre-stimulus negativity associated with differences in preparation for easy and difficult trials could have contributed to differences in post-stimulus positivities in our data, since amplitudes were measured with respect to a pre-stimulus baseline.

Parasuraman *et al.* (1982) used both detection and recognition. Subjects had to determine whether a near threshold tone was present (detection) and identify its frequency (recognition—two frequencies were used in one experiment and four frequencies were used in a second experiment). An early N100 component was larger for more confident detections, but did not relate to recognition accuracy. P3 was larger both for increased confidence in the detection and when the tone was correctly recognized. A later, centro-parietal slow wave was more positive for correctly recognized tones but did not vary significantly with detection confidence. The ERP data of Parasuraman *et al.* indicate that detection and recognition are overlapping processes, and also suggest that the further processing reflected by the slow waves when there is high perceptual demand (near threshold stimuli) is in the domain of recognition of stimulus features.

These studies showed that positive slow wave activity in the epoch following P3, maximal over centro-parietal scalp, was associated with difficult perceptual processing. At the time of these investigations the extent to which such slow wave activity might occur for other types of task demand was not clear. There were studies in which transitions in the rules governing generation of the stimulus sequence elicited late posterior positivities in the post-P3 epoch (Stuss and Picton, 1978; Johnson and Donchin, 1982). The task in these latter investigations was in the domain of conceptual operations rather than perceptual processing, yet late positive activity also appeared as task demand increased. It was an open question as to whether the late positivities observed in other types of experiments were the same as the slow waves observed under high perceptual demand.

Negative Slow Waves—Arithmetic and Mental Rotation

In order to examine the degree of generality or specificity of slow wave activity to type of task demand, we conducted an experiment in which two different types of task demand (perceptual—pattern recognition, and conceptual—arithmetic) were varied orthogonally (Ruchkin, Johnson, Mahaffey and Sutton, 1988). Conceptual difficulty was manipulated by requiring subjects to either memorize a number (easy), use it in mental subtraction (less easy), or use it in mental division (difficult). The stimuli were presented on a cathode ray tube screen, and perceptual difficulty was manipulated by varying the discriminability of the number. It was either intact (easy) or degraded (difficult).

As expected, when the number was degraded, there was a centro-parietal positive slow wave, similar to the slow wave observed by Kok and De Jong (1980) in their degraded stimulus condition. We had postulated that similar positive slow wave activity would occur when arithmetic was difficult, but that it would be at longer latencies than the slow wave associated with perceptual difficulty, since the arithmetic operations would follow perceptual processing. Much to our surprise, we obtained a somewhat different pattern of results. There was a distinct increase in slow wave activity for division, and its onset was clearly at longer latencies than the "perceptual demand" slow wave. However, over posterior scalp, the slow wave activity for division was negative. Positive slow wave activity was only obtained at a pre-frontal site. On the basis of differences in timing, the posterior and pre-frontal slow waves appeared to be different components. (Rösler and Heil [1991] have reported further evidence for a two-component interpretation of the posterior and pre-frontal slow waves, based upon differences in functional behavior.)

We took these findings to mean that the slow wave phenomenon had more variety than previously thought. Given the clear differences in slow wave activity as a function of type of task demand in the experiment of Ruchkin *et al.* (1988), it seemed possible that there were several different types of slow waves, each associated with a different type of sustained mental activity. If this were the case, then slow wave activity might provide ways for conducting detailed ERP studies of specific sustained cognitive operations.

A challenge to this view of slow wave diversity was provided by two studies in which posterior negative slow waves, similar to our finding in arithmetic, were also obtained in mental rotation tasks. The first, by Stuss, Sarazin, Leech and Picton (1983), was reported shortly before we started our perceptual/conceptual slow wave experiment. However, Sam and I did not fully appreciate the possibility that the negative slow wave of Stuss *et al.* was the same as our arithmetic negativity until we completed our experiment. Stuss *et al.* contrasted ERP activity recorded during linguistic decision-making and mental rotation. The mental rotation task elicited a posterior negative slow wave which was larger when subjects gave a higher confidence rating to their judgment about the outcome of the rotation. There was no such slow wave in the linguistic tasks. Peronnet and Farah (1988) found posterior negative slow wave activity during mental rotation whose amplitude was proportional to the angle of rotation (*i.e.*, increased with task demand). The topographic and functional resemblances between the arithmetic and mental rotation negativities raised questions as to whether these negativities were the same. If they were, this would be a sign that the diversity of slow wave activity might be limited, since rather different types of cognitive processing would have produced the same kind of slow wave activity.

Furthermore, the functional significance of the arithmetic and mental rotation slow waves was uncertain, since a delayed response was used in all of the above studies. Thus the time at which the subject completed the task was not known, and hence the specificity of the slow wave activity to the arithmetic and mental rotation operations was unknown. It was possible that these slow waves reflected, at least in part, nonspecific processes (*e.g.*, arousal, executive functions) which, with suitable advance warning, might start before the computations and/or might appear after the computations were completed. Consequently an experiment was conducted intended to clarify the task-specific and nonspecific nature of the slow wave activity (Ruchkin, Johnson, Canoune and Ritter, 1991). Whether the posterior negativities observed in arithmetic and mental rotation were the same was addressed by using both tasks on a random trial-by-trial basis. Whether the slow waves reflected processes specific to the computations or nonspecific processes was addressed by preceding the task stimulus with a warning stimulus and using an immediate response. Use of a warning stimulus indicating the task and difficulty level prior to the task stimulus provided information on whether onset of the negativities was too early to reflect the computational activity, while the immediate response would show whether the negativities were too late.

A major finding of this experiment was that the posterior negative slow wave activity in arithmetic and mental rotation was not a unitary phenomenon. Rather, there were a number of late posterior negativities with different topographies which appeared to be associated with different aspects of the tasks. For both arithmetic and mental rotation, there was a centro-parietal negative slow wave which was larger when task difficulty increased. It was task specific in a temporal sense, since it occurred well after stimulus onset and peaked early enough with respect to the response so that it could have been associated with the computations. At this time we still cannot be completely certain whether this negative slow wave was different in arithmetic and mental rotation, since the anterior-posterior topographies were quite similar for the two types of tasks. However, there was some difference in their lateral topographies, and their morphology and timing differed. The arithmetic negativity was clearly synchronized to the response, while the rotation negativity was more closely synchronized to the stimulus.

There also were an earlier, occipital and a later, parietal negativity. The occipital negativity occurred after the interval between the warning and task stimuli, and hence may have been specifically related to early task operations. In contrast, the parietal negativity followed the response. Hence the late parietal negativity was not related to production of the response, although it was influenced by the difficulty of the prior task. For both arithmetic and mental rotation its amplitude was larger when the preceding task was more difficult.

These various studies indicate that the late negativities constitute a complex group of slow waves. The differences in timing and topography indicate that the various negativities relate to different aspects of the task. While the data are not definitive, as discussed above, they do suggest that the centro-parietal negativity, which is temporally most closely linked to the interval in which the arithmetic and mental rotation operations presumably occur, differs as a function of type of task. To various extents, the amplitudes of these negativities increase with task demand. However, the meaning of this covariation of amplitude with task demand in arithmetic is controversial. Rösler and Heil (1991) claim that the increased negativity in the first arithmetic study by Ruchkin *et al.* (1988) was due to anticipation for the imperative stimulus to which the subject responded (see Ruchkin and Johnson, 1991, for a reply). Since there was no imperative stimulus in the second arithmetic study (Ruchkin *et al.*, 1991), the interpretation of Rösler and Heil does not appear to be viable.

Slow Waves in Learning and Memory

There have been several ERP studies concerned with memory operations. The earliest studies used the well-known Sternberg paradigm for investigating scanning of information in working memory. These studies focussed on relations between P3 latency, reaction time and memory load (Marsh, 1975; Gomer, Spicuzza and O'Donnell, 1976; Adam and Collins, 1978; Ford, Roth, Mohs, Hopkins and Kopell, 1979; Pratt, Michalewski, Patterson and Starr, 1989). Also using P3, investigators have studied ERP activity associated with storage in long term memory. These studies have shown that P3 and/or other late positivities that overlapped with P3 varied as a function of subsequent recognition/recall performance (Sanquist, Rohrbaugh, Syndulko and Lindsley, 1980; Karis, Fabiani and Donchin, 1984; Johnson, Pfefferbaum and Kopell, 1985; Fabiani, Karis and Donchin, 1986; Neville, Kutas, Chesney and Schmidt, 1986; Paller, Kutas and Mayes, 1987; Paller, 1990; Friedman, 1990). In most of these studies the relevant ERP activity was in the 300–1200-msec latency range, and thus these components were associated with relatively short-duration memory operations.

Recently there have been a number of slow wave studies of memory. These experiments have focused upon longer duration processing than in P3 studies. In the slow wave experiments the main experimental maneuver has been to manipulate the difficulty of the memory operation. Thus the focus has been more on the degree or type of effort expended by the subjects rather than on accuracy of subsequent recall or recognition. The earliest studies were by investigators at the Neurological Clinic, University of Vienna. These experiments were concerned with storage in long-term memory and utilized associative learning paradigms (M. Lang, W. Lang, Uhl, A. Kornhuber, Deeke and H. Kornhuber, 1987; W. Lang, M. Lang, Uhl, A. Kornhuber, Deeke and H. Kornhuber, 1988; Uhl, Lang, Lindinger and Deeke, 1990a). Difficulty of learning and elaboration strategies was manipulated. Rösler, Glowalla and Heil (1990a) and Rösler, Heil and Glowalla (1990b) have studied slow waves during retrieval of information from long term memory in which search time was manipulated. Uhl, Goldenberg, Lang, Lindinger, Steiner and Deeke (1990b) reported an experiment involving both recall from long-term memory and subsequent retention of the recalled information in working memory. The type of information (spatial versus visual) was manipulated in the study by Uhl *et al.* (1990b). Ruchkin, Johnson, Canoune and Ritter (1990) and Ruchkin, Johnson, Grafman, Canoune and Ritter (in press) have studied ERP activity during encoding and retention of information in working memory. Information load and type of information (verbal-phonological versus spatial) were manipulated in the studies by Ruchkin *et al.*

The long-term memory storage studies by Lang *et al.* (1987; 1988) were conducted within the context of paired associates learning paradigms. They were especially concerned with the involvement of the left frontal lobe in verbal-cognitive learning, since patients with lesions in this region are impaired in such learning. The investigations of Lang *et al.* were intended to provide ERP information about left frontal lobe activity during verbal-cognitive learning in normal, intact subjects.

In the first study (Lang *et al.*, 1987), subjects learned the Morse codes for a set of letters. On each trial subjects were presented with a letter (S1) to which they tapped the Morse code after presentation of an imperative stimulus (S2). The S1–S2 interval was five seconds. Throughout the S1–S2 interval there was slow negative activity widely distributed over the scalp. The negativity over frontal scalp in the early part of the S1–S2 epoch was clearly lateralized to the left. Furthermore, as learning progressed over trials, the amplitude of the early negativity increased. This increase was confined to frontal and mid-line central scalp electrodes. It was not observed at electrodes placed to overlay the supplementary motor area. There also was a control condition in which subjects were presented with letters for which they had previously learned the Morse codes. There was a negativity in the control condition, but it was smaller than in the learning condition. Overall, the negativity early in the S1–S2 interval was greater in the learning than control condition over frontal, anterior midline and parietal scalp. Moreover, there was no lateralization of the early negativity in the control condition. Lang *et al.* (1987) attributed these differences in slow negativity to differences in information processing associated with different stages of learning—specifically, that subjects were able to utilize more information from prior trials as learning progressed. However, it is conceivable that some of the early slow negativity was ERP activity which was contingent upon preparation for the imperative stimulus, and the authors did use the terminology "CNV" when discussing these slow negativities (although it does not seem likely that the increase with learning was such contingent activity).

Lang *et al.* (1988) further studied slow negative waves in verbal associative learning. There were two factors: 1) degree of learning required (easy or hard) and 2) semantic content (meaningful words or nonsense words). In the easy learning condition the subjects learned to associate pairs of meaningful words which were well-known antonyms (*e.g.*, "hill" and "valley") or pairs of nonsense words which had previously been learned as an associated pair. In the hard learning condition the pairs consisted of either two meaningful words that were not commonly associated or two nonsense words for which there was no previously established association. In the hard learning condition a negative slow wave, largest over left-frontal scalp, followed the second word in the pair in the 1500–2500 msec latency range. The slow negative wave was larger for meaningful words than for nonsense words. There was no such negative wave in the easy learning condition. Since there was no stimulus to be anticipated following the second word, a CNV explanation for the frontal negativity is most unlikely. In view of the differences between ERPs for the meaningful and nonsense words in the hard condition, Lang *et al.* (1988) suggested that the frontal negative slow wave was related to elaborative encoding of verbal material.

The effect of type of elaboration memorization strategy upon slow waves was further investigated by Uhl *et al.* (1990a). Their design was similar to the studies by Lang *et al.* (1987, 1988). Subjects learned word pairs using either of two strategies: 1) semantic, form a sentence linking the two words, or 2) visual imagery, create a visual image combining the two items. There also was a control condition in which there was no learning. During learning there were slow negative waves that were largest over left frontal scalp, while no such negativity occurred in the control condition. The frontal negativity was not affected by the subjects' strategy. Strategy did affect posterior temporal slow wave activity. In the semantic condition there was more posterior negativity over the left hemisphere while in the visual imagery condition the posterior negativity was laterally symmetric.

The amplitudes and topographies of these slow waves appear to be sensitive to task demand and the subject's memorization strategy. However, there appears to be an inconsistency between the interpretation of Lang *et al.* (1988) of the left frontal negative slow wave and the results of the Uhl *et al.* (1990a). In Lang *et al.* its amplitude varied according to whether subjects processed unassociated meaningful or nonsense words. Lang *et al.* conjectured that this difference was due to differences in elaboration during learning of the meaningful and nonsense word pairs. However, the frontal negativity did not vary with semantic versus visual imagery elaboration strategies in the experiment of Uhl *et al.* Clues as to what may actually underlie the variation in frontal negativity as a function of meaningful versus nonsense words may have been provided by the earlier Morse code experiment of Lang *et al.* (1987) and by the study by Rösler *et al* (1990a; 1990b) of retrieval from long-term memory.

In Lang *et al.* (1987) subjects had to retrieve past experiences from long-term memory in order to accomplish the learning. The subjects in Rösler *et al.* learned lists of words and the following day were presented with pairs of words. The task was to decide whether the concepts represented by the two words were learned in the same list. In the latter study, in all conditions there was a prominent long-duration negative wave over left-frontal scalp, which Rösler *et al.* interpreted as a reflection of retrieval operations. The difficulty of the decision was manipulated by varying the number of different lists in which a specific concept appeared (one, two or three lists). The more lists in which a specific concept appeared, the more difficult it was to reach a decision. Overlapping with the long-duration frontal negativity were more rapidly varying slow negativities with broader scalp distributions. The amplitude of a bilaterally distributed frontal negativity varied with the number of lists in which the test words appeared—larger amplitude for the greater number of lists.

With respect to the frontal negativity in the Lang *et al.* (1988) verbal associative learning experiment, when learning the word pairs, the meaningful words contacted semantic memory, and when the two words in the pairs were not related, there may have been a retrieval of more information from long-term memory to aid in the learning process. Thus, some of the frontal negativity in Lang *et al.* (1988) may have been similar to the negativity that Rösler *et al.* (1990a; 1980b) associated with retrieval. The contact with long-term memory stores may also have been different for the meaningful and nonsense words, thus contributing to the differences in frontal negative slow wave activity.

Uhl *et al.* (1990b) have also studied slow wave activity during recall (from long-term memory) and imaging (in working memory) operations. They were concerned with the topography of slow negativities as a function of whether spatial or visual imagery was employed. Upon command, subjects imagined either the faces of well-known politicians (visual), a color (visual) or the relative locations of objects in a previously learned diagram (spatial). It cannot be definitively determined at what point recall was complete and primarily imaging operations were being performed. Early in the recall interval there was a negative slow wave which was maximal over left-frontal scalp. At longer latencies the frontal negativity abated and was followed by broadly distributed slow negative activity with a topography which differed as a function of the type of imagery. This later negativity was larger over parietal scalp for the spatial diagram and larger over temporal and occipital scalp for the faces and colors. Conceivably the early, frontal negativity reflected retrieval from long-term memory (which would be consistent with Rösler *et al.*, 1990a, 1990b), while the late posterior negativity reflected retention in working memory.

Ruchkin *et al.* (1990) studied slow wave activity during encoding and retention of information in working memory. In a remember condition subjects had to retain in memory a string of consonants for an interval of approximately 2.5 seconds. Information load was varied by varying the number of different consonants in the string (one, three or six). In order to distinguish between ERP effects associated with the extraction and immediate processing of information from the stimulus and effects specific to retention of stimulus information over the entire interval, there was a condition in which subjects immediately searched the string of consonants for a match with a previously presented consonant. Thus, in the search condition, for all three load levels, only one piece of information (match or mismatch) had to be remembered until the end of the retention interval.

The amplitude of a posterior positive slow wave, maximally active over the 1000–2000 msec latency range, increased with information load in the remember condition and was negligible in the search condition. A frontal negative slow wave overlapped and followed the posterior positive wave. The frontal negativity was largest over the left hemisphere and was active to the end of the retention interval. It occurred only at the highest load level in the remember condition and was totally absent in the search condition. These two slow waves appear to be specifically associated with encoding and retention of information in working memory over an extended period of time. The relatively early, phasic character of the posterior positivity suggests that it reflects an initial encoding operation. The later, frontal negativity may have been associated with retention processing *per se*.

In a subsequent working memory experiment, Ruchkin *et al.* (in press) compared slow wave activity associated with encoding-retention of either verbal-phonological or spatial information. Subjects were presented with displays of either pronounceable nonsense words or spatial patterns. The sound of the nonsense word or the distances between the elements of the spatial pattern had to be retained in working memory over a five-second interval. Memory load was varied by using either three, four or five syllables or elements in the word or pattern. The early positive posterior wave and the later left-frontal negative wave observed when consonants had to be remembered (Ruchkin *et al.*, 1990) were replicated in the verbal memorization condition. In contrast, the slow wave activity in the visual-spatial task consisted of a large posterior negativity with an early onset and a lower amplitude frontal negativity with a later onset. Both slow waves lasted until at least the end of the recording epoch. The amplitude of the posterior negativity was directly related to memory load—more negative as the number of elements in the pattern increased.

These various results suggest that the posterior positive wave observed in the two studies by Ruchkin *et al.* is specifically related to encoding of verbal information into working memory. The posterior negative wave in the spatial condition, whose topography resembled the posterior negativity associated with imaging spatial information in the study by Uhl *et al.* (1990b), may be associated with maintaining spatial information in working memory.

Frontal negative slow wave activity, often lateralized to the left hemisphere, appears to be a ubiquitous phenomenon in slow wave memory experiments, whether they involve storage to or retrieval from long-term memory or retention in working memory. It may be that a number of different slow waves are involved, each related to a different operation. More than one may be active in a given experiment. For example, we have suggested that the left-frontal negativity in the verbal associative learning experiment of Lang *et al.* (1988) could reflect a mix of both retrieval from and storage in long-term memory. It may also be that the frontal negativity in the studies by Ruchkin *et al.* (1990; in press) reflects a mix of retention in working memory and storage in long-term memory operations. (For example, we sometimes learn new telephone numbers by repeatedly holding them in working memory.)

Conclusion

Taking together the various slow wave studies reviewed above, it should be clear that it makes no sense to refer to the slow wave phenomenon as "the slow wave" any more than it makes sense to refer to event-related potentials as "the event-related potential." There is a variety of slow waves, with different types of slow waves associated with different types of task demand.

At present, slow waves have not been cataloged as components with a set of properties that define each component, as has been done for other ERP phenomena (*e.g.*, P3, CNV, etc.). This is because slow waves have been systematically studied in relatively few, disparate experiments. Thus there is not enough information to make generalizations as to whether a given slow wave phenomenon is the same across different types of studies. The better strategy at this time may be to view slow waves as a means of imaging time intervals and spatial locations of brain activity when subjects are engaged in a specified type of mental processing. This approach may be especially apt when source localization analysis is applied to the slow wave data (Scherg, 1990).

RETROSPECTIVE OVERVIEW

The present paper has been divided into two parts, the first dealing with components that have short latencies and brief durations, and the second dealing with components that occur late in time and have long durations. This was a convenient way of dividing the material covered and fits most ERP investigators' notions of these being two distinct domains of research. Indeed, most investigations of short-latency components have usually paid little attention to slow waves and vice versa.

Initially, it was thought that there is one slow wave, being positive in posterior regions and negative frontally. Since the slow wave occurred after P3, and the experimental designs were such that the slow waves may have been subsequent to the processes that determined the subjects' behavior, its functional significance was conceptualized in terms of further processing associated, for example, with a re-examination of the events which had just occurred. This interpretation was applied to diverse experimental tasks, such as signal detection, discrimination between stimuli, prediction situations, etc. Two major events in slow wave research, described above, have changed matters considerably. One is that a variety of slow waves have been found which have different scalp distributions and are associated with different psychological processes. The other is that the slow waves have more recently been studied in tasks where the subject's performance depended upon very-long-duration processes. Thus, these slow waves may well have been associated with processing that determined the behavioral response. Consequently, the features that used to separate the two domains of research are not as relevant as they once were. Moreover, the division of components into those of short- and long-lasting durations has become arbitrary. On the one hand, the so-called slow waves can be long or short in their duration, depending upon the nature and complexity of the experimental task employed. For example, in mental arithmetic, a centro-parietal negative slow wave peaked prior to the subjects' response. When subjects responded rapidly (*e.g.*, 2 sec), the slow wave's duration was of the order of several hundred msec. When the response was slow (*e.g.*, 4 sec), its duration was of the order of a few thousand msec (Ruchkin *et al.*, 1991).

The early latency components also have a very large range of durations that actually overlap with those of the slow waves. Woldorff *et al.* (1987) have reported a selective attention effect with auditory stimuli that lasts for only 20 msec (from about 30- to 50-msec latency). Selective attention effects associated with spatial selection of visual stimuli typically last for about 100 msec (*i.e.*, the duration of the P1 and N1 components). The MMN usually lasts for about 200 msec. The P3 component can last for 500 or more msec, and the Nd can persist for many hundreds of msec.

There does not appear to be a discontinuity, then, in the duration of ERP components such that a meaningful division into long and short durations can be established. For these various reasons, we believe that it is no longer useful to think of ERPs in terms of a dichotomy of early, phasic components and late, tonic slow waves. Consequently, whereas the manner in which this paper was divided was convenient with regard to the expertise of the authors, upon reflection the distinction between the two lines of research has little theoretical ground.

REFERENCES

ADAM, N. & G. I. COLLINS. 1978. Late components of the visual evoked potential to search in short-term memory. Electroencephalogr. Clin. Neurophysiol. **44:** 147–156.

ALHO, K., P. PAAVILAINEN, K. REINIKAINEN, M. SAMS & R. NÄÄTÄNEN. 1986. Separability of different negative components of the event-related potential associated with auditory stimulus processing. Psychophysiology **23:** 613–623.

ALHO, K., M. SAMS, P. PAAVILAINEN, K. REINIKAINEN & R. NÄÄTÄNEN. 1989. Event-related brain potentials during selective listening. Psychophysiology **26:** 514–528.

ALHO, K., K. SAINIO, N. SAJANIEMI, K. REINIKAINEN & R. NÄÄTÄNEN. 1990. Event-related brain potential of human newborns to pitch change of an acoustic stimulus. Electroencephalogr. Clin. Neurophysiol. **77:** 151–155.

ALHO, K., D. L. WOODS, A. ALGAZI & R. NÄÄTÄNEN. Intermodal selective attention II. Effects of attentional load on processing of auditory and visual stimuli in central space. Electroencephalogr. Clin. Neurophysiol. In press.

BENTIN, S., G. MCCARTHY & C. C. WOOD. 1984. ERP evidence for semantic priming in a memory recognition paradigm. Paper presented at the Third International Conference on Cognitive Neuroscience. Bristol, England.

BENTIN, S., G. MCCARTHY & C. C. WOOD. 1985. Event-related potentials, lexical decision and semantic processing. Electroencephalogr. Clin. Neurophysiol. **60:** 343–355.

BRETON, F., W. RITTER, R. SIMSON & H. G. VAUGHAN JR. 1988. The N2 component elicited by stimulus matches and multiple targets. Biol. Psychol. **27:** 23–44.

CAMPELL, K., I. BELL & C. BASSTIEN. Evoked potential measures of information processing during natural sleep. *In* Sleep, Performance and Arousal. R. Broughton & R. Ogilvie, Eds. Birkhauser. Cambridge, MA. In press.

COURCHESNE, E., S. A. HILLYARD & R. Y. COURCHESNE. 1977. P3 waves to the discrimination of targets in homogeneous and heterogeneous stimulus sequences. Psychophysiology **14:** 590–597.

CSEPE, V., G. KARMOS & M. MOLNAR. 1987. Evoked potential correlates of stimulus deviance during wakefulness and sleep in cat: animal model of mismatch negativity. Electroencephalogr. Clin. Neurophysiol. **66:** 571–578.

DEACON, D., F. BRETON, W. RITTER & H. G. VAUGHAN, JR. 1991. The relationship between N2 and N400: scalp distribution, stimulus probability, and task relevance. Psychophysiology **28:** 185–200.

DONCHIN, E., W. RITTER & W. C. MCCALLUM. 1978. Cognitive psychophysiology: the endogenous components of the ERP. *In* Event-Related Brain Potentials in Man. E. Calloway, P. Tueting & S. H. Koslow, Eds. 349–441. Academic Press. New York.

DUNCAN-JOHNSON, C. C. & E. DONCHIN. 1977. On quantifying surprise: the variation of event-related potentials with subjective probability. Psychophysiology **14:** 456–467.

FABIANI, M., D. KARIS & E. DONCHIN. 1986. P300 and recall in an incidental memory paradigm. Psychophysiology **23:** 298–308.

FISCHLER, I., P. A. BLOOM, D. G. CHILDERS, S. E. ROUCOS & N. W. PERRY, JR. 1983. Brain potentials related to stages of sentence verification. Psychophysiology **20:** 400–409.

FORD, J. M., W. T. ROTH, R. C. MOHS, W. F. HOPKINS & B. S. KOPELL. 1979. Event-related potentials recorded from young and old adults during a memory retrieval task. Electroencephalogr. Clin. Neurophysiol. **47:** 450–459.

FRIEDMAN, D., W. RITTER & R. SIMSON. 1978. Analysis of nonsignal evoked cortical potentials in two kinds of vigilance tasks. *In* Multidisciplinary Perspectives in Event-Related Brain Potential Research. D. Otto, Ed. 194–197. U. S. Government Printing Office. Washington, DC.

FRIEDMAN, D. 1984. P300 and slow wave: the effects of reaction time quartile. Biol. Psychol. **18:** 49–71.

FRIEDMAN, D. 1980. ERPs during continuous recognition memory for words. Biol. Psychol. **30:** 61–87.

GOMER, F. E., R. J. SPICUZZA & R. D. O'DONNELL. 1976. Evoked potential correlates of visual item recognition during memory-scanning tasks. Physiol. Psychol. **4:** 61–65.

HAIDER, M., R. SPONG & D. B. LINDSLEY. 1964. Attention, vigilance, and cortical evoked-potentials in humans. Science **145:** 180–182.

HAMBERGER, M. J. & D. FRIEDMAN. 1990. Age-related changes in semantic activation: evidence from ERPs. *In* Psychophysiological Brain Research. C. H. M. Brunia, A. W. K. Gaillard & A. Kok, Eds. Vol. 1: 279–284. Tilburg University Press. Tilburg.

HARI, R., M. HÄMÄLÄINEN, R. ILMONIEMI, E. KAUKORANTA, K. REINIKAINEN, J. SALMINEN, K. ALHO, R. NÄÄTÄNEN & M. SAMS. 1984. Responses of the primary auditory cortex to pitch changes in a sequence of tone pips: neuromagnetic recordings in man. Neurosci. Lett. **50:** 31–43.

HENIK, A., M. SPEER & W. A. KELLOGG. 1983. The dependence of semantic relatedness upon prime processing. Mem. Cognit. **11:** 366–373.
HANSEN, J. & S. A. HILLYARD. 1980. Endogenous brain potentials associated with selective auditory attention. Electroencephalogr. Clin. Neurophysiol. **49:** 277–290.
HANSEN, J. & S. A. HILLYARD. 1983. Selective attention to multidimensional auditory stimuli. J. Exp. Psychol. Hum. Percept. Perform. **9:** 1–19.
HILLYARD, S. A. & T. W. PICTON. 1987. Electrophysiology of cognition. *In* Higher Functions of the Brain. Part 2. V. B. Mountcastle, F. Plum & S. R. Geiger, Eds. 519–584. American Physiological Society. Baltimore.
HILLYARD, S. A., R. F. HINK, V. L. SCHWENT & T. W. PICTON. 1973. Electrical signs of selective attention in the human brain. Science **182:** 177–180.
JAVITT, D., C. E. SCHROEDER, M. STEINSCHNEIDER, J. C. AREZZO & H. G. VAUGHAN, JR. Demonstration of mismatch negativity in the monkey. Electroencephalogr. Clin. Neurophysiol. In press.
JOHNSON, R., JR. & E. DONCHIN. 1982. Sequential expectancies and decision making in a changing environment: an electrophysiological approach. Psychophysiology **19:** 183–200.
JOHNSON, R., JR., A. PFEFFERBAUM & B. S. KOPELL. 1985. P300 and long-term memory: latency predicts recognition performance. Psychophysiology **22:** 497–507.
KARIS, D., M. FABIANI & E. DONCHIN. "P300" and memory: individual differences in the Von Restorff effect. Cognit. Psychol. **16:** 177–216.
KLINKE, R., H. FURHSTORFER & P. FINKENZELLER. 1968. Evoked responses as a function of external and stored information. Electroencephalogr. Clin. Neurophysiol. **25:** 119–122.
KOK, A. & H. L. DE JONG. 1980. Components of the event-related potential following degraded and undegraded stimuli. Biol. Psychol. **11:** 117–133.
KUTAS, M., & S. A. HILLYARD. 1980. Reading senseless sentences: brain potentials reflect semantic incongruity. Science **207:** 161–163.
KUTAS, M., & S. A. HILLYARD. 1989. An electrophysiological probe of incidental semantic association. J. Cognit. Neurosci. **1:** 38–49.
LANG, M., W. LANG, F. UHL, A. KORNHUBER, L. DEECKE & H. H. KORNHUBER. 1987. Slow negative potential shifts indicating verbal cognitive learning in a concept formation task. Hum. Neurobiol. **6:** 183–190.
LANG, W., M. LANG, F. UHL, A. KORNHUBER, L. DEECKE & H. H. KORNHUBER. 1988. Left frontal lobe in verbal associative learning: a slow potential study. Exp. Brain Res. **70:** 99–108.
MÄNTYSALO, S. & R. NÄÄTÄNEN. 1987. The duration of a neuronal trace of an auditory stimulus as indicated by event-related potentials. Biol. Psychol. **24:** 183–195.
MARSH, G. R. 1975. Age differences in evoked potential correlates of a memory scanning process. Exp. Aging Res. **1:** 3–16.
MCCALLUM, W. C., K. BARRETT & P. V. POCOCK. 1989. Late components of auditory event-related potentials to eight equiprobable stimuli in a target detection task. Psychophysiology **26:** 683–694.
MCCALLUM, W. C., S. F. FARMER & P. V. POCOCK. 1984. The effects of physical and semantic incongruities on auditory event-related potentials. Electroencephalogr. Clin. Neurophysiol. **59:** 477–488.
NÄÄTÄNEN, R. 1975. Selective attention and evoked potentials in humans—a critical review. Biol. Psychol. **2:** 237–307.
NÄÄTÄNEN, R. 1990. The role of attention in auditory information processing as revealed by event-related potentials and other brain measures of cognitive function. Behav. Brain Sci. **13:** 201–288.
NÄÄTÄNEN, R. & A. W. K. GAILLARD. 1983. The orienting reflex and the N2 deflection of the event-related potential (ERP). *In* Tutorials in Event-Related Potential Research: Endogenous Components. A. W. K. Gaillard & W. Ritter, Eds. North-Holland. Amsterdam.
NÄÄTÄNEN, R., A. W. K. GAILLARD & S. MÄNTYSALO. 1978. Early selective-attention effect on evoked potential reinterpreted. Acta Psychol. **42:** 313–329.

NÄÄTÄNEN, R., P. PAAVILAINEN, K. ALHO, K. REINIKAINEN & M. SAMS. 1987. Interstimulus interval and the mismatch negativity. *In* Evoked Potentials III. C. Barber & T. Blum, Eds. Butterworth. London.

NÄÄTÄNEN, R., P. PAAVILAINEN, H. TIITINEN, D. JIANG & K. ALHO. Attention and mismatch negativity. Submitted.

NÄÄTÄNEN, R., M. SIMPSON & N. E. LOVELESS. 1982. Stimulus deviance and evoked potentials. Biol. Psychol. **14:** 53–98.

NEVILLE, H. J., M. KUTAS, G. CHESNEY & A. L. SCHMIDT. 1986. Event-related brain potentials during initial encoding and recognition memory of congruous and incongruous words. J. Mem. Lang. **25:** 75–92.

NOVAK, G. P., W. RITTER & H. G. VAUGHAN, JR. Mismatch detection and the latency of temporal judgments. Psychophysiology. In press a.

NOVAK, G. P., W. RITTER & H. G. VAUGHAN, JR. The chronometry of attention-modulated processing and automatic mismatch detection. Psychophysiology. In press b.

NOVAK, G. P., W. RITTER, H. G. VAUGHAN, JR. & M. L. WIZNITZER. 1990. Differentiation of negative event-related potentials in an auditory discrimination task. Electroencephalogr. Clin. Neurophysiol. **75:** 255–275.

NYMAN, G., K. ALHO, P. LAURINEN, P. PAAVILAINEN, T. RADIL, K. REINIKAINEN, M. SAMS & R. NÄÄTÄNEN. 1990. Mismatch negativity (MMN) for sequences of auditory and visual stimuli: evidence for a mechanism specific to the auditory modality. Electroencephalogr. Clin. Neurophysiol. **77:** 436–444.

PALLER, K. A. 1990. Recall and stem-completion priming have different electrophysiological correlates and are modified differentially by directed forgetting. J. Exp. Psychol. Learn. Mem. Cognit. **16:** 1021–1032.

PALLER, K. A., M. KUTAS & A. R. MAYES. 1987. Neural correlates of encoding in an incidental learning paradigm. Electroencephalogr. Clin. Neurophysiol. **67:** 360–371.

PARASURAMAN, R., F. RICHER & J. BEATTY. 1982. Detection and recognition: concurrent proceses in perception. Percept. Psychophys. **31:** 1–12.

PERONETT, F. & M. J. FARAH. 1989. Mental rotation: an event-related potential study with a validated mental rotation task. Brain Cognit. **9:** 279–288.

PICTON, T. W., K. B. CAMPBELL, J. BARIBEAU-BRAUN & G. B. PROULX. 1978. The neurophysiology of human attention: a tutorial review. *In* Attention and Performance VII. J. Requin, Ed. 429–467. Wiley. New York.

PRATT, H., H. J. MICHALEWSKI, J. V. PATTERSON & A. STARR. 1989. Brain potentials in a memory scanning task. I. Modality and task effects on potentials to the probes. Electroencephalogr. Clin. Neurophysiol. **72:** 407–421.

RENAULT, B. & N. LESEVRE. 1979. A trial-by-trial study of the visual omission response in reaction time situations. *In* Human Evoked Potentials. D. Lehmann & E. Callaway, Eds. 317–329. Plenum Press. New York.

RITTER, W. 1978. Latency of event-related potentials and reaction time. *In* Multidisciplinary Perspectives in Event-Related Brain Potential Research. D. Otto, Ed. 173–174. U. S. Government Printing Office. Washington, DC.

RITTER, W., P. PAAVILAINEN, J. LAVIKAINEN, K. REINIKAINEN & R. NÄÄTÄNEN. Event-related potentials to repetition and change of auditory stimuli. Submitted.

RITTER, W., R. SIMSON & H. G. VAUGHAN, JR. 1972. Association cortex potentials and reaction time in auditory discrimination. Electroencephalogr. Clin. Neurophysiol. **33:** 547–555.

RITTER, W., R. SIMSON & H. G. VAUGHAN, JR. 1983. Event-related potential correlates of two stages of information processing in physical and semantic discrimination tasks. Psychophysiology **20:** 168–179.

RITTER, W., R. SIMSON, H. G. VAUGHAN, JR. & D. FRIEDMAN. 1979. A brain event related to the making of a sensory discrimination. Science **203:** 1358–1361.

RÖSLER, F., U. GLOWALLA & M. HEIL. 1990. Slow negative potentials during retrieval from long-term memory. *In* Psychophysiological Brain Research. C. H. M. Brunia, A. W. K. Gaillard & A. Kok, Eds. Vol. 1: 244–247. Tilburg University Press. Tilburg.

RÖSLER, F., M. HEIL & U. GLOWALLA. 1990. Monitoring retrieval from long-term memory by means of slow event-related brain potentials. Psychophysiology **27:** 4A, S61.

RÖSLER, F. & M. HEIL. 1991. Toward a functional categorization of slow waves: taking into account past and future events. Psychophysiology **28:** 344–358.

ROTH, W. T., J. M. FORD & B. S. KOPELL. 1978. Long latency evoked potentials and reaction time. Psychophysiology **15:** 17–23.

RUCHKIN, D. S. & R. JOHNSON, JR. 1991. Complexities related to cognitive slow wave experiments: a reply to Rosler and Heil. Psychophysiology **28:** 359–362.

RUCHKIN, D. S., R. JOHNSON, JR., D. MAHAFFEY & S. SUTTON. 1988. Toward a functional categorization of slow waves. Psychophysiology **25:** 339–353.

RUCHKIN, D. S., R. JOHNSON, JR., H. CANOUNE & W. RITTER. 1990. Short-term memory storage and retention: an event-related brain potential study. Electroencephalogr. Clin. Neurophysiol. **76:** 419–439.

RUCHKIN, D. S., R. JOHNSON, JR., H. CANOUNE & W. RITTER. 1991. Event-related potentials during arithmetic and mental rotation. Electroencephalogr. Clin. Neurophysiol. **79:** 473–487.

RUCHKIN, D. S., R. JOHNSON, JR., J. GRAFMAN, H. L. CANOUNE & W. RITTER. Distinctions and similarities among working memory processes: an event-related potential study. Cognit. Brain Res. In press.

RUCHKIN, D. S., S. SUTTON & M. STEGA. 1980a. Emitted P300 and slow wave event-related potentials in guessing and detection tasks. Electroencephalogr. Clin. Neurophysiol. **49:** 1–14.

RUCHKIN, D. S., S. SUTTON, M. L. KIETZMAN & K. SILVER. 1980b. Slow wave and P300 in signal detection. Electroencephalogr. Clin. Neurophysiol. **50:** 35–47.

RUGG, M. D., J. FURDA & M. LORIST. 1988. The effects of task on the modulation of event-related potentials by word repetition. Psychophysiology **25:** 55–63.

SAMS, M., K. ALHO & R. NÄÄTÄNEN. 1983. Sequential effects in the ERP in discriminating two stimuli. Biol. Psychol. **17:** 41–58.

SAMS, M., P. PAAVILAINEN, K. ALHO & R. NÄÄTÄNEN. 1985. Auditory frequency discrimination and event-related potentials. Electroencephalogr. Clin. Neurophysiol. **62:** 437–448.

SANQUIST, T. F., J. W. ROHRBAUGH, K. SYNDULKO & D. B. LINDSLEY. 1980. Electrocortical signs of levels of processing: perceptual analysis and recognition memory. Psychophysiology **17:** 568–576.

SCHERG, M. 1990. Fundamentals of dipole source potential analysis. *In* Auditory Evoked Electric and Magnetic Fields. Topographic Mapping and Functional Localization. Advances in Audiology. F. Grandori, G. L. Romani & M. Hoke, Eds. Vol. 6: 40–69. Karger. Basel.

SCHERG, M., J. VAJSAR & T. W. PICTON. 1989. A source analysis of the late human auditory evoked potentials. J. Cognit. Neurosci. **1:** 336–355.

SIMSON, R., H. G. VAUGHAN, JR. & W. RITTER. 1976. The scalp topography of potentials associated with missing visual and auditory stimuli. Electroencephalogr. Clin. Neurophysiol. **40:** 33–42.

SIMSON, R., H. G. VAUGHAN, JR. & W. RITTER. 1977. The scalp topography of potentials in auditory and visual discrimination tasks. Electroencephalogr. Clin. Neurophysiol. **42:** 528–535.

SMITH, M. C. 1979. Contextual facilitation in a letter search task depends on how the prime is processed. J. Exp. Psychol. Hum. Percept. Perform. **5:** 239–251.

SQUIRES, N. K., K. SQUIRES & S. A. HILLYARD. 1975. Two varieties of long-latency positive waves evoked by unpredictable stimuli in man. Electroencephalogr. Clin. Neurophysiol. **38:** 387–401.

STUSS, D. T. & T. W. PICTON. 1978. Neurophysiological correlates of human concept formation. Behav. Biol. **23:** 135–162.

STUSS, D. T., F. F. SARAZIN, E. E. LEECH & T. W. PICTON. 1983. Event-related potentials during naming and mental rotation. Electroencephalogr. Clin. Neurophysiol. **56:** 133–146.

SUTTON, S., M. BRAREN, J. ZUBIN & E. R. JOHN. 1965. Evoked potential correlates of stimulus uncertainty. Science **150:** 1187–1188.

SUTTON, S., P. TUETING, J. ZUBIN & E. R. JOHN. 1967. Information delivery and the sensory evoked potential. Science **155:** 1436–1439.

TREISMAN, A. M. & G. GELADE. 1980. A feature integration theory of attention. Cognit. Psychol. **12:** 97–136.

TUETING, P. 1978. Event-related potentials, cognitive events, and information processing.

In Multidisciplinary Perspectives in Event-Related Brain Potential Research. D. Otto, Ed. 159–169. U. S. Government Printing Office. Washington, DC.

TUETING, P., S. SUTTON & J. ZUBIN. 1970. Quantitative evoked potential correlates of the probability of events. Psychophysiology **7:** 385–394.

WIJERS, A. A., L. J. OTTEN, S. FEENSTRA, G. MULDER & L. J. M. MULDER. 1989. Brain potentials during selective attention, memory search, and mental rotation. Psychophysiology **26:** 452–466.

WINKLER, I., K. REINIKAINEN & R. NÄÄTÄNEN. Event-related brain potentials reflect traces of echoic memory in humans. Submitted.

WOLDORFF, M. G., S. A. HACKLEY & S. A. HILLYARD. 1991. The effects of channel-selective attention on the mismatch negativity wave elicited by deviant tones. Psychophysiology **28:** 30–42.

WOLDORFF, M., J. C. HANSEN & S. A. HILLYARD. 1987. Evidence for effects of selective attention in the mid-latency range of the human auditory event-related potential. *In* Current Trends in Event-Related Potential Research. R. Johnson, Jr., J. W. Rohrbaugh & R. Parasuraman, Eds. 146–154. Elsevier. Amsterdam.

UHL, F., W. LANG, G. LINDINGER & L. DEECKE. 1990a. Elaborative strategies in word pair learning—DC-potential correlates of differential frontal and temporal lobe involvement. Neuropsychologia **28:** 707–717.

UHL, F., G. GOLDENBERG, W. LANG, G. LINDINGER, M. STEINER & L. DEECKE. 1990b. Cerebral correlates of imagining colours, faces and a map—II. Negative cortical DC potentials. Neuropsychologia **28:** 81–93.

Event-Related Potential Investigations of Cognitive Development and Aging[a]

DAVID FRIEDMAN

Departmentof Medical Genetics
Annex Room 308
New York State Psychiatric Institute
Box 58
722 West 168th Street
New York, New York 10032

Sam Sutton's discovery in 1965 of the P300 wave of the event-related potential (ERP) enabled two generations of cognitive scientists to pursue a quest that is still thriving today—to relate the brain's time-locked electrical activity to behavioral phenomena. Current ERP research has two primary goals: 1) to more precisely identify those cognitive events (*e.g.*, memory storage, memory retrieval, stimulus classification, and response selection) that are reflected in the electrical perturbations recorded at the scalp; and 2) to identify where in the brain those electrical phenomena are generated. Since 1965, discoveries on both fronts have been made.

As Sutton and Ruchkin (1984) have argued, we need a "dictionary" of ERP components that relates a particular mental process within the behavioral domain to a particular component within the physiological domain. Knowledge of the physiological sources of the components could help identify those areas of the brain that are dysfunctional in specific clinical syndromes and could serve as converging evidence for the kinds of cognitive activity the components reflect.

Consistent with this strategy, one possibility for understanding where in the brain ERP components originate is to study brain-injured subjects (*e.g.*, Knight, 1990). For example, if a particular ERP component is generated in frontal cortex, then one would not expect to record an ERP with normal morphology in patients with frontal lesions (see Knight, 1984 for an example). One problem with the application of the ERP technique to these kinds of problem areas is its exquisite sensitivity to task demands, instructions and subjective strategies (*cf.*, Donchin & Coles, 1988; Donchin, Ritter & McCallum, 1978). Thus, the effect that these variables have on the cognitive ERPs must be understood before meaningful interpretation of ERP findings in clinical samples can be made.

In order to achieve this goal it is necessary to employ tasks for which the underlying cognitive structure is well understood. In this manner, changes in the ERP waveform can be anchored in the behavioral domain and thus related to systematic differences in the cognitive demands of the task. Using the parallel strategy briefly described above one could then study subject groups whose cognitive skill and information processing capabilities on the task in question are known

[a] The research reviewed here was supported in part by Grants HD14959 and AG005213 from the United States Public Health Service and by the New York State Department of Mental Hygiene. The Computer Center at New York State Psychiatric Institute is supported in part by a grant (MH-30906) from the National Institute of Mental Health. Dr. Friedman is supported in part by Research Scientist Development Award #K02 MH00510.

to differ so that between-group differences in the endogenous components could be related to between-group differences in a particular cognitive function. This convergent approach has been used previously to assess changes in the ERP waveform as cognitive development unfolds (*e.g.*, Courchesne, 1978; Friedman, Sutton, Putnam, Brown & Erlenmeyer-Kimling, 1988), and as humans age (*e.g.*, Ford & Pfefferbaum, 1980; Friedman, Putnam & Hamberger, 1990; Hamberger & Friedman, 1990). It relies upon the fact that, for example, as children mature, their information processing capabilities improve (*cf.*, Bruner, Olver & Greenfield, 1966), presenting the possibility for differences between age groups to be anchored in such (specifiable) age-related changes in cognitive skill. Similarly, the increase in problems with memory as humans age (see, for example, the volume edited by Poon, 1980) provides a "natural" experiment for furthering our understanding of those ERP components that reflect processes that occur during memory storage and retrieval, as well as for obtaining new knowledge on age-related changes in memory function. Moreover, using groups of subjects with well-defined cognitive deficits should yield critical information on the cognitive functions reflected in the ERP since these deficits should be reflected in abnormal endogenous ERP components. This strategy is the basis, for example, for the application of ERP methodology to the study of schizophrenia (*e.g.*, Pfefferbaum, Ford, White & Roth, 1989) and Alzheimer's disease (*e.g.*, Friedman, Hamberger, Stern & Marder, in press).

In this paper I shall very briefly review the cognitive development and aging ERP literatures, dealing only with the "cognitive" or endogenous ERP components. This will be followed by a selective review of work from my laboratory covering a fairly large age range, that will focus primarily on ERP correlates of memory function across the life span. Before briefly reviewing this cognitive development and aging ERP literatures, it will be necessary to describe the ERP findings during tasks that tap memory function, with which this paper is primarily concerned.

ERPs during Memory Tasks

Encoding

The ERP provides complimentary data to traditional behavioral measures at the time of encoding. For example, during incidental (*e.g.*, Paller, Kutas & Mayes, 1987a) as well as intentional (*e.g.*, Friedman, 1990b) learning, it has been demonstrated that the amplitude of electrical activity elicited by to-be-remembered items predicts the accuracy of subsequent recall or recognition, suggesting that this activity may provide a measure of encoding of the to-be-remembered item. This has been termed the ERP "memory effect" (Friedman, 1990b), or "Dm" (difference in subsequent memory—Paller et al., 1987a). More information about this ERP activity is needed, since robust effects are not always obtained (*e.g.*, Friedman, 1990a; Johnson, Pfefferbaum & Kopell, 1985; see Paller, 1990 for an overview). Nevertheless, the range of parameters that modulate this activity need to be explored from a life span perspective, since this might shed light not only on age-related differences in memory function, but also on the underpinnings of the "memory effect" itself. Friedman (1990b; see also Karis, Fabiani & Donchin, 1984) has hypothesized that this activity may reflect elaboration (on which both children and older adults have been shown to be deficient—see reviews below), so that the interpretation of life span behavioral differences in memory performance is directly relevant to this hypothesis. However, aside from our studies during continuous recognition (reviewed below), there are, to my knowledge, no life span studies of the ERP "memory effect."

Retrieval

The ERP also provides complimentary information to that inferred from behavioral measures during the test phases of memory experiments. Recent behavioral studies of memory suggest that there may be two types of memory, termed "implicit" and "explicit," each possibly mediated by a unique set of brain and cognitive structures (see, for example, Squire, 1987). Implicit memory is exhibited by an increase in a subject's skill in processing a previously presented stimulus, for example, during repetition priming within a lexical decision task, while explicit memory is demonstrated by a subject's awareness that he/she encountered that stimulus in a previous episode, for example, during free recall or recognition memory testing. The two kinds of memory are thus functionally distinguished by the type of retrieval test. I use the terms implicit and explicit to identify two types of memory test and not theoretical forms of memory, thus avoiding the use of the same terms to refer to a type of test and a type of memory (Richardson-Klavehn & Bjork, 1988). Amnesic patients perform poorly on explicit memory tasks, but appear normal on implicit memory tasks (Graf, Squire & Mandler, 1984). This fact has been cited as one of the main pieces of evidence for suggesting that performance on the two types of memory test is mediated by different brain systems, although this inference is as controversial as the idea that the two types of test tap two distinct forms of memory (see, for example, Richardson-Klavehn and Bjork, 1988). Nevertheless, the dissociation of performance on these two kinds of test is important, since it may provide a means of distinguishing a variety of memory disorders and ultimately understanding their biological bases (*e.g.*, Squire, 1986).

A consistent ERP finding during explicit recognition testing and repetition priming in an implicit task is that the ERP elicited by the second presentation of an item is more positive than the ERP evoked by its first presentation (the ERP "repetition effect;" for examples of adult studies, see Bentin & Moscovitch, 1990; Fabiani, Karis & Donchin, 1986; Friedman, 1990b; Neville, Kutas, Chesney & Schmidt, 1986; Rugg & Nagy, 1989; Rugg, Furda & Lorist, 1988; for studies of children, see Berman, Friedman & Cramer, 1990; Stelmack, Saxe, Noldy-Cullum, Campbell & Armitage, 1988). The repetition effect appears to span two components, a negativity at about 400 msec (N400), larger to new items, and a subsequent positivity, P3b (*i.e.*, classical P300), smaller to new items. Since both N400 and P3b amplitudes are modulated by repetition in similar fashion (*i.e.*, to new items, respectively, greater negativity and less positivity; to old items, respectively, less negativity and greater positivity), the data suggest that an overlapping negativity (or positivity) could mediate the effect. However, it is currently unclear whether repetition modulates one, or multiple ERP processes (see Friedman, 1990b; Rugg and Nagy, 1989; Smith and Halgren, 1989; and Rugg and Doyle, in press for discussions; see also the section below dealing with aging, ERPs and memory). Moreover, explicit and implicit ERP repetition effects have not been directly compared, nor has their developmental course been charted. To my knowledge, there are only a few explicit/implicit behavioral studies in which a wide age range has been used (*e.g.*, Graf, 1990; Greenbaum & Graf, 1989; Parkin & Streete, 1988) and, other than our own studies discussed below, no ERP investigations exist.

Brief Review of ERPs and Cognitive Development

Surprisingly few ERP investigations exist that have been designed specifically from a developmental perspective. Much of our early developmental ERP knowledge was gleaned from studies of abnormal groups of subjects (*e.g.*, dyslexia, attention deficit disorder), in which nondysfunctional young children served as normal controls (*e.g.*, Buchsbaum & Wender, 1973; Ornitz, Ritvo, Panman, Lee, Carr & Walter, 1968). However, when studies were designed to answer developmental questions (*e.g.*, Courchesne, 1978; Friedman, Brown, Sutton & Putnam, 1983; Kok & Rooijakkers, 1985), the evidence indicated that there were both qualitative and quantitative age-related changes. The most consistent of the quantitative changes was the systematic variation in P3b latency from young childhood through old age (*e.g.*, Goodin, Squires, Henderson & Starr, 1978; Mullis, Holcomb, Diner & Dykman, 1985). The decrease in P3b latency from the age of 6 through 15 and then its steady increase through old age followed the well-known reaction time pattern, especially at the older ages, where slowing was a key characteristic reported in studies of performance and aging (*e.g.*, Botwinick, 1984; Cerella, 1985). Qualitative changes were more difficult to discern, although evidence from both cross-sectional (*e.g.*, Courchesne, 1978; 1983; Goodin *et al.*, 1978; Friedman, Brown, Vaughan, Cornblatt & Erlenmeyer-Kimling, 1984; Friedman *et al.*, 1988; Kok & Rooijakkers, 1985) and longitudinal (*e.g.*, Friedman, Putnam & Sutton, 1990; Kurtzburg, Vaughan & Kreutzer, 1979) studies suggest that there may be differences in cognitive processing mode between age groups.

This latter kind of evidence has usually been inferred from the differences in ERP configuration or morphology found between age groups (*cf.*, Courchesne, 1983; Friedman *et al.*, 1983; Kok & Rooijakkers, 1985; for reviews, see Friedman, 1991; and Kurtzberg, Vaughan, Courchesne, Friedman, Harter & Putnam, 1984). For example, Courchesne (1978) reported that a negative-positive complex (labelled "Nc-Pc") elicited by "novel" stimuli was present in the ERPs of young children, but was dramatically reduced in amplitude and/or absent in the ERPs of older children and young adults. Courchesne (1978) speculated that this age-related difference in waveform morphology reflected differences in mode of processing the novel stimuli by the young compared to the older children and adults and related the ERP changes to behavioral "stages" of the kind hypothesized by Bruner *et al.* (1966) and Inhelder and Piaget (1964). While the existence of such ERP-stage relationships can be debated, some data do suggest that there are ERP morphological differences between adults and children for the "cognitive" ERPs. Specifying just what such differences reflect has been and continues to be difficult.

Since cognitive development depends upon increases in memory function (Kail, 1984), the use of tasks designed to tap all aspects of memory function and their maturation seems especially appropriate for developmental ERP investigations. Thus, studies of repetition and semantic priming, as well as recognition memory, all seem like excellent candidates for developmental study from an ERP vantage point. This is a relatively new area for ERP investigation, and few studies exist. I shall review our studies designed from this perspective in a later section of this paper. First, however, a brief review of what we have learned from behavioral studies of cognitive development, priming and memory will be necessary.

Behavioral Studies of Priming and Memory during Cognitive Development

Despite the fact that young children benefit more than older children from semantic information in on-line tasks, such as lexical decision (*e.g.*, Schvanaveldt,

Ackerman & Semlear, 1977), other evidence points to less elaboration and use of context by young children to aid in later recall or recognition of stimuli presented during the on-line series. This apparent paradox is understandable, however, on the basis of Graf and Mandler's (1984) distinction between activation, an automatic process that takes place whenever a word is presented and is necessary for priming to occur, and elaboration, an effortful process conducive to the later retrieval of an event. Activation appears to be responsible for on-line word recognition during such tasks as lexical decision, whereas elaboration seems to be necessary for differences in delayed memory performance, *i.e.*, recognition and/or recall of items based on context presented during the on-line task.

During recognition memory, both behavioral components are thought to underly performance (Mandler, 1980), with recognition a function of the familiarity of the item (represented by intraevent integrative processes, *e.g.*, activation) and retrieval mechanisms (dependent upon interevent elaborative processes). Consonant with a deficit in elaborative processing, Chi (1976; see also Case, 1984) has concluded that one source of the child's decreased memory performance may be poorer encoding efficiency, and several experimenters (*e.g.*, Emmerich & Ackerman, 1979; Ghatala, Carbonari & Bobele, 1980; Perlmutter, 1980) have reported results consistent with this interpretation. Coupled with the results of studies that point to increasing benefit from semantic strategies on memory performance as age increases (Ghatala *et al.*, 1980; Kau & Winer, 1987; Melkman & Deutsch, 1977), in particular elaboration (*e.g.*, Pressley, 1982), a review of the developmental behavioral data with respect to the above-described recognition processes suggests that when orienting activity is unconstrained (*e.g.*, during intentional learning) the young child may not engage in between-item elaboration to facilitate subsequent memory performance.

The results of several behavioral investigations suggest that preschool and elementary school age children are quite capable of automatically extracting semantic information (*e.g.*, Geis & Hall, 1976; Means & Rohwer, 1976; Perlmutter, Schork & Lewis, 1982). However, it is still possible that the young child's episodic memory traces, relative to those of the adult, are more heavily weighted by sensory compared to conceptual (*i.e.*, semantic) features. Several studies have supported this idea (*e.g.*, Ackerman, 1981; Cramer, 1976; Sophian & Hagen, 1978), especially when the child's encoding operations are unconstrained (*cf.*, Emmerich & Ackerman, 1979). However, the evidence is not unequivocal (*e.g.*, Rosinski, Golinkoff & Kukish, 1975).

Negative ERP Deflections and Cognitive Development

Although preliminary and speculative, ERP evidence consistent with this latter notion has come from our laboratory (Berman *et al.*, 1990; Berman, Friedman & Cramer, 1991), as well as from the ERP data of others (*e.g.*, Noldy, Stelmack & Campbell, 1990; Robertson, Mahesan & Campbell, 1988). These data are consonant with a substantial body of behavioral and theoretical work (*cf.*, Paivio, 1986) suggesting that the underlying representations for pictures and words may be different from one another. Moreover, the data also lend support, albeit preliminary and sketchy, to the hypothesis that there are different representational systems for pictures and words that may undergo developmental change (Berman *et al.*, 1990; Robertson *et al.*, 1988).

Several investigators working with children have recorded negative components within the 300–500 msec latency range (*e.g.*, Courchesne, 1978; Friedman *et al.*, 1988) that are similar in latency and scalp distribution to the N400 originally reported by Kutas and Hillyard (1980). The young adult data suggest that N400 indexes the degree of semantic relatedness between stimuli, since its amplitude varies inversely with the degree of semantic association between a prime and its target (Kutas and Hillyard, 1989). Using pictorial stimuli, we (Friedman, Putnam, Ritter, Berman & Hamberger, in press) recently demonstrated that the amplitude of a negativity at 400 msec reflected the degree of semantic and/or physical relationship between two sequentially presented pictures of common objects for all age groups between 6 and young adult. Since this occurred even though subjects were not specifically instructed to assess the degree of relationship, the data were taken as evidence that by the age of 6 semantic and/or physical relationships among stimuli can be automatically extracted. Moreover, these data suggest that, in similar fashion to the N400s elicited by verbal stimuli, pictures also elicit a negativity that appears to be responsive to the semantic relationships between pictorial concepts.

These behavioral and ERP studies briefly reviewed in the last two sections, led us to embark upon a series of investigations of memory function and cognitive development. The tasks we designed were modifications (for ERP recording) of the continuous recognition memory paradigm originally described by Shepherd and Teghtsoonian (1967), in which the number of intervening items (*i.e.*, lag) between first and second presentations of an item could be manipulated. This allowed us to assess the effect of lag on the size of the ERP (and reaction time) repetition effect during an explicit memory task. We also designed a version of this task that allowed us to study the ERP repetition effect during implicit task conditions.

Developmental Studies of ERPs during Continuous Recognition Memory

These studies were performed in collaboration with Dr. Steven Berman of my laboratory. The experiments were designed to assess the modulation of ERP repetition effects by explicit and implicit task instructions using pictorial and lexical equivalents of the same concepts.

The subjects of these investigations were children (N = 14; age range = 7–10; mean = 9.2); adolescents (N = 13; range = 14–16; mean = 14.8); and young adults (N = 14; range = 20–30; mean = 24.5). All subjects had IQs within the normal range and all children were reading at the 50th percentile or better for their grade level. In each block of continuous recognition trials, one-third of the items were new and were never repeated (*i.e.*, these were control items), one-third were new and were subsequently repeated and, thus, one-third were old. During these explicit memory blocks, subjects were asked to identify all old (seen previously in the block) and new (not seen previously) items via speeded, choice button press responses. Pictures and words were presented via MacIntosh Plus microcomputer (300 msec duration; 2-sec ISI) in separate blocks (total of 8 blocks; 85 stimuli per block), and old items followed their first presentation counterparts after either 2, 8 or 32 intervening items (*i.e.*, lags) following their initial presentation (equiprobable). In order to assess ERP repetition effects during implicit task instructions, we administered 2 additional blocks of 85 trials in which items were repeated (separate picture and word blocks with order of these counterbalanced across subjects). The implicit memory task was designed to unconfound the effect of button pressing from the effect of repetition on the ERPs by using an orienting question for both pictures and words that yielded a large proportion of negative responses (*e.g.*, identify "things that you can read;" *e.g.*, book, magazine). Thus, for the implicit blocks, subjects generated reaction times (RT) in response to the instructed items, but these were never any of the items that repeated.

The behavioral results indicated that, across age groups, more accurate and faster responding occurred to new than to old items. There were developmental effects on the behavioral measures: RT was slowest in children, faster in adolescents and fastest in young adults, with a similar increment in percent correct as a function of age. For all three age groups, picture RTs and accuracy were respectively faster and more accurate than word RTs and accuracy.

FIGURE 1 depicts the ERPs elicited by old and new pictures and words for explicit and implicit instructions at the midline electrode sites. Because I am not certain that the negativity (peaking between about 350 and 400 msec) preceding the large-amplitude positivity (*i.e.*, classical P300 or P3b) in FIGURE 1 is synonymous with the N400 recorded by Kutas and Hillyard (1980), I shall refer to this deflection as Neg400 (see also Friedman *et al.*, in press). The ERP results indicated that, across words and pictures, new items were associated with greater Neg400s and smaller P3bs than old items, replicating previous findings (*e.g.*, Friedman, 1990b; Neville *et al.*, 1986; Rugg & Nagy, 1989). Slow wave, which was maximal at the anterior electrodes, and may be synonymous with "frontal positive slow wave," was larger to new than to old items for all age groups. Ruchkin and Sutton (1983) have suggested that this activity reflects "further processing." Such processing would be required of new items, in order for them to be encoded and stored for subsequent retrieval. The fact that all age groups showed this relationship for slow wave suggests that this activity may be functionally homologous across this age range.

Although all age groups showed larger frontal positive slow wave when elicited by new compared to old items, when the ERPs elicited by new items were decomposed into those ERPs elicited by new items that were subsequently recognized and those new items that were subsequently unrecognized there was an Age by Subsequently Recognized interaction for P3b, and a marginally significant interaction for positive slow wave. The adolescents and adults showed the expected effect (subsequently recognized > subsequently unrecognized for P3b), while the adults showed it for both P3b and positive slow wave. In contrast, the children showed the opposite effect.

As can be seen in the explicit panel of FIGURE 1, pictures produced larger Neg400s than words at the anterior electrodes, but this was modulated by a Picture/Word by Age interaction, indicating that this effect was larger for children than for the adolescents or adults. Since the children produced much larger amplitude ERPs than the adolescents or adults, it was necessary to normalize the data (*cf.*, McCarthy & Wood, 1985). However, the interaction remained significant after scaling. The Age by Picture/Word interaction was also significant during the implicit blocks, and survived scaling, suggesting developmental differences in the processing of words and pictures that are independent of whether the instructions are implicit or explicit.

Although preliminary and difficult to understand at this stage of our knowledge, the subsequent memory data suggest that, although the children appear to be encoding the items for subsequent recognition (as evidenced by new > old for slow wave), the kind of processing they are performing may be different than that of the adolescents and adults. Evidence concerning the ERP "memory effect" in adults suggests that semantic orienting tasks elicit larger amplitudes than orthographic tasks (*cf.*, Paller *et al.*, 1987a). However, even orthographic

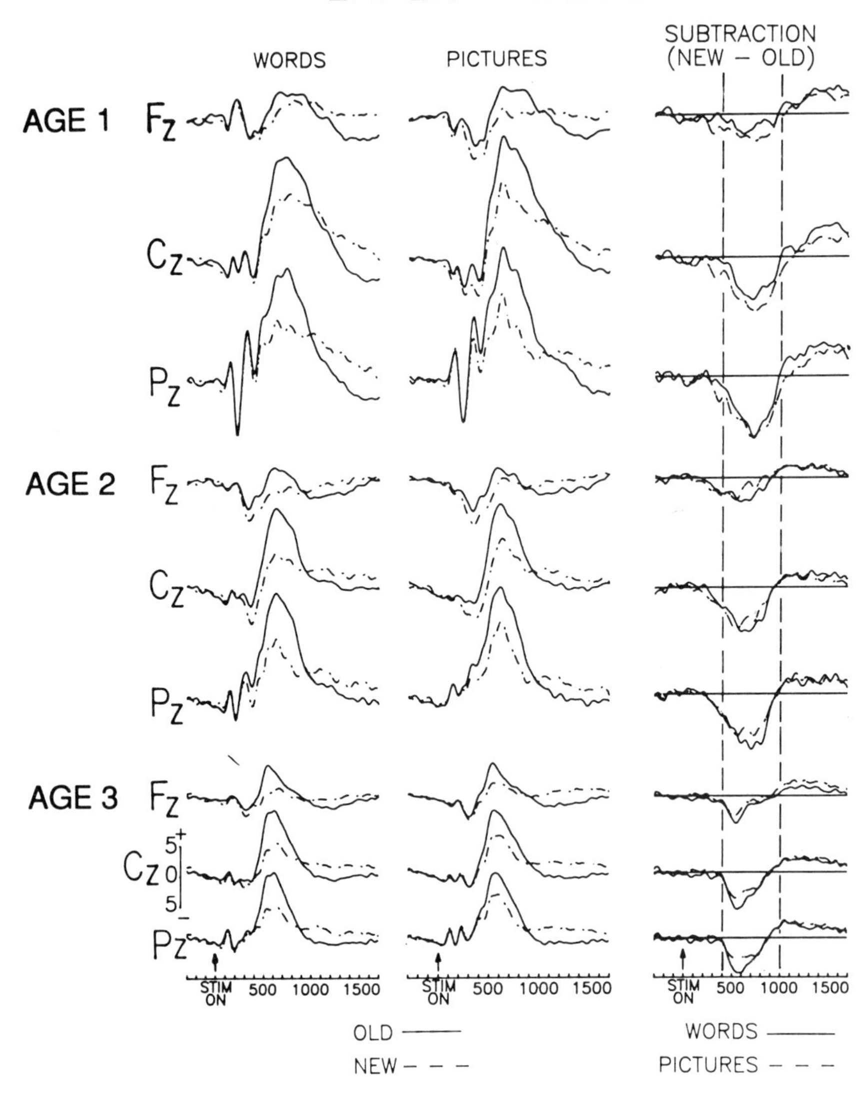

FIGURE 1. *Left*. Grand mean ERPs averaged across subjects within each age group elicited by first presentation items (new), and by these same items presented for the second time (old) under explicit memory instructions. *Right*. Grand mean ERPs averaged across subjects within each age group elicited by first presentation items (new), and by these same items presented for the second time (old) under implicit memory instructions. The data are from Berman *et al.* (1990). *Arrows* mark stimulus onset, with time lines every 100 msec. Age 1 = children; Age 2 = adolescents; Age 3 = young adults. (From Berman *et al.*, 1990. Reprinted by permission from the *International Journal of Psychophysiology*.)

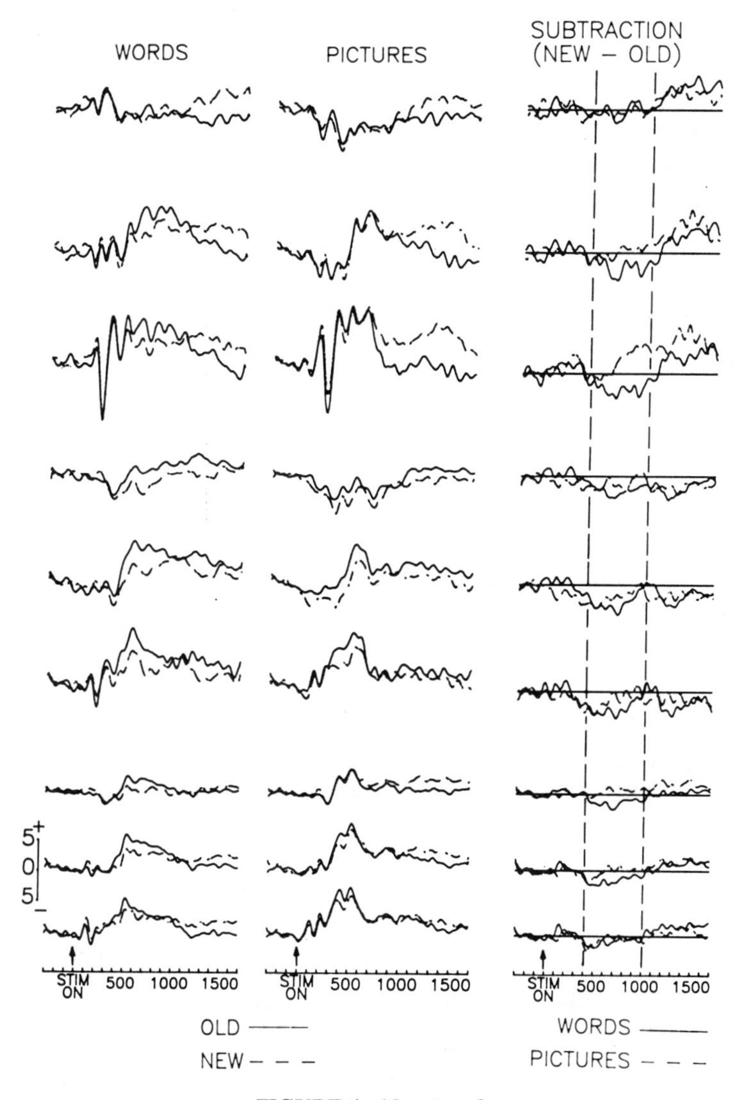

FIGURE 1. (*Continued*)

orienting instructions elicited some "Dm," suggesting that other kinds of elaboration may modulate the amplitude of this electrical activity. As stated previously, one hypothesis for the occurrence of the ERP "memory effect" is that it reflects elaborative processes that may facilitate subsequent recall and/or recognition (Friedman, 1990b; Karis *et al.*, 1984). In the data of Karis *et al.*

(1984) for subjects who used "elaborative strategies," the ERP "memory effect" was reflected in a "frontal positive slow wave." Similarly, in Friedman's data (1990b), the "memory effect" was synonymous with a frontal positive slow wave. Since children are known to increase their use of elaborative strategies as they mature (e.g., Pressley, 1982), one speculative interpretation is that the reversal of the "memory effect" for the children is due to the young child's inability to spontaneously utilize this kind of rehearsal strategy. However, more definitive evidence for this hypothesis awaits a better understanding of the "Dm" effect as well as experiments in which type of encoding is manipulated in a developmental setting.

The larger Neg400s to pictures than words in children compared to adolescents and adults (see the unsubtracted ERPs at Fz in the explicit panel of FIG. 1) is similar to recent findings by Noldy *et al.* (1990), who found an Fz N400 to be larger to words than pictures for adults, whereas in another study by this same group (Robertson *et al.*, 1988), a frontal N400 was larger to objects than words in children, but not adults. The larger negativity to pictures than words in children in the current study could be due to the less automatic association between the picture and its name in this age group. Alternatively, this finding might be explained on the basis alluded to above (Behavioral Studies of Priming and Memory) that children tend to encode pictures schematically, employing sensory features, whereas adults tend towards conceptual or semantic encoding (*cf.*, Ackerman, 1981). Since pictures have very distinct sensory features, whereas words do not (Nelson *et al.*, 1977), this would have been consonant with the child's inherent tendency, possibly leading to larger Neg400s to pictures than words in children. For the adults, whose tendency is toward conceptual encoding, the picture's sensory features would have been less salient leading to smaller Neg400 differences for adolescents and adults (in fact, note the *smaller* picture than word difference means of the explicit panel of FIG. 1 for the two older groups compared to children). Moreover, one question that arises is to what extent these word/picture differences result from using modality-homogeneous blocks, possibly permitting independent strategies for each surface form. In planned experiments, words and pictures will be intermixed, thus allowing us to comment on whether differences in performance and ERP measures between pictures and words are a function of differential encoding of individual items, as opposed to differential preparation for modality-specified blocks.

In showing an effect of repetition of a prior event on the ERP without requiring an overt response of previous occurrence (during the implicit condition—see the implicit panel of FIG. 1), these data represent the first demonstration of "implicit" effects on ERPs recorded in a developmental study. The data suggest that for Neg400 and P3b, the effect of implicit and explicit instructions are relatively invariant across the age range tested.

To summarize, the results of these studies suggest that the major differences between children and adults lies in encoding of items for subsequent retrieval from memory, since the age-related effect on slow wave activity occurred only during intentional (*i.e.*, explicit) learning conditions. In contrast, during both explicit and implicit blocks, the new/old repetition effect (comprising Neg400 and P3b) was highly similar in all age groups.

Our studies of continuous recognition and cognitive development were followed by investigations using similar tasks designed to assess ERP and behavioral changes during the adult years, thus allowing us to obtain a life-span perspective on the cognitive ERPs and memory function. Before reviewing the resultant data, I turn first to brief reviews of behavioral and ERP studies of aging and memory.

Behavioral Studies of the Effects of Aging on Memory Function

It is, by now, axiomatic that older adults show poorer performance during explicit memory tests compared to younger adult controls (for a recent review see Light, 1991). It has been more difficult to specify, in information processing terms, the locus of the memory performance difference (Salthouse, 1980; 1982). However, most evidence suggests that older adults perform as well as younger controls in tasks which do not require the conscious recollection of previously presented material, *i.e.*, on implicit memory paradigms such as repetition priming, performance on which is assumed to be mediated via relatively automatic processes, such as activation (Burke, White & Diaz, 1987; Howard, 1983; but see Rose, Yesavage, Hill & Bower, 1986 and Howard, Shaw & Heisey, 1986a).

Despite the fact that the majority of evidence suggests that older relative to younger adults show equivalent priming facilitation for a variety of stimulus types, the older adult appears to be at a disadvantage when conscious recollection of these same stimuli is required, whether the material is well- or newly-learned associations (*e.g.*, Howard, Heisey & Shaw, 1986b; Mitchell & Perlmutter, 1986; Light & Singh, 1987; Light, Singh & Capps, 1986). The results of some studies suggest that the elderly are deficient in contextual encoding and the "elaboration" of processing for storage of items in and retrieval of items from memory (see, for examples, Botwinick, 1984), deficits similar to those proposed to account for deficiencies in the memory function of young children (see above). Since lack of this elaborative strategy might be expected to result in less "elaborated" memory traces for the elderly, this could be one explanation of the memorial performance differences (favoring young adults) so often reported in the behavioral literature on aging.

ERPs, Aging and Memory

Several studies of the cognitive ERPs in elderly subjects have been published (*e.g.*, Bashore, Osman & Heffley, 1989; Ford, Pfefferbaum, Tinklenberg & Kopell, 1982; Picton, Stuss, Champagne & Nelson, 1984; Pfefferbaum, Ford, Wenegrat, Roth & Kopell, 1984). The main findings are: P3b latencies are longer in the elderly relative to younger adults (*e.g.*, Brown *et al.*, 1983; Goodin *et al.*, 1978a; Picton *et al.*, 1984; Pfefferbaum *et al.*, 1984), consistent with the behavioral literature on age changes in the speed of information processing (for examples, see Botwinick, 1984). P3b amplitude for the elderly is reduced, and its scalp distribution is more equipotential across a midline electrode montage, compared to younger adult controls (*e.g.*, Beck, Swanson & Dustman, 1980; Friedman, Putnam & Sutton, 1989; Mullis *et al.*, 1985; Pfefferbaum *et al.*, 1984; Picton *et al.*, 1984; Strayer, Wickens & Braune, 1986). The significance of this topographic shift is not yet understood; in the tasks studied to date it appears unrelated to changes in performance (*e.g.*, Pfefferbaum *et al.*, 1984; Strayer *et al.*, 1986), although we (Friedman, Berman & Hamberger, in preparation) have some preliminary evidence suggesting that the P3b shift is correlated with memory performance.

For verbal stimuli, Waugh and Norman (1965) have distinguished between two memory stores, primary (or immediate) memory, as typically assessed by digit span and characterized by limited capacity and fast decay, and secondary memory (the repository of newly learned information), characterized as a more stable memory store with a slow decay. According to the model, items enter the temporary or primary store as they are presented and are then transferred to the more permanent stores or are lost over time. Studies designed specifically to investigate memory and ERPs in elderly subjects have assessed primary memory using the Sternberg memory scanning paradigm (*e.g.*, Ford *et al.*, 1982).

With the exception of the study by Marsh (1975), the remaining studies of primary memory by Ford and Pfefferbaum and their colleagues (Ford, Hopkins, Pfefferbaum & Kopell, 1979; Ford *et al.*, 1982a; Pfefferbaum, Ford, Roth & Kopell, 1980a; see review by Ford & Pfefferbaum, 1985) have been consistent in showing longer mean reaction times and steeper reaction time slopes in the elderly compared to the young, indicating longer scan times per item in the Sternberg memory paradigm. Although as previously stated, P3b latencies are consistently longer in the elderly, the slope of P3b latency and memory set size does not differ between the elderly and the younger adult, suggesting that response processing (*i.e.*, organization and execution of the motor response) might account for the between-age group reaction time difference. Other than our own studies to be described below, there are (to my knowledge) no studies of normal aging in which ERPs have been recorded during tasks which tap "secondary" memory (*i.e.*, newly learned information). Older adults are known to be more impaired on recall or recognition of material from secondary compared to primary memory (*e.g.*, Poon & Fozard, 1980). Further, to my knowledge, only two teams investigating ERPs and aging (Harbin, Marsh & Harvey, 1984; Friedman *et al.*, 1989) have used tasks designed to elicit N400, a component implicated in tasks which tap memory function. Thus, our studies of normal aging (described below), were designed to elicit this negativity, as well as other scalp electrical phenomena during tasks specifically constructed to assess memory function in the normally aging adult.

ERP Studies of Memory in Normal Young, Middle-Aged and Older Adults

In similar fashion to our studies of cognitive development described in the initial sections of this paper, we began to pursue the question, using ERP techniques, of where in the sequence of memory-related events, the older adult's difficulty might lie. We also employed the continuous recognition paradigm as the explicit task, as well as an implicit task designed to allow a more direct comparison of repetition effects during explicit and implicit conditions than was possible using the design described in the section on cognitive development above.

Our experiments were motivated by two general findings in the cognitive aging behavioral literature: 1) that older adults perform worse in the recall of material intentionaly learned compared to material incidentally learned, suggesting that encoding difficulties are responsible for the performance differential (*e.g.*, Perlmutter, 1978); and 2) that older adults are thought to have poorer recall of memorized material because they do not initially process the material to "deep" enough (*e.g.*, semantic) levels (see, for example, Botwinick, 1984; Craik & Simon, 1980; and Salthouse, 1982 for reviews).

Participants in this study were females, all of whom had IQs and modified Mini-Mental-State (MMS; Mayeux, Stern, Rosen & Leventhal, 1981) scores within the normal range. In addition, all older adults (65 and older) were free of dementia, depression and limitation in the activities of daily living as assessed by the Short

CARE (Gurland, Golden, Teresi & Challop, 1984), which is a structured interview, standardized on large samples of community dwelling older individuals. Subjects were recruited into 3 age groups: young (N = 18; mean age = 24.94); middle-aged (N = 15; 48.86); and Older (N = 18; 70.11) adults.

Each subject participated in two sessions on the same day, one in the morning (referred to as AM), the other after lunch (referred to as PM). In the morning session, subjects were presented with 8 blocks of word trials (108 stimuli/block; randomized separately for each subject). Four blocks were continuous recognition (*i.e.*, intentional learning), and 4 were "elaboration of processing" blocks (incidental learning). For 2 of the elaboration blocks, subjects were asked to identify "animal" and "non-animal" words (*i.e.*, a semantic classification, a relatively "deep" level of processing) and, in the other 2 blocks of trials, words which were presented in upper case letters versus those presented in lower case letters (an orthographic task, a relatively "shallow" level of processing). The ability of subjects of different ages to perform in the continuous recognition and elaboration tasks was assessed in the morning, but the comparison of the two levels of elaboration (semantic versus orthographic-incidentally learned) and continuous recognition (intentional) as they affected recognition memory, were assessed on performance and ERPs in the afternoon session. For the afternoon, a subset of the items presented in the morning (from both the continuous recognition and elaboration of processing blocks) were presented for recognition, with an equal number of foil or new items. Only the AM session ERP data will be presented here, although the reaction time indices from the PM session will be described.

The design of the continuous recognition paradigm was highly similar to that detailed above for our studies of cognitive development, with the exception that only words were used as stimuli. Two-thirds of the items were new (half repeated and half did not), and, thus, one-third were old. The "elaboration of processing" blocks were constructed in identical manner to those for continuous recognition. For each condition (continuous recognition, semantic, orthographic), subjects were required to make a choice, speeded button press reaction time response (orthographic = upper case versus lower case words; semantic = animal versus non-animal words; continuous recognition = old versus new words). The words (300 msec on time; 2-sec inter-stimulus-interval) were presented on a video monitor. Word stimuli were taken from the 925 nouns for which Paivio, Yuille, and Madigan (1968) had originally collected norms. All word characteristics that could have affected their memorability were balanced across conditions of the experiment. The design of the AM portion of the experiment enabled the comparison of the old/new ERP repetition effect under explicit and implicit conditions. For the former, the subject was required to consciously identify old and new words; for the latter, the subject's task was incidental to the repetition of items within the stimulus series, and no explicit old/new judgement was required.

The continuous recognition data (Friedman & Hamberger, 1989; Friedman *et al.*, in preparation) and the elaboration data (Hamberger & Friedman, in revision) recorded in the AM session, and the PM data were obtained in collaboration with Drs. Steven Berman and Marla Hamberger of my laboratory.

FIGURE 2 presents the ERP data elicited by first and second presentations of the same item in the AM session. The depicted data were recorded from the Pz scalp site from the three age groups during continuous recognition and elaboration blocks. TABLE 1 presents the reaction time data corresponding to the ERP data depicted in FIGURE 2. The table also presents the RTs elicited by control items that were only presented once in the AM ("not repeated" in TABLE 1). Using these items allowed a comparison of the effect on RT of two- versus three-word

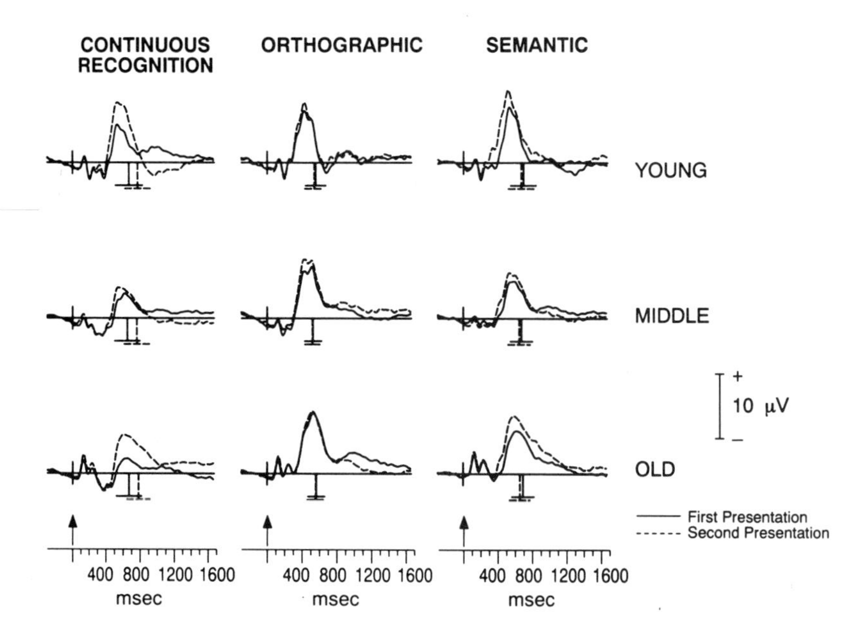

FIGURE 2. Grand mean ERPs averaged across subjects within each age group during the AM session. The data depict the effect of repetition during continuous recognition and elaboration of processing (semantic and orthographic) blocks. *Arrows* and *hash marks* indicate stimulus onset, with time lines every 100 msec. *Vertical bars* indicate mean reaction time, and *horizontal bars* indicate the mean within-subject standard deviation of reaction time.

presentations when the second (for those items not repeated in the AM) and third (for those items repeated in the AM) presentation occurred in the PM. The data in TABLE 1 are collapsed across age groups, since there were no significant age-related differences for RT, nor were there any interactions with Age for the variables manipulated during the AM and PM phases. The effects on RT of repetition for each age group in the AM session can also be seen in FIGURE 2.

As can be seen in FIGURE 2, for all age groups for both semantic and continuous recognition conditions, repetition (*i.e.*, second presentation) in the AM led to greater positivity relative to the first presentation, spanning negative-going activity

TABLE 1. Reaction Time in AM and PM Sessions

	AM Session Presentation Number			PM Session Number of Times Seen		
AM Condition	Not Repeated	First	Second	Foil	Two	Three
Semantic	695.17	698.23	671.72	1080.00	968.38	952.87
Orthographic	560.58	561.31	554.74		1075.53	1062.71
ConRec	677.84	676.98	780.25		1075.84	1019.94
Grand Mean					1039.92	1011.84

at about 400 msec and a subsequent positivity with a peak latency of approximately 600 msec. This is the well-known "ERP repetition effect" (for other examples, see Bentin & Moscovitch, 1990; Rugg *et al.*, 1988; Rugg & Nagy, 1989; Smith & Halgren, 1989). In contrast to the semantic and continuous recognition data, the repetition effect does not appear to be present or is much less marked for the ERPs elicited during orthographic blocks. Note, however, that for the oldest age group, the ERP repetition effect elicited during the AM appears to be prolonged relative to the younger age groups.

Another aspect of these data is of interest. Note that, although greater positivity is associated with the repeated AM item for elaboration as well as continuous recognition blocks (but much smaller for orthographic blocks, as indicated by a Condition by Repetition interaction), as can be seen in FIGURE 2 as well as in TABLE 1, repetition produces RT prolongation during continuous recognition, but facilitation during elaboration blocks (also much less marked for orthographic blocks, as indicated by a Condition by Repetition interaction for RT). Highly similar behavioral findings have been reported by Moscovitch (1982), who repeated items after lags of 0, 7 or 29 intervening items following their initial presentation during lexical decision and continuous recognition memory tasks. Although we did not use lexical decision to assess implicit effects, we also manipulated lag (2, 8 or 32 intervening items) during our elaboration of processing and continuous recognition blocks. Similar to Moscovitch's (1982) findings, lag did not affect reaction time during elaboration blocks, but had highly significant effects (*i.e.*, RT increased as lag increased) during continuous recognition memory blocks (Friedman, 1990b; Friedman *et al.*, in preparation). Thus, when conscious recollection is required (*i.e.*, explicit memory), repetition produces RT prolongation and lag has a dramatic effect on RT, while during an implicit test, repetition produces RT facilitation and lag has little effect on RT. Moreover, lag did not affect ERP parameters during the elaboration of processing blocks (Hamberger & Friedman, in revision), but there was a trend for it to affect ERP parameters during continuous recognition (Friedman, 1990b; Friedman *et al.*, in preparation). Thus, both the ERP and RT data suggest a dissociation between implicit and explicit tests of retention, supporting a larger array of previous findings (for a review see Richardson-Klaven & Bjork, 1988).

Although the ERPs recorded from the middle-aged and oldest age groups appear smaller with prolonged P3b latencies, the effect of repetition and condition (*i.e.*, continuous recognition, semantic, orthographic) on ERP amplitude and RT in the AM and on RT in the PM were remarkably similar for all age groups. These visual impressions were supported by a lack of Age Group by Condition or Repetition interactions either for the ERPs or the behavioral data, whether in AM or PM.

Repetition priming during these incidental conditions (*i.e.*, orthographic and semantic blocks) can be considered an implicit task (*e.g.*, Graf & Mandler, 1984), since the subject is not required to overtly indicate the item's previous occurrence, and performance enhancement to the second presentation (*i.e.*, decreased RT) demonstrates the unintentional effect of previous experience. Moreover, since both the ERP and RT differentiated first from second presentation primarily during semantic blocks, word presentation may not automatically induce access to word meaning if it is not required for task performance, suggesting that the word repetition effect is dependent upon the level of processing required by the task (see also Rugg *et al.*, 1988). The data suggest that repetition effects are similar in all age groups, both from a behavioral and physiological perspective. Further, to the extent that the ERP repetition effect reflects implicit retention, the data suggest that there are few age-related differences when memory is tested in this fashion.

However, there may be differences in the duration of the repetition effect, as suggested by the data in FIGURE 2. To assess this, we (Hamberger & Friedman, 1990) computed difference waveforms (new-old) and calculated sequential measures of 40 msec of averaged ERP activity (relative to pre-stimulus baseline). *T* tests were then computed to determine if these new-old difference scores differed reliably from zero. These appear in FIGURE 3 for the semantic data at the Cz and Pz electrode sites for ERP activity beginning at 300 msec post-stimulus and ending at 1000 msec post-stimulus. A *t* value below the shaded area indicates that the new ERP was significantly more negative than the old ERP. As can be seen, the onset of the repetition effect does not appear to differ as a function of age. However, compared to the younger age groups, the oldest age group shows a longer duration effect that lasts until about 1000 msec post-stimulus.

Most theories of semantic organization identify access to word meaning and semantic priming as automatic processes. It has also been suggested that automatic processes, unlike more effortful processes such as rehearsal, are not susceptible to disruption by the aging process (*e.g.*, Hasher & Zacks, 1979). To the extent that the ERP repetition effect during these semantic tasks reflects some of the processes involved in activation, these data suggest that the processes underlying such activation may change with increasing age (see Hamberger & Friedman, 1990 for more complete details). This change may take the form of a decrement in the speed of activation, supporting recent behavioral findings (*cf.*, Howard *et al.*, 1986a; but, see Burke *et al.*, 1987 for contrary findings). Salthouse (1988) has suggested that slower speed of activation might result in the activation of fewer nodes in semantic memory, resulting in less "elaborated" memory traces for the elderly. Thus, this might be one mechanism accounting for the conclusion (previously discussed) that the elderly do not spontaneously employ "elaborative" strategies when storing items in memory for later retrieval.

ERP Studies of Repetition Priming during Abnormal Aging

One problem with the data described above is that we required a task-relevant decision (indicated via a reaction time response) on each trial, which undoubtedly led to large-amplitude P3bs and, therefore, the possibility that the true nature of the repetition effect was masked by this overlapping positivity. Moreover, because during semantic blocks RT was shorter to new than old items and P3b latency also tended to peak earlier to old items, the difference waveforms (on which FIG. 3 is based) cannot be unambiguously interpreted as only reflecting the onset and duration of a negativity that is larger to new items. We (Friedman *et al.*, in press), therefore, decided to replicate this experiment again using semantic and orthographic conditions, employing the methodology described by Rugg (1987). Instead of requiring a task-relevant decision to every item, we required RTs to low-probability targets (which did not repeat), while frequently occurring non-targets (which did not require a response) were repeated. P3b is dramatically smaller to high probability events (*e.g.*, Duncan-Johnson & Donchin, 1982). The use of high probability non-targets that did not require a task-relevant decision should have dramatically reduced or eliminated the presence of overlapping P3b activity, thus dissociating the repetition effect from P3b. We again compared young and older controls, and also added a sample of patients with probable Alzheimer's disease (PAD), since there is some question in the behavioral literature as to whether patients with presumed Alzheimer's disease show evidence of implicit retention (*e.g.*, Heindel, Salmon & Butters, 1990; Salmon, Shimamura, Butters & Smith, 1988).

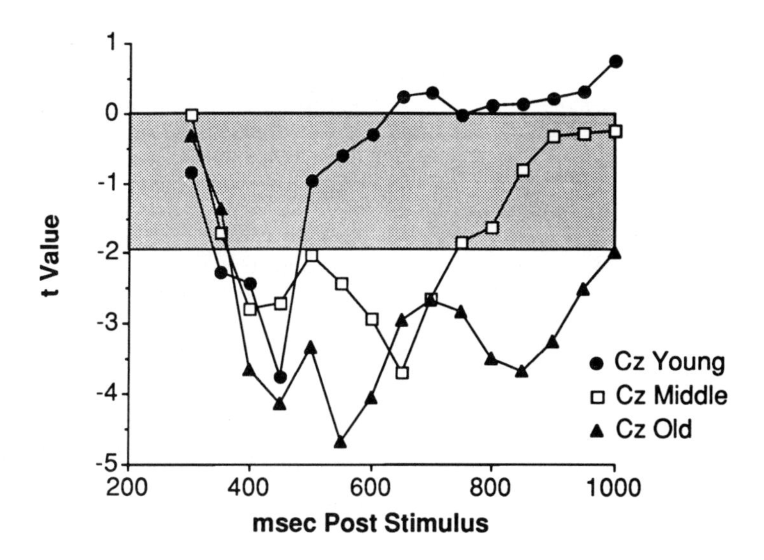

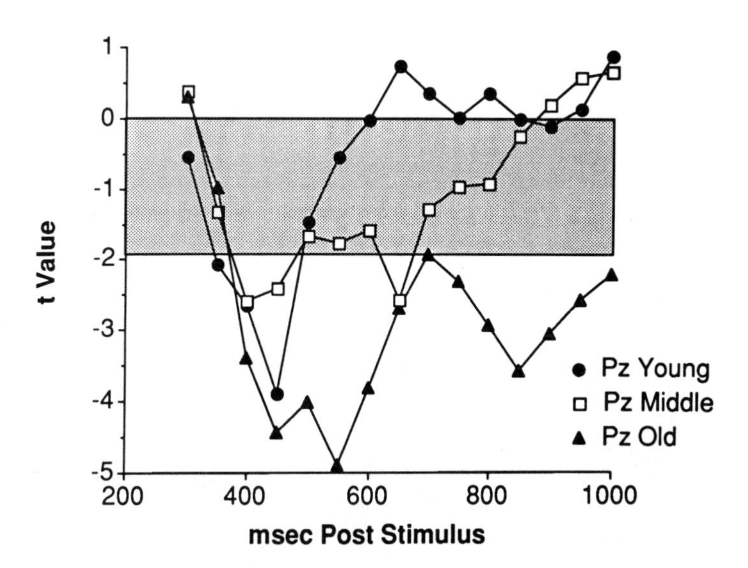

FIGURE 3. Results of t tests computed on first minus second presentation subtraction waveforms (depicted in FIGURE 2) for the semantic condition at the Cz (*above*) and Pz (*below*) electrode sites during elaboration of processing blocks. Negative values below the *shaded area* are significant at the 0.05 level or better.

For this study, patients with PAD were diagnosed using NINCDS-ADRDA criteria (McKann, Drachman, Folstein, Katzman, Price & Stadian, 1984), and were required to have had the illness for between 6 months and three years at study entrance. For the 10 patients all test results (*e.g.*, computed tomography, serum B_{12}), except for the Modified Mini-Mental State examination (MMS—Mayeux *et al.*, 1981; max = 57) were required to be within normal limits. Young (mean age of 24.1) and older (mean age of 69.8) controls (age-matched to the PAD patients—mean age of 70.6) had IQs (greater than 90) and MMS scores within the normal range (50 or better; young = 56.6; older = 54.4). The mean MMS score for the PAD sample was 38.2. Older control subjects were free of dementia, and depression, and were not limited in the activities of daily living as assessed by the Short CARE.

Four blocks each of semantic (targets = animal words) and orthographic (targets = words printed in upper case letters) conditions were administered (order of presentation counterbalanced across subjects within each group). In each block there were 18 targets to which the subject made a speeded reaction time response, and 72 non-targets (either non-animal or lower case words), which did not require a response. Only non-target words in each condition were repeated (average lag of 14 items between first and second presentations). Of the 72 non-target words, 24 were seen only once, and 24 were seen twice. Following the 8 blocks, an unexpected free recall test was administered.

The grand average ERPs at midline and O1 and O2 sites for all three groups are depicted in FIGURE 4. This figure depicts the results from the semantic blocks only, as repetition effects were not reliable for any group during the orthographic task. FIGURE 4 shows that both young and older control subjects display large-amplitude repetition effects during semantic blocks, *i.e.*, greater positivity to the second (old) relative to the first (new) presentation of the item, onsetting at about 250 msec and lasting until approximately 1000 msec post-stimulus (depending upon the group). The grand average ERPs of the PAD subjects (rightmost column of FIG. 4), although in the same general direction as the controls, display markedly smaller repetition effects. These visual impressions were supported by the statistical analyses.

Similar to FIGURE 3, FIGURE 5 depicts the t values computed on the new-old difference waveforms for each of 17, 100-msec post-stimulus periods of averaged ERP activity. A t value below the shaded area indicates a significant difference (from zero), in which the new ERP was more negative than its old counterpart. As can be seen, there are many fewer significant differences for the PAD group than for either the young or older controls. Moreover, as in FIGURE 3, these data also suggest a difference between young and older controls in the duration of the repetition effect (longer in older than younger controls) but, in addition, suggest a difference in onset (longer in older controls). Since RT was not a factor here, these data suggest that the difference in onset between older and younger subjects may have been masked in our previous study (FIG. 3).

FIGURE 6 depicts the distributions of the midline electrode summary measures of averaged ERP activity (from 300–600 and 700–1000 msec post-stimulus) for each of the 30 participants. As can be seen for the 300–600-msec measure, many of the PAD patients are clustered around zero, indicating little difference between old and new ERPs. For the 700–1000-msec index, 6 of the 10 patients display a clear ERP repetition effect, whereas the new-old difference was close to zero or reversed (*i.e.*, new more positive than old) for the remaining patients. By contrast, for both young and older controls, the majority of subjects in each group showed unequivocal repetition effects.

Although the PAD group produced longer reaction times and poorer accuracy measures than the young or older controls, they showed highly similar between-condition effects as compared to the young and older controls (*i.e.*, for all three groups, RTs were longer in semantic than orthographic blocks). These data suggest, therefore, that the processing of PAD subjects can be directed by the appropriate encoding task. The recall data were as expected: the patients were markedly impaired (mean of 1.8 items correctly recalled), and the young (20.9) recalled more items than the older (16.8) controls (this difference was not significant).

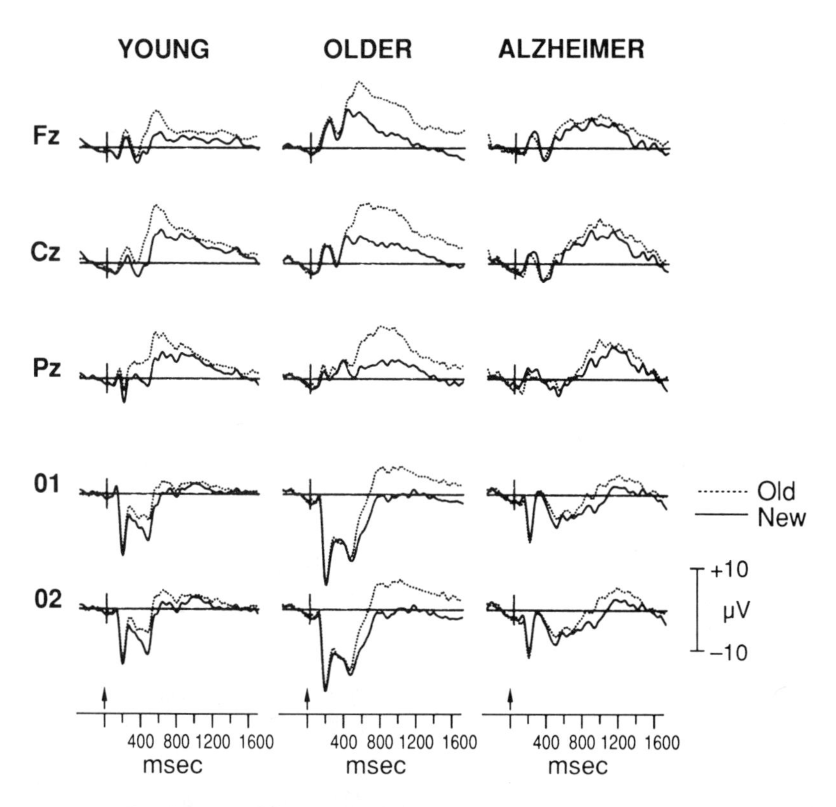

FIGURE 4. Grand mean ERPs, averaged across subjects within each group, elicited by first (new) and second (old) presentations of words seen during the semantic condition. *Vertical hash marks* and *arrows* indicate stimulus onset, with time lines every 200 msec. (Adapted from Friedman *et al.*, in press.)

These data suggest (again, to the extent that the ERP repetition effect reflects implicit retention) that this kind of retention is intact in at least some PAD patients in the early stages of the disease. However, it is not possible to know, with certainty, whether the repetition effect recorded here reflects only implicit memory. For example, several of our subjects on post-experimental inquiry, mentioned that they had noticed some of the repetitions. Thus, explicit recognition may also have occurred.

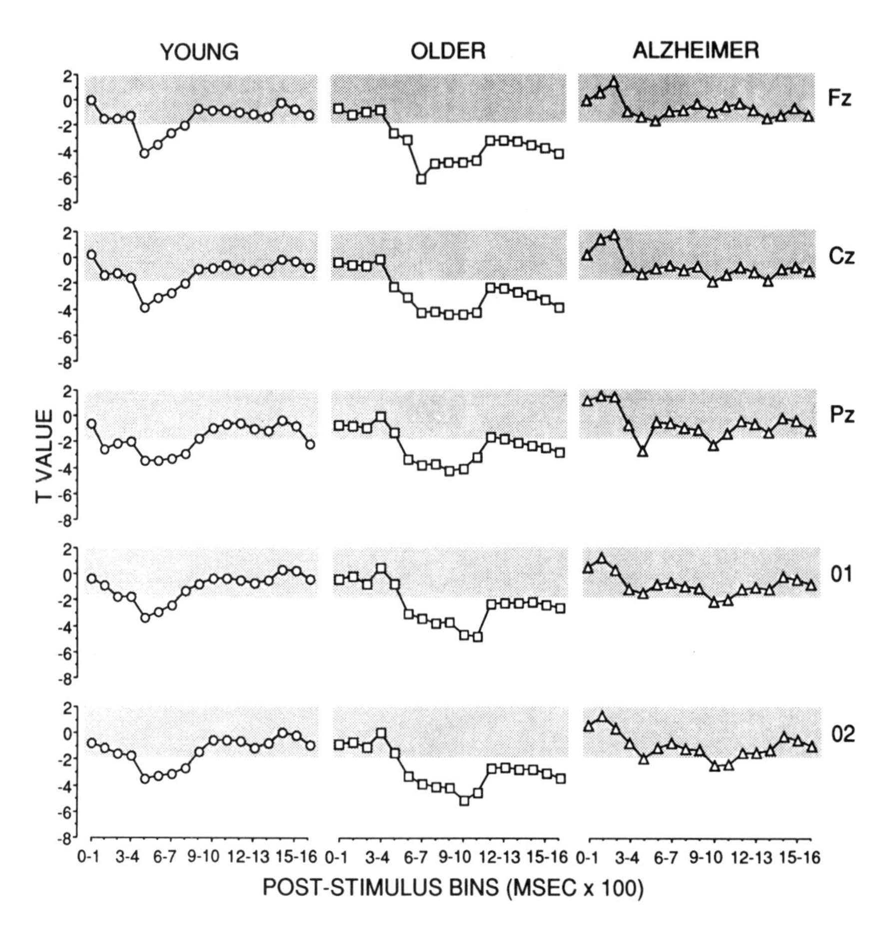

FIGURE 5. *T* values computed on the subtraction of new-old semantic ERPs depicted in FIGURE 4 for each of the 17 100-msec post-stimulus averaged voltages. Negative *t* values exceeding the *shaded areas* are significant at $p < 0.05$. (From Friedman *et al.*, in press. Reprinted by permission from the *Journal of Clinical and Experimental Neuropsychology*.)

This is a distinct possibility, since the recent finding of relative impairment of elderly subjects on indirect memory tasks (*e.g.*, Howard *et al.*, 1986a; Rabbitt, 1982—at a long repetition delay only), may have been due to the use of intentional memory strategies by the younger controls (Chiarello & Hoyer, 1988; Graf, 1990). For example, in a stem completion task, young subjects may use word stems as cues to generate words from a prior study phase, thus leading to relative impairment of older relative to younger subjects on the indirect memory task.

Rugg and colleagues (1991) have recently shown that during an explicit recognition task, the old/new ERP repetition effect was dramatically reduced or absent in patients who had undergone unilateral anterior temporal lobectomy. By contrast, in a task similar to that used here, in which item repetition was incidental to the subject's primary task and did not require an overt recognition response, the ERP repetition effect was of normal amplitude in these same patients. These data suggest that the ERP repetition effect, elicited during tasks in which attention is directed away from item repetition, is dependent upon structures outside the anterior temporal lobe. Thus, in addition to likely damage to brain structures that mediate conscious recollection of recent experiences (probably located in the medial temporal lobe—Squire, 1987), as evidenced by dramatically reduced free recall performance in all of our probable Alzheimer's disease patients, brain regions outside the anterior temporal lobe may also be damaged in those patients who do not show normal ERP repetition effects.

Consonant with the results of the current study, Knopman and Nissen (1987) reported that of 28 PAD patients in their sample, 19 showed intact implicit memory for procedural knowledge, while 9 did not. One recent report suggests that stem completion priming (a form of repetition priming) is impaired in PAD (*e.g.*, Salmon *et al.*, 1988), while another suggests it is intact (*e.g.*, Grosse, Wilson & Fox, 1990). Moreover, the results of other studies support the notion of variability in clinical symptomatology and affected brain areas in PAD (*e.g.*, Pfefferbaum, Sullivan, Jernigan, Zipursky, Rosenbloom, Yesavage & Tinkleberg, 1990). Taken in conjunction with these reports, our data recorded from patients with probable Alzheimer's disease suggest that one potentially fruitful avenue of research will be to attempt to determine if there are any characteristics that distinguish those patients who do from those who do not show the ERP repetition priming effect.

Processes Reflected in the ERP Repetition Effect

During implicit repetition priming experiments, such ERP repetition effects have been interpreted as reflecting an increase in negativity to new items that is

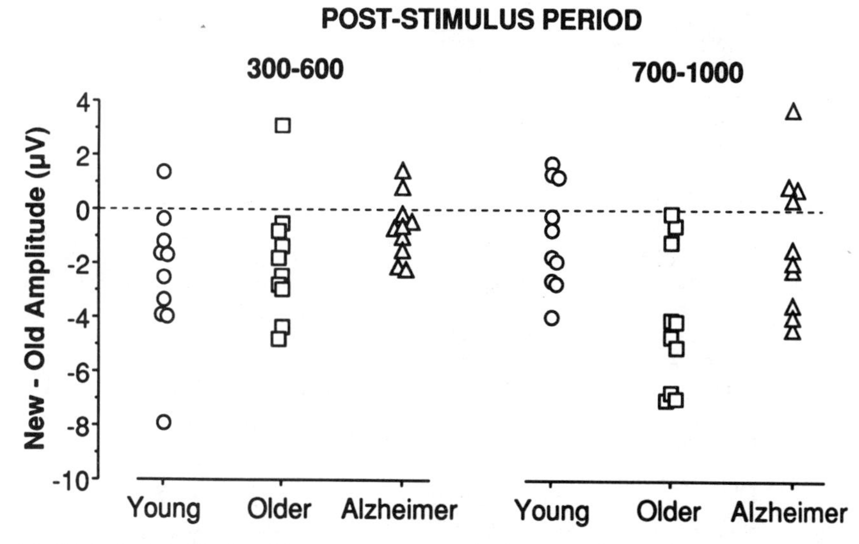

FIGURE 6. Individual midline summary scores in the semantic condition for the 300–600 and 700–1000 msec periods of ERP activity.

reduced or absent to old items (*cf.*, Berman *et al.*, 1991; Rugg, 1987; Rugg *et al.*, 1988), presumably because whatever behavioral process the negativity reflects is not as detailed, or is not required, on second presentation. The size of the repetition effect could thus reflect the consistency or overlap in the processing of the two (or more) presentations of the item (Scarborough, Gerard & Cortese, 1979). However, recent evidence obtained during lexical decision experiments (Rugg, 1990; reviewed by Rugg & Doyle, in press), suggests that a late positive component (P600), subsequent to the negativity, but with a different functional role, is also modulated during implicit paradigms. During explicit recognition memory experiments, where an overt recognition response is required, two ERP components are also thought to occur, an earlier negativity larger to new items (perhaps synonymous with the N400), and a subsequent positivity (consistent with a P3b interpretation) larger to old items (*e.g.*, Berman *et al.*, 1991; Friedman *et al.*, in preparation; Rugg & Nagy, 1989; Smith & Halgren, 1989). Smith and Halgren (1989) have argued that the N400 could reflect "associative activation," while P3b might reflect "contextual closure" (Halgren & Smith, 1987), similar, respectively, to the familiarity (Mandler, 1980) or perceptual fluency (Jacoby & Dallas, 1981), and the episodic retrieval (Mandler, 1980) components that are thought to underlie recognition performance (*cf.*, Mandler, 1980). Smith and Halgren (1989) claimed that their ERP repetition effect reflected the episodic retrieval component and not the familiarity or perceptual fluency component (but see Rugg & Nagy, 1989 for opposing arguments). In the lexical decision study mentioned above, Rugg (1990) found that P600 was larger to repeated low frequency words after a fifteen-minute study-test delay, whereas the preceding negativity, N400, was not modulated by repetition after this relatively long interval, thus suggesting that the two components are unique with respect to functional correlates. On this basis, and in contrast to Smith and Halgren (1989), Rugg and Doyle (in press) have argued that their late positive wave (most likely P3b) could reflect the "familiarity" component in such two-process theories of recognition memory. More recently, Bentin and Moscovitch (1990) have suggested that the ERP repetition effect could reflect the "memory strength" of a trace. Their conclusion was based on the finding that the size of the ERP repetition effect was larger the more times an item was repeated and was larger the shorter the lag since the last presentation, consistent with a "familiarity" or "fluency" interpretation of the magnitude of the ERP repetition effect. However, although Bentin and Moscovitch's repetition effect spanned a negativity and a subsequent positivity (*i.e.*, P3b), they made no attempt to differentiate the two on the basis of their putative behavioral correlates.

The current data also suggest the existence of at least two ERP components active during repetition. In the continuous recognition data presented in FIGURE 2, we (Friedman *et al.*, in preparation) were able to differentiate the negativity (300–500 msec) from the overlapping and subsequent positivity (500–900 msec) on the basis of scalp distribution, and relationship to the experimental variables. A similar dissociation based on scalp distribution was reported by Berman *et al.* (1990). It is not possible, however, on the basis of the current data, to come to any firm conclusion as to whether these two distinct brain potential components are associated with the two unique behavioral processes thought to subserve recognition performance.

Relationship of these Findings to Age-Related Memory Performance Differences

Since the young, middle-aged, and older adults all produced similar old/new effects on both the N400 and P3b components (as evidenced by the lack of age by old/new interactions during the AM tasks described above), in light of the above-reviewed findings, and to the extent that the relevant processes are reflected in the ERPs, the data could be interpreted as suggesting that the older adults engaged similar processing mechanisms when retrieving items from memory.

If the ERPs suggest similar retrieval mechanisms, where can the oft-reported age-related explicit memory difference be localized? Some evidence comes from our studies of continuous recognition. As can be seen in FIGURE 2 (top and middle rows), both young and middle-aged adults show an old/new crossover during continuous recognition beginning at about 700 msec post-stimulus (new > old). By contrast, the older adults do not. In our continuous recognition studies, subjects were informed that an item would not be presented more than twice. Since slow wave has been linked to "further processing," of the kind that might be involved in encoding a new item for subsequent retrieval, the greater slow wave to new items is consistent with this kind of activity. Such processing of new items might in some way enhance the probability of subsequent recognition. Thus, in a previous paper dealing with continuous recognition (Friedman, 1990b) with only young adults as subjects, this "crossover" effect was interpreted as indicating a lack of "positive slow wave" in the second presentation ERPs, since "further processing" of the repeated item would not be necessary. This interpretation was bolstered by the finding that, in our age-related study of continuous recognition memory, the amplitudes of new-old subtraction indices for the slow wave region of the waveform were negatively correlated with age, but positively correlated with the signal detection measure of sensitivity. Although preliminary, requiring interpretive caution, the current data suggest the oldest group is not performing (or is doing less of) this "further processing," suggesting less efficient encoding of items for subsequent retrieval. Thus, the oldest group's poorer performance during continuous recognition (d prime for elderly of 4.51; for young adults, 5.11) may be due to poorer encoding during the input stage of processing.

Some additional support for this notion comes from data in which we averaged the new ERPs according to subsequent performance on second presentation. Those data showed a trend (highly similar to that described above from our developmental study), for the young adult group to show the expected "memory effect" (subsequently recognized > subsequently unrecognized), with no difference for the two older groups.

Recent data concerned with the ERP subsequent "memory effect" or "Dm," has shown greater amplitudes when the memory test requires free recall rather than recognition (Paller, McCarthy & Wood, 1988). This is shown in FIGURE 7. As suggested by Paller *et al.* (1988), this could be due to the fact that sorting ERPs on the basis of subsequent recall provides a criterion that is more sensitive to differences in encoding strength between words, whereas recognition is more subject to extraneous factors such as guessing. This is consistent with the fact that a robust subsequent memory effect is not always obtained in studies of recognition memory (*e.g.*, Friedman, 1990a; Johnson *et al.*, 1985; Paller *et al.*, 1987b), although the reported effects were usually in the right direction (subsequently recognized > subsequently unrecognized). In one such study of continuous recognition memory (Friedman, 1990b), in which a robust "Dm" was obtained, it was speculated that the ERP memory effect might reflect elaborative processes, since the to-be-remembered words were highly abstract, possibly precluding the use of a distinctive attribute to aid subsequent recognition. In this case, the memory effect was reflected in a "frontal positive slow wave," spanning several ERP deflections, including the P3b. Similar to these data, Karis *et al.* (1984) reported that a "frontal positive slow wave" was related to recall performance in a Von Restorff memory paradigm (see also Sanquist, Rohrbaugh, Syndulko & Lindsley, 1980). This aspect of their slow wave was larger to words that were recalled compared to those that were not, and was largest in a group of subjects whose memory strategies were characterized as elaborative rather than rote. In a very recent follow-up experiment, Fabiani *et*

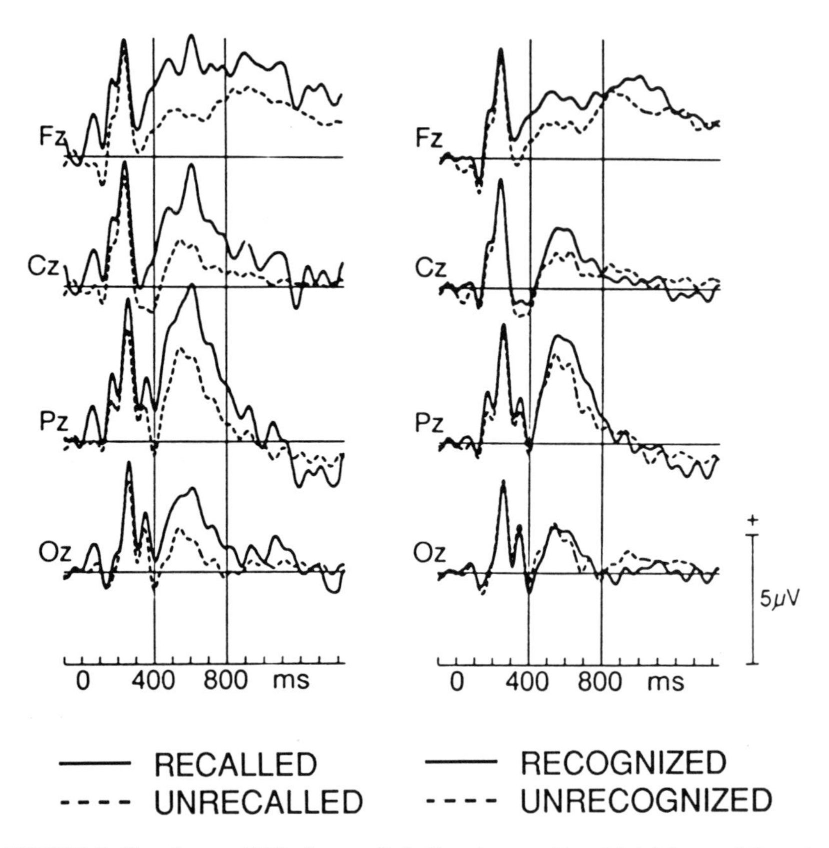

FIGURE 7. Grand mean ERPs for recall (*left*) and recognition (*right*) from midline sites averaged according to subsequent memory performance. Time lines every 100 msec. (Modified from Paller *et al.*, 1988.)

al. (1990) replicated this finding on a within-subjects basis by inducing either rote or elaborative strategies. In their data, frontal positive slow wave predicted subsequent recall under an elaborative strategy, whereas P3b predicted subsequent performance when those same subjects were instructed to use a rote strategy. Although Paller *et al.* (1987) reported a "Dm" with a different scalp distribution than the frontal positive slow waves of Friedman (1990b), Karis *et al.* (1984) and Fabiani *et al.* (1990), their Dm was largest when subjects processed words for their semantic content rather than for orthographic content, consistent with an "elaborative" interpretation of the subsequent memory effect.

The lack of a robust subsequent "memory effect" in the older age group (described above) could be due to their failure to use elaborative strategies to aid subsequent recognition (see, for examples, Botwinick, 1984). However, the fact that the memory effect was not robust in any age group in this sample of subjects argues for interpretive caution and necessitates further experimentation in which more direct control over strategy is achieved, and/or type of processing is manipulated (*e.g.*, semantic; orthographic) with an elderly sample.

SUMMARY AND CONCLUSIONS

Comparison of the results of the studies of cognitive development and normal aging suggests a large degree of commonality in both behavioral and ERP effects across a wide age range. Whether measured in young children, adolescents, young, middle-aged or elderly adults, the size of the ERP repetition effect did not differ among the various age groups. This was true whether memory was tested directly during continuous recognition or indirectly during variants of semantic categorization tasks. Similarly, in the studies of adult aging, the degree of RT facilitation during the semantic task did not differ with age and, in both the studies of cognitive development and aging, the degree of RT prolongation during the explicit tasks did not appear to differ as a function of age. Moreover, in the studies of adult aging, the effects of three versus two exposures of a word assessed in the PM session (TABLE 1), modulated RT similarly in all three age groups. These data argue for continuity of information processing across a very wide age range during both direct and indirect memory tasks, when retention is assessed during the recognition (for explicit testing) and repetition (for implicit testing) phases of the task.

Since ERP and RT modulation do not appear to differ with age during the retrieval phases of these experiments, how can the performance differential seen in young children and older adults be explained? Some evidence comes from the ERP data recorded during the study phases of our explicit tasks. During continuous recognition, both young children and elderly adults did not show the typical subsequent "memory effect." In the case of the children, the subsequently unrecognized ERP was larger than the subsequently recognized ERP, whereas for the older adults, there was no difference between these two ERPs. Moreover, during these same tasks, the young children did show the "crossover" (new > old) pattern for slow wave activity, whereas the older adults did not. Since these ERP findings were obtained during the acquisition phase (*i.e.*, to new items that had to be encoded for subsequent retrieval), the data argue for encoding difficulties as one means of explaining the performance differences seen at the two ends of the age spectrum. However, since the older adults displayed a different new/old pattern for slow wave activity, the two age groups may differ qualitatively in the strategies employed to encode items for subsequent retrieval. Such strategic differences are known to exist (*cf.*, Kail, 1984; but, see Light, 1991 for a different conclusion), but the current data are too preliminary to even attempt speculation about the source of such differences. Nevertheless, taken together, these results argue for continued study of memory function across a wide age range to more precisely pinpoint where in the information processing sequence the age-related differences lie.

ACKNOWLEDGMENTS

The author thanks Mr. Charles L. Brown for computer programming and data reduction, Ms. Concetta DiCaria, and Ms. Margaret Cramer for aid in the construction of figures. The work reported in this review was greatly furthered by collaborations with Drs. Steve Berman, Lois Putnam, Joan G. Snodgrass, and Marla Hamberger. Dr. Samuel Sutton was involved in the initial discussions of the rationale and experimental design for most of the studies reported here. His input, as always, was invaluable in the conduct of the research and the interpretation of the results.

REFERENCES

ACKERMAN, B. P. 1981. Encoding specificity in the recall of pictures and words in children and adults. J. Exp. Child Psychol. **31:** 193–211.

BASHORE, T. R., A. OSMAN & E. F. HEFFLEY III. 1989. Mental slowing in elderly persons: a psychophysiological analysis. Psychol. Aging **4:** 235–244.

BECK, E. C., C. SWANSON & R. E. DUSTMAN. 1980. Long latency components of the visually evoked potential in man: effects of aging. Exp. Aging Res. **6:** 523–545.

BENTIN, S. & M. MOSCOVITCH. 1990. Neurophysiological indices of implicit memory performance. Bull. Psychon. Soc. **28:** 346–352.

BERMAN, S., D. FRIEDMAN, M. CRAMER & L. PUTNAM. 1988. Event-related potentials (ERPs) in continuous recognition memory for pictures and words., Psychophysiology **25:** 435 (abstract).

BERMAN, S., D. FRIEDMAN & M. CRAMER. 1990. A developmental study of event-related potentials to pictures and words during explicit and implicit memory. Int. J. Psychophysiol. **10:** 191–198.

BERMAN, S., D. FRIEDMAN & M. CRAMER. 1991. ERPs during continuous recognition memory for words and pictures. Bull. Psychon. Soc. **29:** 113–116.

BOTWITNICK, J. 1984. Aging and Behavior. 3rd Edit. Springer. New York.

BRUNER, J. S., R. S. OLVER & P. M. GREENFIELD. 1966. Studies in Cognitive Growth. John Wiley and Sons. New York.

BUCHSBAUM, M. & P. WENDER. 1973. Average evoked responses in nornal and minimally brain dysfunctioned children treated with amphetamine. Arch. Gen. Psychiatry **29:** 764–770.

BURKE, D. M., H. WHITE & D. L. DIAZ. 1987. Semantic priming in young and older adults: evidence for age constancy in automatic and attentional processes. J. Exp. Psychol. Hum. Percept. Perform. **13:** 79–88.

CASE, R. 1984. The process of stage transition: a neo-Piagetian view. *In* Mechanisms of Cognitive Development. R. J. Sternberg, Ed. 19–44. W. H. Freeman. New York.

CERELLA, J. 1985. Information processing rates in the elderly. Psychol. Bull. **98:** 67–83.

CHI, M. T. H. 1976. Short-term memory limitations in children: capacity or processing deficits? Mem. Cognit. **4:** 559–572.

CHIARELLO C. & W. J. HOYER. 1988. Adult age differences in implicit and explicit memory: time course and encoding effects. Psychol. Aging **3:** 358–366.

COURCHESNE, E. 1978. Neurophysiological correlates of cognitive development: changes in long-latency event-related potentials from childhood to adulthood. Electroencephalogr. Clin. Neurophysiol. **45:** 468–482.

COURCHESNE, E. 1983. Cognitive components of the event-related potential: changes associated with development. *In* Tutorials in Event-Related Potential Research: Endogenous Components. A. W. K. Gaillard & W. Ritter, Eds. 329–344. North Holland. Amsterdam.

CRAIK, F. I. M. & E. SIMON. 1980. Age differences in memory: the roles of attention and depth of processing. *In* New Directions in Memory and Aging: Proceedings of the George A. Talland Memorial Conference. L. W. Poon, J. L. Fozard, L. S. Cermak, D. Arenberg & L. W. Thompson, Eds. 95–112. Laurence Erlbaum. Hillsdale.

CRAMER, P. 1976. Changes from visual to verbal memory organization as a function of age. J. Exp. Child Psychol. **22:** 50–57.
DONCHIN, E., & M. G. H. COLES. 1988. Is the P300 component a manifestation of context updating? Behav. Brain Sci. **11:** 357–374.
DONCHIN, E., W. RITTER & C. MCCALLUM. 1978. Cognitive psychophysiology: the endogenous components of the ERP. *In* Event-Related Brain Potentials in Man. E. Callaway, P. Tueting & S. Koslow, Eds. 349–411. Academic Press. New York.
DUNCAN-JOHNSON, C. C., & E. DONCHIN. 1982. The P300 component of the event-related brain potential as an index of information processing. Biol. Psychol. **14:** 1–52.
EMMERICH, H. J. & B. P. ACKERMAN. 1979. The effect of orienting activity on memory for pictures and words in children and adults. J. Exp. Child Psychol. **28:** 499–515.
FABIANI, M., D. KARIS & E. DONCHIN. 1986. P300 and recall in an incidental memory paradigm. Psychophysiology **23:** 298–308.
FABIANI, M., D. KARIS & E. DONCHIN. 1990. Effects of mnemonic strategy manipulation in a Von Restorff paradigm. Electroencephalogr. Clin. Neurophysiol. **75:** 22–35.
FORD, J. M., W. F. HOPKINS, A. PFEFFERBAUM & B. S. KOPELL. 1979. Age effects on brain responses in a selective attention task. J. Gerontol. **34:** 388–395.
FORD, J. M., & A. PFEFFERBAUM. 1980. The utility of brain potentials in determining age-related changes in central nervous system and cognitive functioning. *In* Aging in the 1980s. L. W. Poon, Ed. 115–124. American Psychological Association. Washington, DC.
FORD, J. M. & A. PFEFFERBAUM. 1985. Age-related changes in event-related potentials. *In* Advances in Psychophysiology. P. K. Ackles, J. R. Jennings & M. G. H. Coles, Eds. 301–339.
FORD, J. M., A. PFEFFERBAUM, J. R. TINKLENBERG & B. S. KOPELL. 1982. Effects of perceptual and cognitive difficulty on P3 and RT in young and old adults. Electroencephalogr. Clin. Neurophysiol. **54:** 311–321.
FRIEDMAN, D. 1990a. Endogenous event-related electrical activity during continuous recognition memory for pictures. Psychophysiology **27:** 136–148.
FRIEDMAN, D. 1990b. Endogenous event-related brain potentials during continuous recognition memory for words. Biol. Psychol. **30:** 61–87.
FRIEDMAN, D. 1991. The endogenous scalp-recorded brain potentials and their relationship to cognitive development. *In* Handbook of Cognitive Psychophysiology: Central and Autonomic Nervous System Approaches. J. R. Jennings & M. G. H. Coles, Eds. 621–656. John Wiley and Sons. Chichester.
FRIEDMAN, D. & M. HAMBERGER. 1989. ERPs during continuous recognition memory in young, middle-aged and elderly adults. Psychophysiology **27:** 4A, S27 (abstract).
FRIEDMAN, D., C. BROWN, S. SUTTON & L. E. PUTNAM. 1983. Cognitive potentials in a picture matching task: comparison of children and adults. *In* Event-Related Potentials in Children: Basic Concepts and Clinical Application. A. Rothenberger, Ed. 325–336. North Holland. Amsterdam.
FRIEDMAN, D., C. BROWN, H. G. VAUGHAN, JR., B. CORNBLATT & L. ERLENMEYER-KIMLING. 1984. Cognitive brain potential components in adolescents. Psychophysiology **21:** 83–96.
FRIEDMAN, D., M. HAMBERGER & L. M. PUTNAM. 1990. The "Frontal lobe dysfunction" hypothesis in the elderly: evidence from event-related potentials. *In* Psychophysiological Brain Research. C. H. M. Brunia, A. W. K. Gaillard & A. Kok, Eds. 50–54. Tilburg University Press. Amsterdam.
FRIEDMAN, D., S. BERMAN & M. HAMBERGER. Recognition memory and ERPs: age-related changes in young, middle-aged and elderly adults. In preparation.
FRIEDMAN, D., L. E. PUTNAM & M. HAMBERGER. 1990. Cardiac deceleration and E-wave brain potential components in young, middle-aged, and elderly adults. Int. J. Psychophysiol. **10:** 185–190.
FRIEDMAN, D., L. PUTNAM & S. SUTTON. 1989. Event-related potentials in children, young adults and senior citizens: homologous components and scalp distribution changes. Dev. Neuropsychol. **5:** 33–60.
FRIEDMAN, D., L. PUTNAM & S. SUTTON. 1990. Longitudinal and cross-sectional comparisons of young children's cognitive ERPs and behavior in a picture-matching task: preliminary findings. Int. J. Psychophysiol. **8:** 213–221.

FRIEDMAN, D., S. SUTTON, L. PUTNAM, C. BROWN & L. ERLENMEYER-KIMLING. 1988. ERP components in picture matching in children and adults. Psychophysiology **25:** 570–590.

FRIEDMAN, D., L. PUTNAM, W. RITTER, S. BERMAN & M. HAMBERGER. A developmental event-related potential study of picture matching in children, adolescents and adults: a replication and extension. Psychophysiology. In press.

FRIEDMAN, D., M. HAMBERGER, Y. STERN & K. MARDER. Event-related potentials (ERPs) during repetition priming in Alzheimer's patients and young and older controls. J. Clin. Exp. Neuropsychol. In press.

GHATALA, E. S. & J. R. LEVIN. 1981. Children's incidental memory for pictures: item processing versus list organization. J. Exp. Child Psychol. **31:** 231–244.

GHATALA, E. S., J. P. CARBONARI & L. Z. BOBELE. 1980. Developmental changes in incidental memory as a function of processing level, congruity and repetition. J. Exp. Child Psychol. **29:** 74–89.

GOODIN, D. C., K. C. SQUIRES, B. H. HENDERSON & A. STARR. 1978. Age-related variations in evoked potentials to auditory stimuli in normal human subjects. Electroencephalogr. Clinical Neurophysiol. **44:** 447–458.

GRAF, P. 1990. Life-span changes in implicit and explicit memory. Bull. Psychon. Soc. **28:** 353–358.

GRAF, P., & G. MANDLER. 1984. Activation makes words more accessible, but not necessarily more retrievable. J. Verb. Learn. Verb. Behav. **23:** 553–568.

GRAF, P. & D. L. SCHACTER. 1985. Implicit and explicit memory for new associations in normal and amnesic subjects. J. Exp. Psychol. Learn. Mem. Cognit. **11:** 501–518.

GRAF, P., L. R. SQUIRE & G. MANDLER. 1984. The information that amnesic patients do not forget. J. Exp. Psychol. Learn. Mem. Cognit. **10:** 164–178.

GREENBAUM, J. L. & P. GRAF. 1989. Preschool period development of implicit and explicit remembering. Bull. Psychon. Soc. **27:** 417–420.

GROSSE, D. A., R. S. WILSON & J. H. FOX. 1990. Preserved word-stem-completion priming of semantically encoded information in Alzheimer's disease. Psychol. Aging **5:** 304–306.

GURLAND, B., R. R. GOLDEN, J. A. TERESI & J. CHALLOP. 1984. The SHORT CARE: An efficient instrument for the assessment of depression, dementia and disability. J. Gerontol. **39:** 166–169.

HALGREN, E. & M. E. SMITH. 1987. Cognitive evoked potentials as modulatory processes in human memory formation and retrieval. Hum. Neurobiol. **6:** 129–140.

HAMBERGER, M. & D. FRIEDMAN. 1990. Age-related changes in semantic activation: evidence from event-related potentials. *In* Psychophysiological Brain Research. C. H. M. Brunia, A. W. K. Gaillard & A. Kok, Eds. 279–284. Tilburg University Press. Amsterdam.

HAMBERGER, M. & D. FRIEDMAN. ERP correlates of repetition priming and stimulus classification in young, middle-aged and older adults. J. Gerontol. Psychol. Sci. *In* revision.

HARBIN, T. J., G. R. MARSH & M. T. HARVEY. 1984. Differences in the late components of the event-related potential due to age and to semantic and non-semantic tasks. Electroencephalogr. Clin. Neurophysiol. **59:** 489–496.

HASHER, L. & R. T. ZACKS. 1979. Automatic and effortful processes in memory. J. Exp. Psychol. Gen. **108:** 356–388.

HEINDEL, W. C., D. P. SALMON & N. BUTTERS. 1990. Pictorial priming and cued recall in Alzheimer's disease and Huntington's disease. Brain Cognit. **13:** 282–295.

HOWARD, D. V. 1983. The effects of aging and degree of association on the sementic priming of lexical decisions. Exp. Aging Res. **9:** 145–151.

HOWARD, D. V., R. J. SHAW & J. G. HEISEY. 1986a. Aging and the time course of semantic activation. J. Gerontol. **41:** 195–203.

HOWARD, D. V., J. G. HEISEY & R. J. SHAW. 1986b. Aging and the priming of newly learned associations. Dev. Psychol. **22:** 78–85.

INHELDER, B. & J. PIAGET. 1964. The Early Growth of Logic in the Child: Classification and Seriation. Harper and Row. New York.

JACOBY, L. L., & M. DALLAS. 1981. On the relationship between autobiographical memory and perceptual learning. J. Exp. Psychol. Gen. **110:** 306–340.

JOHNSON, R., JR. 1988. Scalp-recorded P300 activity in patients following unilateral temporal lobectomy. Brain **111:** 1517–1529.

JOHNSON, R., JR., A. PFEFFERBAUM & B. S. KOPELL. 1985. P300 and long-term memory: latency predicts recognition performance. Psychophysiology **22:** 497–507.

KAIL, R. 1984. The Development of Memory in Children. W. H. Freeman. New York.

KARIS, D., FABIANI, M. & DONCHIN, E. 1984. "P300" and memory: individual differences in the Von Restorff effect. Cognit. Psychol. **16:** 177–216.

KAU, A. S. M. & G. A. WINER. 1987. Incidental learning in young children tested with words or words plus pictures as stimuli. J. Exp. Child Psychol. **43:** 359–366.

KNIGHT, R. T. 1984. Decreased response to novel stimuli after prefrontal lesions in man. Electroencephalogr. Clin. Neurophysiol. **59:** 9–20.

KNIGHT, R. T. 1990. Neural mechanisms of event-related potentials: evidence from human lesion studies. *In* Event-Related Brain Potentials: Basic Issues and Applications. J. Rohrbaugh, R. Johnson, Jr. & R. Parasuraman, Eds. 3–18. Oxford University Press. New York.

KNIGHT, R. T., D. SCABINI, D. L. WOODS & C. C. CLAYWORTH. 1989. Contributions of temporal-parietal junction to the human auditory P3. Brain Res. **502:** 109–116.

KNOPMAN, D. S. & M. J. NISSEN. 1987. Implicit learning in patients with probable Alzheimer's disease. Neurology **37:** 784–788.

KOK, A. & J. A. J. ROOIJAKKERS. 1985. Comparison of event-related potentials of young children and adults in a visual recognition and word reading task. Psychophysiology **22:** 11–23.

KURTZBERG, D., H. G. VAUGHAN, JR. & J. KREUTZER. 1979. Task-related cortical potentials in children. *In* Cognitive Components in Cerebral Event-Related Potentials and Selective Attention, Progress in Clinical Neurophysiology. J. Desmedt, Ed. Vol. 6: 216–223. Karger. Basel.

KURTZBURG, D., H. G. VAUGHAN, JR., E. COURCHESNE, D. FRIEDMAN, M. R. HARTER & L. E. PUTNAM. 1984. Developmental aspects of event-related potentials. *In* Brain and Information: Event Related Potentials. R. Karrer, J. Cohen & P. Tueting, Eds. Ann. N. Y. Acad. Sci. **425:** 300–318.

KUTAS, M. & S. A. HILLYARD. 1980. Reading senseless sentences: brain potentials reflect semantic incongruity. Science **207:** 203–205.

KUTAS, M. & S. A. HILLYARD. 1989. An electrophysiological probe of incidental semantic association. J. Cognit. Neurosci. **1:** 38–49.

KUTAS, M., & C. VAN PETTEN. 1988. Event-related brain potential studies of language. *In* Advances in Psychophysiology. R. Jennings, M. G. H. Coles & P. Ackles, Eds. 139–187. JAI Press Inc. Greenwich.

LIGHT, L. L. 1991. Memory and aging: four hypotheses in search of data. Ann. Rev. Psychol. **42:** 333–376.

LIGHT, L. L. & A. SINGH. 1987. Implicit and explicit memory in young and older adults. J. Exp. Psychol. Learn. Mem. Cognit. **13:** 531–541.

LIGHT, L. L., A. SINGH & J. L. CAPPS. 1986. Dissociation of memory and awareness in young and older adults. J. Clin. Exp. Neuropsychol. **8:** 62–74.

MANDLER, G. 1980. Recognizing: the judgement of previous occurrence. Psychol. Rev. **87:** 252–271.

MARSH, G. R. 1975. Age differences in evoked potential correlates of a memory scanning process. Exp. Aging Res. **1:** 3–16.

MAYEUX, R., Y. STERN, J. ROSEN & J. LEVENTHAL. 1981. Depression, intellectual impairment and Parkinson's disease. Neurology **31:** 645–650.

MCKHANN, G., D. DRACHMAN, M. FOLSTEIN, R. KATZMAN, D. PRICE & E. M. STADIAN. 1984. Clinical diagnosis of Alzheimer's disease: report of the NINCDS-ADRDA Work Group under the auspices of Department of Health and Human Services Task Force on Alzheimer's disease. Neurology **34:** 939–944.

MCCARTHY, G. & C. C. WOOD. 1985. Scalp distributions of event-related potentials: an ambiguity associated with analysis of variance models. Electroencephalogr. Clin. Neurophysiol. **62:** 203–208.

MCCARTHY, G., C. C. WOOD, P. D. WILLIAMSON & D. D. SPENCER. 1989. Task-dependent field potentials in human hippocampal formation. J. Neurosci. **9:** 4253–4268.

Means, B. M. & W. D. Rohwer. 1976. Attribute dominance in memory development. Dev. Psychol. **12:** 411–417.
Melkman, R. & C. Deutsch. 1977. Memory functioning as related to developmental changes in bases of organization. J. Exp. Child Psychol. **23:** 85–97.
Mitchell, D. B. & M. Perlmutter. 1986. Semantic activation and episodic memory: age similarities and differences. Dev. Psychol. **22:** 86–94.
Moscovitch, M. 1982. A neuropsychological approach to perception and memory in normal and pathological aging. *In* Aging and Cognitive Processes. F. I. M. Craik & S. Trehub, Eds. 55–78. Plenum. New York.
Mullis, R. J., P. J. Holcomb, B. C. Diner & R. A. Dykman. 1985. The effects of aging on the P3 component of the visual evoked potential. Electroencephalogr. Clin. Neurophysiol. **62:** 141–149.
Nelson, D. L., V. S. Reed & C. L. McEvoy. 1977. Learning to order pictures and words: a model of sensory and semantic encoding. J. Exp. Psychol. Hum. Learn. Mem. **3:** 485–497.
Neville, H., M. Kutas, G. Chesney & A. L. Schmidt. 1986. Event-related brain potentials during initial encoding and recognition of congruous and incongruous words. J. Mem. Lang. **25:** 75–92.
Noldy, N. E., R. M. Stelmack & K. B. Campbell. Event-related potentials and recognition memory for pictures and words: the effects of intentional and incidental learning. Psychophysiology. In press.
Ornitz, E. M., E. R. Ritvo, L. E. Panman, Y. H. Lee, E. M. Carr & R. D. Walter. 1968. The auditory evoked response in normal and autistic children during sleep. Electroencephalogr. Clin. Neurophysiol. **25:** 221–230.
Paivio, A. 1986. Mental Representations: a Dual Coding Approach. Oxford University Press. New York.
Paivio, A., J. C. Yuille & S. A. Madigan. 1968. Concreteness, imagery, and meaningfulness values for 925 nouns. J. Exp. Psychol. Monogr. Suppl. **76:** 1–9.
Paller, K. A. 1990. Recall and stem-completion priming have different electrophysiological correlates and are differentially modified by directed forgetting. J. Exp. Psychol. Learn. Mem. Cognit. **16:** 1021–1032.
Paller, K. A., M. Kutas & A. R. Mayes. 1987. Neural correlates of encoding in an incidental learning paradigm. Electroencephalogr. Clin. Neurophysiol. **67:** 360–371.
Paller, K. A., G. McCarthy & C. C. Wood. 1988. ERPs predictive of subsequent recall and recognition performance. Biol. Psychol. **26:** 269–276.
Parkin, A. J. & S. Streete. 1988. Implicit and explicit memory in young children and adults. Br. J. Psychol. **79:** 361–369.
Perlmutter, M. 1978. What is memory aging the aging of? Dev. Psychol. **14:** 330–345.
Perlmutter, M. 1980. A developmental study of semantic elaboration and interpretation of recognition memory. J. Exp. Child Psychol. **29:** 413–427.
Perlmutter, M., E. Schork & D. Lewis. 1982. Effects of semantic and perceptual orienting tasks on preschool children's memory. Bull. Psychon. Soc. **19:** 65–68.
Pfefferbaum, A., J. M. Ford, W. T. Roth & B. S. Kopell. 1980. Age differences in P3-reaction time associations. Electtroencephalogr. Clin. Neurophysiol. **49:** 257–265.
Pfefferbaum, A., J. M. Ford, B. G. Wenegrat, W. T. Roth & B. S. Kopell. 1984. Clinical application of the P3 component of the ERP I: normal aging. Electroencephalogr. Clin. Neurophysiol. **59:** 85–103.
Pfefferbaum, A., J. M. Ford, P. M. White & W. T. Roth. 1989. P3 in schizophrenia is affected by stimulus modality, response requirements, medication, and negative symptoms. Arch. Gen. Psychiatry **46:** 1035–1044.
Pfefferbaum, A., E. V. Sullivan, T. L. Jernigan, R. B. Zipursky, M. J. Rosenbloom, J. A. Yesavage & J. R. Tinkleberg. 1990. A quantitative analysis of CT and cognitive measures in normal aging and Alzheimer's disease. Psychiatry Res. Neuroimag. **35:** 115–136.
Picton, T. W., D. T. Stuss, S. C. Champagne & R. F. Nelson. 1984. The effects of age on human event-related potentials. Psychophysiology **21:** 312–325.

Poon, L. W. (Ed.) 1980. Aging in the 1980s: Psychological Issues. American Psychological Association. Washington, DC.
Poon, L. W. & J. L. Fozard. 1980. Age and word frequency effects in continuous recognition memory. J. Gerontol. **35:** 77–86.
Pressley, M. 1982. Elaboration and memory development. Child Dev. **53:** 296–309.
Rabbitt, P. M. A. 1982. How do old people know what to do next? *In* Aging and Cognitive Processes. F. I. M. Craik & S. Trehub, Eds. 79–98. Plenum Press. New York.
Richardson-Klavehn, A. & R. A. Bjork. 1988. Measures of memory. Annu. Rev. Psychol. **39:** 475–543.
Robertson, A. L., K. Mahesan & K. B. Campbell. 1988. Developmental differences in event-related potentials during a lexical and object decision task. Paper presented at the Society for Psychophysiological Research, October. San Francisco.
Rose, T. L., J. A. Yesavage, R. D. Hill & G. H. Bower. 1986. Priming effects and recognition memory in young and elderly adults. Exp. Aging Res. **12:** 31–37.
Rosinski, R. R., R. M. Golinkoff & K. S. Kukish. 1975. Automatic semantic processing in a picture-word interference task. Child Dev. **46:** 247–253.
Ruchkin, D. S. & S. Sutton. 1983. Positive slow wave and P300: association and dissociation. *In* Tutorials in ERP Research: Endogenous Components. A. W. K. Gaillard & W. Ritter, Eds. 233–250. North Holland. Amsterdam.
Rugg, M. D. 1987. Dissociation of semantic priming, word and non-word repetition effects by event-related potentials. Quart. J. Exp. Psychol. **39A:** 123–148.
Rugg, M. D. 1990. Event-related brain potentials dissociate repetition effects of high- and low-frequency words. Mem. Cognit. **18:** 367–379.
Rugg, M. D. & M. C. Doyle. Event-related potentials and stimulus repetition in direct and indirect tests of memory. *In* Cognitive Electrophysiology. H. J. Heinze, T. Munte & G. R. Mangun, Eds. Birkhauser. Boston. In press.
Rugg, M. D. & M. E. Nagy. 1989. Event-related potentials and recognition memory for words. Electroencephalogr. Clin. Neurophysiol. **72:** 395–406.
Rugg, M., D.-J. Furda & M. Lorist. 1988. The effects ot task on the modulation of event-related potentials by word repetition. Psychophysiology **25:** 55–63.
Rugg, M. D., R. C. Roberts, D. D. Potter, C. D. Pickles & M. E. Nagy. 1991. Event-related potentials related to recognition memory: effects of temporal lobectomy and unilateral temporal lobe epilepsy. Brain **114:** 2313–2332.
Salmon, D. P., A. P. Shimamura, N. Butters & S. Smith. 1988. Lexical and semantic priming deficits in patients with Alzheimer's disease. J. Clin. Exp. Neuropsychol. **10:** 477–494.
Salthouse, T. A. 1982. Adult Cognition: an Experimental Psychology of Human Aging. Springer-Verlag. New York.
Salthouse, T. A. 1988. Initiating the formalization of theories of cognitive aging. Psychol. Aging **3:** 3–16.
Sanquist, T. F., J. W. Rohrbaugh, K. Syndulko & D. B. Lindsley. 1980. Electrocortical signs of levels of processing: perceptual analysis and recognition memory. Psychophysiology **17:** 568–576.
Scarborough, D. L., L. Gerard & C. Cortese. 1979. Accessing lexical memory: the transfer of word repetition effects across task and modality. Mem. Cognit. **7:** 3–12.
Schvaneveldt, R., B. P. Ackerman & T. Semlear. 1977. The effect of semantic context on children's word recognition. Child Dev. **48:** 612–616.
Smith, M. E. & E. Halgren. 1989. Dissociation of recognition memory components following temporal lobe lesions. J. Exp. Psychol. Learn. Mem. Cognit. **15:** 50–60.
Smith, M. E., E. Halgren, M. Sokolik, P. Baudena, A. Musolino, C. Liegeois-Chauvel & P. Chauvel. 1990. The intracranial topography of P3 event-related potential elicited during auditory oddball. Electroencephalogr. Clin. Neurophysiol. **76:** 235–248.
Smith, M. E., J. M. Stapleton & E. Halgren. 1986. Human medial temporal lobe potentials evoked in memory and language tasks. Electroencephalogr. Clin Neurophysiol. **63:** 145–159.
Snodgrass, J. G. & J. Corwin. 1988. Pragmatics of measuring recognition memory: applications to dementia and amnesia. J. Exp. Psychol. Gen. **116:** 34–50.

Squire, L. R. 1986. Mechanisms of memory. Science **232:** 1612–1619.

Squire, L. R. 1987. Memory and Brain. Oxford University Press. New York.

Stelmack, R. M., B. J. Saxe, N. Noldy-Cullum, K. B. Campbell & R. Armitage. 1988. Recognition memory for words and event-related potentials: a comparison of normal and disabled readers. J. Clin. Exp. Neuropsychol. **10:** 185–200.

Sternberg, S. 1966. High speed scanning in human memory. Science **153:** 652–654.

Strayer, D. L., C. D. Wickens & R. Braune. 1987. Adult age differences in the speed and capacity of information processing: 2. An electrophysiological approach. Psychol. Aging **2:** 99–110.

Sutton, S. & D. S. Ruchkin. 1984. The late positive complex—advances and new problems. *In* Brain and Information: Event Related Potentials. R. Karrer, J. Cohen & P. Tueting, Eds. Ann. N. Y. Acad. Sci. **425:** 1–23.

Waugh, N. C. & D. A. Norman. 1965. Primary memory. Psychol. Rev. **72:** 89–104.

Wijker, W., P. C. M. Molenaar & M. W. van der Mollen. 1989. Age changes in scalp distribution of cognitive event-related potentials elicited in an oddball task. J. Psychophysiol. **3:** 179–189.

Attention and Para-Attentional Processing

Event-Related Brain Potentials as Tests of a Model

KARL H. PRIBRAM[a] AND DIANE McGUINNESS[b]

[a]Center for Brain Research and Informational Sciences
Radford University
P.O. Box 5867, R.U. Station
Radford, Virginia 24142

[b]University of South Florida at Fort Myers
811 College Parkway
Fort Myers, Florida 33907

INTRODUCTION

Event-related electrical brain potentials are unique in providing a "window" or "lens" with a resolving power of milliseconds through which input processing can be assessed. They are therefore ideal in tracking the rapid sequence of brain responses that immediately occur when a sensory input is processed. These responses make up the orienting reaction. Those components of orienting which are accessible to awareness are "attended;" the remaining components comprise para-attentional processes.

The aim of the review is to relate, to the extent possible, the operations of various brain systems to the attentional and para-attentional components of orienting. With repetition of an input pattern there is a shift from attentional to para-attentional processing: processing which is automatic but directly influences attention. The shift is called habituation, the pattern becomes familiar. With any change of input pattern or the context in which it occurs, the orienting reaction recurs, it is dishabituated. Dishabituation reflects the response to the novel configuration which has been produced by the change.

The processes of habituation and dishabituation are disturbed when certain parts of the brain are resected. The disturbances are selective; some of the components of orienting are vulnerable to one site of brain resection, other components to other sites. It is therefore necessary to review first the evidence which furnishes the basis for a component analysis of orienting. This evidence comes mainly from psychophysiology, recordings of visceroautonomic indicators of orienting. Next, the relationships between these indicators and brain systems are reviewed. The relationships are established by neurobehavioral and neurochemical studies.

These psychophysiological, neurobehavioral and neurochemical analyses yield a model of orienting which is "tested" below. The tests consist of relating brain electrophysiological evidence to the model. This evidence allows specification of the neural processes involved in the various phases of orienting. In addition, the model is updated with respect to the delineation of an extralemniscal processing system involved in targeted awareness. Thus, a lemniscal automatic para-attentional process becomes defined upon which a set of generalized and targeted attentional control processes operate in a top-down fashion.

Historical Overview

Initially, behavioral and physiological manifestations of central processing of input—attention—led neuroscientists to the view that attention could be ordered along a quantitative continuum; an organism is more or less "aroused" or "activated" and that this occurs because of the way certain brain systems are functioning.

In early research one source of activation or arousal (terms at that time used synonymously) was found in the mesencephalic reticular activating system of the brain stems. Lesions of this network of fibers led to somnolence, coma or even death (Moruzzi & Magoun, 1949) while stimulation of the same system produced alertness and behavioral activity (Lindsley, 1961). Further research produced similar results following electrical stimulation of the hypothalamus of the diencephalon (Abrahams, Hilton & Zbrozna, 1964) and also from the amygdala of the forebrain (Gastaut, 1954). A continuum of behavior from orienting to rage and attack occurred as the amount of stimulation increased. Certain peripheral physiological indices such as the Galvanic skin response (GSR) and heart rate were found to correspond to these levels of arousal. The question fundamental to this line of inquiry was the *degree* or amount to which an organism is aroused. Research of this sort asks questions such as: How *much* attention can be paid?; How *long* can attention be paid?; and can attention be *maintained* in the face of distraction?

In experimental psychology the focus has been different. It has been accepted that, in general, animals and humans attend to something. If they did not, no experiments could be carried out. Interest was generated in the study of attention only when it appeared to break down. Animals were found to notice some features of a compound stimulus but not others, such as color but not shape. Humans and animals alike behaved as if their processing capabilities were limited. They produced responses that indicated they had not processed all the available sensory input.

Thus the initial question arose, spearheaded by Broadbent's classic text in 1958: Are the limits on attention due to filters on the input side, or because of limitation on the organization of behavior? In effect this asked: *Where* is the bottleneck?

Bottleneck models carry the implicit assumption that the brain is indeed like a bottle and that input from the environment is the substance that flows into it, always in one direction. The question "where is the bottleneck?" has never been answered satisfactorily for the simple reason that the brain is not built like a bottle. In a sophisticated review of the problem, Erdelyi (1974) concludes on the basis of 20 years of data, that the limits on processing (selection) are "ubiquitous throughout the nervous system" and need not occur with conscious awareness. In other words, there are a multitude of overlapping systems, some parallel in operations (a wide bottleneck—or several bottles), some sequential (narrow bottleneck), and those intermediate between them.

A beginning was made toward a neuropsychological analysis of this problem when we realized that the issues raised by the hypothesis of a bottleneck of limited capacity might be more productively phrased in terms of a central *competency* (Pribram, 1971, 1974, 1976).

Competencies can be multiple, both on the input and output sides, and the ultimate capability, rather than being conceived of as the capacity of a box of finite limits, can be better construed as a flexible matrix of interlocking competencies. Evidence has confirmed that with extensive practice a formerly limited "capacity" becomes less and less restricted (Logan, 1979; Hirst, Spelke, Reaves, Caharack & Neisser, 1980).

Earlier studies reviewed by Garner (1962) demonstrate that with appropriate experimental designs almost unlimited processing is revealed. For instance, Anderson and Fitts (1958) showed that subjects could handle 17 alternative bits of information at one time. This represented stimulus parameters of color, shape and location, each with nine alternatives, $9 \times 9 \times 9$ or 729 differentiable signals! One of the mechanisms by which such large amounts of information can be processed is by grouping or chunking the bits into larger categories (Miller, 1956; Simon, 1974). In contrast to the rigid external structure implied by bottleneck of limited capacity models, the evidence on chunking shows that an endoskeleton, an internal structure, can be formed which determines the competence of a processing channel.

Focus on the competence to *organize* information has led to the second question asked by experimental psychologists, which is, in effect: "What is the nature of selectivity in processing?" This is the question that lies at the heart of research on attention *span* which deals with such issues as how much can we attend to at any one time, and what kind of stimuli can be attended to more readily than others? The two questions are intimately interwoven because it has become clear that "what kind" determines "how much."

Problems arise with questions of this type because they move us away from the study of attention as a simple function toward a study of attention as a process based on structure. Because of this, research on attention has come to resemble research on cognitive efficiency. We can attend more readily to stimuli that are comprehensible, or have become so through learning or practice.

The model of attention described in this essay is based on data that go a long way toward integrating the neurophysiological and psychological traditions. The current model extends our earlier model (Pribram and McGuinness 1975) by incorporating neurochemical and neurophysiological data that have accumulated since its initial publication. Furthermore, data are here presented according to technique: this should allow for an easier disciplinary evaluation of the model.

Outline of the Model

A considerable body of evidence is accruing to the effect that the central processing of sensory input proceeds automatically under certain circumstances and more deliberately under others. Posner (1973) has data which indicate that automatic processing proceeds by virtue of activity of the "extrinsic" sensory-motor projection systems. Controlled processing entails activity of the "intrinsic" sensory associated system not only of the frontal but also of the posterior cortical convexity (Bolster & Pribram, in preparation). While automatic processing is para-attentional and in large part parallel in nature, controlled processing involves steps, the serial engagement of several attentional systems which range from orienting to effortful search.

Originally, three classes of neural systems were discerned to involve the control of orienting and the major portion of this review will be devoted to these systems. However, in the section on event-related electrical brain potentials, an additional system will be described, which details the basic automatic process upon which the three control systems are shown to operate.

Initially, the three classes of attentional control systems were identified as dealing with (1) "arousal," (2) "activation," and (3) "effort," but these terms of themselves are not of prime importance. As evidence accrued each of these forms was shown to signify one pole of a dimension. Thus, arousal became paired with familiarization and activation with targeted readiness, and effort was better represented by a comfort-effortful innovation dimension. What is important to retain is that operational definitions of the three dimensions are available and thus the concepts underlying the classification can be subjected to further test.

The defining operations upon which the classification was based center on the orienting reaction. Orienting *per se* was shown (*e.g.*, Sharpless & Jasper, 1956) to consist of two components: a brief reflexive response signalled by psychophysiological indicators and a somewhat more prolonged reaction signalled by behavioral orienting. Furthermore, with repetition of the stimulus, the orienting reaction ordinarily decrements—this is called habituation. The brief reflexive portion of the orienting reaction habituates rapidly while the more selective, target behavioral portion habituates slowly or not at all.

Habituation of the reflex component of the orienting reaction is impaired when the orbitofrontal cortex or the temporal pole including the amygdala are damaged (Kimble, Bagshaw & Pribram, 1965; Bagshaw, Kimble & Pribram, 1956; Bagshaw & Benzies, 1968; Luria, Pribram, & Homskaya, 1964, Grueninger & Grueninger, 1973; Pribram, Reitz, McNeil & Spevack, 1979). Such lesions result in a total absence of the ordinarily present viscerautonomic components of the orienting reaction, and we have suggested that there may be a causal relationship between the occurrence of visceroautonomic responses and the production of habituation. In the absence of visceroautonomic activity the orienting stimulus fails to become familiar with the result that behavioral orienting to the same stimulus continues unabated.

Whereas the behavioral component of the orienting reaction is resistant to orbitofrontal and amygdala lesions, this component is impaired when the nigrostriatal (basal ganglia) system, the cingulate and the cortical convexity become damaged (Heilman & Valenstein, 1972; Wright, 1979, 1980a, 1980b). Such damage leads to "neglect" of the stimulus, a failure to orient within the sensory field affected by the damage, especially when the system is put out of balance by unilateral lesions. In such instances the neglect is limited to the sensory field contralateral to the lesion.

The visceroautonomically reinforced aspects of orienting thus appear to result in generalized "arousal" while more selective "activation" characterizes behavioral readiness to orient.

In the course of our experimental analysis, a third distinction became necessary. Under many circumstances generalized arousal and selective activation appear to reflexively couple input to output and output to input. On other occasions, however, the components of orienting become uncoupled—such uncoupling appears to entail more prolonged chronic "arousal" involving internally controlled dishabituation often experienced as anxiety, "discomfort" or "effort." Prolongation provides an opportunity for innovation. Damage to the hippocampal system of the brain interferes with uncoupling: animals with such lesions are hyperdistractable (*i.e.*, they dishabituate more readily than controls) provided they are not engaged in a task, in which they become highly resistant to distraction (Douglas & Pribram, 1966; Crowne & Riddell, 1969).

Arousal—familiarization, activation—targeted readiness, and comfort-effortful innovation are therefore three separate dimensions of controls on attention initiated by the orienting reaction. These dimensions can be "dissected" by making the appropriate brain lesions. The next section is devoted to portraying more fully the relationships between these three aspects of orienting and to attention more generally.

Arousal—Familiarization: Habituation of the Orienting Reaction

The arousal component of the orienting reaction is said to occur when an input change produces a measurable brief (several seconds) change in a physiological (*e.g.*, GSR) indicator over a baseline. In psychophysiology such brief changes are referred to as phasic. The types of input change that produce arousal have been studied extensively: they are changes in stimulations that are in one way or another relevant to the well being of the organism. They include sudden changes in intensity to which the organism is unaccustomed, changes in timing of inputs, and changes in the context in which a figure appears. In short, arousal results when, in the history of the organism's experience, a relevant input is novel. Inherent in these operations is the inference that the input is matched against some residual in the organism of its past experience, some familiar representation, a neuronal model of iterated inputs, a competence (Bruner, 1957; Miller, Galanter & Pribram, 1960; Pribram, 1971). Without matching there could be no novelty nor even a measure of change.

Any small change in a parameter of the signal will reconstitute the arousal reaction (Sokolov, 1960, 1963). The waning or habituation of the arousal response must therefore be due to the establishment of a residual neuronal model of that event. Further, certain stimuli which have special relevance, such as one's name, produce dishabituation in an appropriate context, suggesting that familiarization is a process that makes the neuronal model readily accessible. Thus, there are two related consequences of arousal, 1) a visceroautonomic reaction and 2) with stimulus repetition, familiarization.

Activation—Readiness: the Maintenance of Targeted Orienting

The interaction between behaving organisms and their environment is not one-sided. The organism is not just a switchboard for incoming stimulation. Rather, the essence of behaving organisms is that they are spontaneously active, generating changes in the environment often by way of highly programmed, *i.e.*, serially ordered responses (Miller, *et al.*, 1960; Pribram, 1960a, 1962, 1963, 1971). These organizations of behavior must involve the construction of neuronal models in at least two ways: 1) control of the somatomotor system which effects the responses, and 2) feedback from the outcomes (reinforcing consequences) of the behavior. Sherrington (1955), in discussing central representations, framed the question: "Is the organism intending to *do* something about the stimulus variables in the situation?" Germana (1968, 1969) in a review of the evidence suggested that any "neuronal model" must include such "demand" characteristics. Thus he proposed that Pavlov's "What is it?" reaction (which we have called "arousal" and the process of familiarizing the input) does not occur in isolation from a "What's to be done?" reaction. As we shall see, our analysis would suggest that both reactions occur and that they can be distinguished: arousal and familiarization indicating "What is it?" and activation of targeted readiness signalling "What's to be done?"

Readiness differs from familiarization, therefore, in selectively targeting possible outcomes of behavior. Maintaining readiness is reflected in an increase in cortical negativity (CNV) (*e.g.*, Walter, Cooper, Aldridge, McCallum & Winter, 1964; Donchin, Otto, Gerbrandt & Pribram, 1971) and heart rate deceleration (Lacey & Lacey, 1970) which is measured over minutes (and therefore referred to in psychophysiology as tonic).

Comfort—Effort: Innovative Attention

Thus the systems involved in familiarization and targeted readiness can be distinguished: arousal defined as a visceroautonomic reaction which is critical to familiarizing the input, activation as a maintenance of targeted readiness to respond. Under many circumstances, the two reactions appear to be yoked. In such situations they share the function of reflexively coupling input to output, stimulus to response. In the absence of control, behaving organisms would be constantly aroused by their movements and moved by arousing inputs. There must be some long range, or sustained, control process that involves both generalized arousal and active selection which allows *uncoupling* and recoupling to take place. As a rule, initiated inputs (the reinforcing consequences of actions) appear to produce more complexly structured neuronal models than repetitions of simple inputs *per se*. This is largely due to the participation of the central motor systems in *generating* input: *i.e.*, in producing the environmental outcomes that reinforce behavior. Thus, it takes longer to form a habit in, than to habituate to, the same situation. The coordinating process, requiring innovative change from primitive input-output (stimulus-response) states, can be experienced as discomfort or *effort*.

The effort accompanying innovative change (during problem solving) is reflected, both centrally and peripherally where isometric muscular contraction (Berdina, Kolenko, Kotz, Kuzetzov, Rodinow, Savtchencko & Thorevsky, 1972) and increased blood flow are accompanied by chronic accelerations of heart rate (Lacey & Lacey, 1970).

Effort is here defined as a measure of the *efficiency* with which energy (metabolic output) is expended in producing a "change of state" in control systems. Our definition of energy is in keeping with the definition of energy in physics (see also McFarland, 1971) as the capacity for doing work, *i.e.*, for innovation, for changing the state of a system (or maintaining a state in the face of changes in external parameters). Effort additionally measures the *cost* of such change and is thus an index of the efficiency, the negentropy, with which the work is accomplished.

Basis for the Model

The Amygdala Circuits and Familiarization

Studies on the behavior of neural systems during arousal in animals have revealed that brief psychophysiological responses to sudden changes in stimulus events are a ubiquitous property of certain portions of the central nervous system. In an extensive series of experiments, reviewed by Groves and Thompson (1970), these authors distinguished a system of "arousal" neurons in the medial portions of the spinal cord. This system of neurons in turn converges with another more laterally placed set of decrementing neurons onto a final common path that habituates and dishabituates much as does the motor behavior in which these neural systems are involved. There is every reason to believe that the rostral extension into the mesencephalic brainstem of the column of medially placed cells accounts for the well documented arousal effects of stimulations of the reticular formation (see Lindsley, 1961; Magoun, 1958 for review). Such effects are obtained even more rostally in the diencephalon in a continuation of this neuron system into the hypothalamus where episodes of general alerting, fighting and fleeing are produced by electrical or chemical stimulation to the so-called "defense" or "stop" region of the hypothalamus.

General alerting is produced as well by electrical stimulation of the orbitofrontal cortex, midline and medial thalamus and amygdala. The reaction closely resembles that produced by stimulation of the hypothalamus and mesencephalic reticular core (Wilcott & Hoel, 1973). Such stimulation also results in visceroautonomic activity and in the activation of the cells of the reticular nucleus of the thalamus, momentarily closing sensory input gates (Skinner, 1989). More on this below.

These effects have been shown to be related to the psychophysiological components of the orienting reaction. Abrahams and Hilton (1958) and Abrahams *et al.* (1964) found that in attempting to produce a defense response by stimulation of the hypothalamus, at first a much lower degree of arousal occurred, indicated by pupil dilation and postural alerting. Only when the level of stimulation was increased and maintained for a few seconds, did hissing, snarling, running and piloerection occur. In the later study, alerting psychophysiological components were measured in greater detail, and during mild stimulations the authors observed changes in pupil dilation, respiration and blood flow to accompany head movements and pricking the ears. These same changes were also recorded during responses to simple auditory, visual or cutaneous stimuli, in the absence of hypothalamic stimulation. Since these physiological changes are the same as those observed in all orienting responses, the defense reaction could therefore be considered in part as due to an increase of arousal.

Converging on these hypothalamic structures are two reciprocally acting circuits regulating arousal. These circuits center on the amygdala. This structure is classified as a basal ganglion and part of the limbic forebrain (for an extensive review see Pribram & Kruger, 1954; and Pribram & McGuinness, 1975). One of these circuits involves the ventrolateral frontal cortex and is excitatory since resections of this structure *invariably* eliminate visceral-autonomic orienting responses. The other, opposite in function, is related to the orbitofrontal cortex which has been shown to be the rostral pole of an extensive inhibitory pathway (Kaada, Pribram & Epstein, 1949; Pribram, 1961, 1987; Sauerland & Clemente, 1973; Skinner & Lindsley, 1973; Wall & Davis, 1951).

Observations of the behavior of amygdalectomized animals (Pribram & Bagshaw, 1953), confirm the opponent nature (Solomon, 1980) of these two systems. Ordinarily amygdalectomy produces monkeys that are tame, unresponsive to threat and nonaggressive. However, the opposite finding has also been occasionally observed (*e.g.*, Rosvold, Mirsky & Pribram, 1954). Studies by Ursin and Kaada (1960) using more restricted lesions and electrical stimulations have identified two reciprocal amygdala systems that account for opponent reciprocity.

Reciprocal innervation allows sensitive modulation (tuning) of the arousal mechanism. This is in accord with evidence from other control functions of the amygdala and related structures. For instance, injections of carbachol into the amygdala have no effect unless the animal is already drinking, in which case the amount of drinking becomes proportional to the amount of carbachol injected (Russell, Singer, Flanagan, Stone & Russell, 1968). The fronto-amygdala influence finely tunes viscero-autonomic arousal initiated by the hypothalamic mechanism. It is as if, in the absence of the fronto-amygdala systems, the animal would fail to control his drinking behavior: once started he would drink under circumstances in which others would stop. This is exactly what happens—and more. Both eating and drinking are controlled in this fashion (Fuller, Rosvold & Pribram, 1957).

A clue to what these controls on arousal accomplish, comes from the finding that despite an essentially normal reactivity to shock, the amygdalectomized subjects have fewer spontaneous GSRs during the shock sessions, suggesting a change in base level (Bagshaw & Pribram, 1968). That baseline changes do occur after amygdala lesions was demonstrated directly in sustained chronic response measures (see below) and indirectly by various studies which showed that although behavioral and some electrocortical responses appeared to be normal during orienting (Schwartzbaum, Wilson & Morrisette, 1961; Bagshaw & Benzies, 1968) the background level of these responses is lower than in controls. Ear flicking is practically absent during interstimulus intervals (Bateson, 1972), and it takes less time for the lesioned animals to attain a criterion of slow wave activity in the EEG (Bagshaw & Benzies, 1968) in the preparatory phase of the experiment. While electromyographic (EMG) responses occur with normal latency, the amplitude of these responses is considerably reduced (Pribram et al., 1979). These results indicate that at the forebrain level, just as at the spinal level in Groves' and Thompson's experiments (1970), arousal and decrementing systems converge to produce orienting, habituation and dishabituation.

Perhaps the most striking chronic psychophysiological change to follow amygdalectomy was the finding of a paradoxically elevated basal heart rate (Bagshaw & Benzies, 1968; Pribram *et al.*, 1979). This puzzled us considerably and made data collection analysis difficult (operated and control monkeys had to be matched for basal rate; it had to be shown that no ceiling effect was operating). We wondered whether "arousal" as a concept was in fact untenable in the face of lack of evidence for orienting coupled with an elevated heart rate. Experimental results obtained by Elliott (Elliott, Bankart & Light, 1970) and their analysis clarified the issues. They expected an elevated heart rate to accompany arousal (defined as a response to collative variables such as surprise, and novelty of input much as we have defined them here) but as they were recording longer lasting rather than brief changes he found the opposite: "These collative variables either have no effect on tonic heart rate or they had an effect (deceleratory) opposite to expectations; but response factors and incentive factors (reinforcing consequences) had strong accelerating effects."

Arousal is ordinarily followed by heart rate *deceleration*, which is indicative of activation. By contrast, the monkeys with absent arousal reactions show an *elevated* heart rate. They thus appear to be working with considerable *effort*. In accord with the psychophysiological data on humans, such elevated heart rate is manifest when the situation demands the concentration of attention. Our observations suggest that without such expenditure of effort the amygdalectomized monkeys tend to fall asleep.

We therefore interpret the effects of amygdalectomy as follows: because the specific controls on arousal are removed, arousal results not in familiarization of the situation by altering the access to the neuronal model, but in immediate reflexive distraction. This increased distractibility evokes a defensive effort to cope with the situation. The defense reaction is characterized by an attempt to shut off further input (see Pribram, 1969), an effect inferred from neurophysiological evidence of control over input. The effort is reflected in an elevated heart rate and other changes in chronic autonomic variables indicative of a continuing defense against impending breakdown in the coordination involved in maintaining a set in the face of distraction.

This interpretation is borne out by the results of an experiment in which infant kittens were raised in isolation. When their orienting behavior was examined after six months of isolation, the kittens' visceroautonomic and endocrine reactivity was essentially that of amygdalectomized subjects: they had not learned to cope with situations (had not built up neuronal models) and thus showed the "defensive" syndrome suggestive of considerable effort (Konrad & Bagshaw, 1970).

In summary, studies relating brain function to the visceroautonomic components of the orienting reaction have identified a system of neurons which familiarize a novel input. This core system of neurons extends from the spinal cord through the brain stem reticular formation, including hypothalamic sites and lies in close proximity to those responsible for the engenderment of visceroautonomic responses to novelty. Forebrain control over this corebrain arousal system is exerted by reciprocal facilitatory and inhibitory circuits centered on the amygdala. These circuits control the onset and duration of arousal by controlling the onset and duration of visceroautonomic responses.

It is the relationship between the lack of visceroautonomic responses to orienting and the failure to habituate behaviorally that indicates that a deficiency is produced in a central process by which organisms become familiar with an input: that is, they have ready access to their neuronal model for updating or orienting (dishabituating). Mild disturbances of this process produce the clinical picture of "déjà" and "jamais vu." More severe disturbances produce the automatisms occurring during psychomotor seizures in the presence of epileptic lesions in the region of the amygdala.

Based on the results of the experiments reviewed here, Mednick and Schulsinger (1968) and Venables (Gruzelier & Venables, 1972) have reported two classes (GSR responders and nonresponders) of patients diagnosed as schizophrenics. Responders have a much better prognosis than nonresponders. In fact, the classification has been successfully used as a screening device to identify children in families with a history of schizophrenia who are at risk. Identification can be made before the children show overt symptoms and can, therefore, be sheltered from being exposed to overly traumatic situations.

The Basal Ganglia and the Maintenance of Targeted Readiness

In structures such as the mesencephalic reticular formation and the hypothalamic region a system can be identified with the familiarization process detailed above: when excited as by a novel input, this system operates to stop behavioral reactions to that input by virtue of habituation and/or satiety. Closely coupled to this "stop" or "interrupt" process is its reciprocal, a process that operates to continue targeted behaviors. This readiness process was discovered in relationship to food appetitive processes: in collaboration with one of us (KHP), Anand and Brobeck (1952) discovered that stereotaxic lesions of the "far-lateral" hypothalamic region produced aphagia (animals who failed to eat and starved to death if left alone). Anand (1963) went on to show electrophysiologically (with unit recordings) the activity in this region was reciprocal to that in the ventromedial nucleus of the hypothalamus; when an animal began eating or drinking, unit recordings in the far lateral hypothalamic region were active and those obtained from the ventromedial nucleus were inactive; when satiety set in due to an increase in blood sugar level (as reflected in the arteriovenous ratios), the cells of the ventromedial nucleus became active, while recordings from the far-lateral region showed diminished activity.

The aphagia produced by far-lateral hypothalamic lesions turned out to be peculiar. Teitelbaum in a long series of studies (Teitelbaum, 1955; Teitelbaum & Epstein, 1962; Teitelbaum & Milner, 1963) showed that animals with such lesions would eat if given food which had proven to be highly attractive to nonlesioned animals—sweets, for instance. It was as if the lesioned animals were "finicky" and simply ignored food because their appetite threshold had been markedly raised.

A similar decrease in responsivity to other forms of stimulation has been classically observed to follow certain lesions in the frontal and parietal regions of the cerebral hemispheres of humans (*e.g.*, Semmes, Weinstein, Ghent & Tueber, 1963) and animals (see below). Ignoring becomes especially manifest after unilateral lesions when both the ipsilateral and the contralateral hemifields are simultaneously stimulated. In such instances the stimulus contralateral to the lesion is routinely ignored. This is the syndrome of "neglect."

Heilman and his group (*e.g.*, Heilman & Valenstein, 1972; Heilman & Watson, 1977) have systematically produced "neglect." These investigators find that certain lesions of the mesencephalic reticular formation and of the far-lateral hypothalamic system interfere with the targeted aspects of orienting. Behavioral orienting to food and water has been shown to follow electrical stimulation of this system. Such orienting is prolonged and maintains readiness. Behaviorally, targeted orienting is markedly different from the generalized alerting produced by stimulation of ventromedial hypothalamic system which interrupts ongoing adaptive behavior even to the point of producing sham rage (Hoebel, 1974, 1976; Hernandez & Hoebel, 1978; Abrahams & Hilton, 1958).

There are no cells in the far-lateral hypothalamic region. Rather, this region consists mainly of the median forebrain bundle connecting the mid- and forebrain. The bundle is crossed with fibers connecting the amygdala with the ventromedial hypothalamic system. Ungerstedt (1974) showed that the dopaminergic fibers originating in the substantia nigra and terminating in the basal ganglia (caudate, putamen and globus pallidus) make up a great portion of the median forebrain bundle as it traverses the far-lateral hypothalamic region. Teitelbaum (1955) and Fibiger, Phillips and Clouston (1973) have established that the food "neglect" syndrome is due to lesions of this tract by using antidopaminergic agents to produce "finickiness' and neglect.

Recall that lesions of the amygdala (and those of the ventromedial nucleus of the hypothalamus which results in excessive eating) produced a failure to habituate and thus a continuation of generalized orienting over repetitions of a sensory input. Contrast this to the effects of lesions of the basal ganglia system coursing through the far-lateral hypothalamic region which produce a failure in targeted orienting, neglect and finickiness. It is these reciprocal effects that provide a strong support for the distinction between a "familiarize" and a "readiness" system.

Studies on animal and human patients with lesions in the basal ganglia (Bowen, 1976) also show this inability to maintain targeted attention. In a series of studies employing multiple small stereotactic lesions in the globus pallidus, putamen and caudate nucleus Denny-Brown and Yanagisawa (1976) report their findings with the following summary: "What then is absent? It would appear to be the activating 'set' or 'pump primer' for a certain act, the preparation of the mechanism preparatory to a motor performance oriented to the environment." They also note a particular type of ramp discharge in electrical activity in putamen neurones (see also DeLong & Strick, 1974) which precedes the motor performance at every stage. They suggest this operates as a facilitatory discharge which establishes a "climate" for performance.

They further suggest ". . . the basal ganglia have all the aspects of a 'clearing house' that accumulates samples of ongoing cortical projected activity and, on a competitive basis, can facilitate any one and suppress all others." This indicates that the part of this system relates to an ability to transfer attention from one type of stimulus to another and maintain that attentional set.

The Hippocampal System and Innovative Effort

Data on animal behavior following hippocampectomy indicate that this structure and its connections are critical in coordinating the familiarization and readiness systems. While orienting, subjects with bilateral hippocampectomy show a greater number of, and a greater amplitude of galvanic skin response than controls—a visceroautonomic reactivity opposite to that observed in nonresponding amygdalectomized monkeys. In addition, brief galvanic skin responses terminate considerably more rapidly in hippocampectomized subjects than in controls. It appears from this that hippocampectomized monkeys restabilize more rapidly than normal subjects whose slower galvanic skin response recovery may indicate a more prolonged processing time.

A further change is that such subjects show delayed or absent orienting reactions when thoroughly occupied in performing some other task (Crowne & Riddell, 1969; Kimble, Bagshaw & Pribram, 1965; Raphelson, Isaacson & Douglas, 1965; Riddell, Rothblat & Wilson, 1969; Wicklegren & Isaacson, 1963). In short, the animals appear to be abnormally undistractible while occupied. But in some situations this apperance of undistractibility is restricted to the overt *responses* of the organism, not to orienting *per se*. Douglas and Pribram (1969) used distractors in a task in which responses had been required to each of two successive signals. Hippocampectomized monkeys initially responded much as did controls by manipulating the distractors which appeared between the two signals, increasing the time between the two responses.

However, the controls began to ignore the distractors and speeded their inter-response time. In the hippocampectomized group the number of manipulations declined but their inter-response time remained slow. In this situation, hippocampectomized monkeys continued to be *perceptually* distractible while becoming behavirorally habituated and undistractible. This result is reminiscent of that obtained in man with medial temporal lesions: instrumental behavior can to some considerable extent be shaped by task experience, but verbal reports of the subjective aspects of experience fail to indicate prior acquaintance with the situation (Milner, 1958).

The dissociation between habituation (familiarization) of perceptual responses and habituation involving somatomotor performance appears to be part of a more general effect of hippocampal lesions. In a discrimination reversal situation, extinction of previously learned behavior and acquisition of new responses was observed. In contrast to their controls, the monkeys with the hippocampal lesion remained at a chance level of performance for an inordinately long time (Pribram, Douglas & Pribram, 1969) despite the fact that their recovery from extinction and the slope of their reversal learning curves was completely normal. This was due to the "capture" of the behavior by the 50% intermittent schedule of reinforcement (Spevak & Pribram, 1973). This result suggested that self-directed "observing" responses (indicative of "attention") were relinquished when the probabilities of reinforcement ranged around the chance level.

Taken together, these experimental results suggest that interference with the hippocampal circuit reduces the organism to a state in which the more effort demanding relationships between perception and action, between observing and instrumental responses, and between stimulus and response are replaced by more primitive relationships in which either input or output captures an aspect of the behavior of the organism without the coordinating intervention of central control. The mechanism by which the hippocampal circuit accomplishes the more complex relationship has been studied by making recordings of electrical activity from the hippocampus, with both micro- and macroelectrodes. Before we come to these studies, however, we need to review the neurochemistry, not only of the hippocampal but also the amygdala and basal ganglia systems.

Extension of the Model: Neurochemical Analysis

The evidence reviewed so far has indicated that the neural systems involved in orienting are composed of sets of reciprocally acting mechanisms. Reciprocity has been analyzed by Fair (1965) as an "answering" process and has been the subject of an extensive series of studies by Solomon and his group (see Solomon, 1980 for review) under the label of "opponent process theory." Pribram (1977) has suggested that reciprocity is based on the action of neurochemical systems that to a considerable extent coincide with the three sets of systems (familiarization, readiness and effort) delineated by psychophysiological and neurobehavioral techniques.

A caveat: Each of the "systems" described are of course sets of systems. As already noted, the amygdala is made up of three groups of nuclei: basolateral, central and corticomedial (see Pribram & Kruger, 1954 for review). The basal ganglia are composed of the caudate nucleus, putmen, nucleus accumbens and pallidum. The hippocampus has, in subprimate mammals, a dorsal and a ventral portion—the dorsal portion becomes a vestigial rudiment in primates, the induseum griseum. Furthermore, different layers of the hippocampal formation have different functions in behavior (see Lindsley & Wilson, 1976; in Isaacson & Pibram, *The Hippocampus*, Vol. II). Thus, when matching neurochemical systems to the sets of systems described so far, this can be done at present only with broad strokes.

Generally speaking, the following scheme can be made out: a serotonergic-adrenergic interaction involving the amygdala systems; a cholinergic-dopaminergic interaction involving mesolimbic (n. accumbens), pallidal and caudate basal ganglia systems; and a cholinergic-aminergic interaction involving the hippocampal systems. These reciprocal interactions are superimposed on or activated within a set of steroid, adrenocortical-adrenocorticotrophic and peptide mechanisms that further modulate processing.

Serotonergic-Adrenergic Interactions

A large amount of research (*e.g.*, reviews by Jouvet, 1974; Barchas, Ciaranello, Stolk & Hamburg, 1972) has related the serotonergic and adrenergic systems to the phases of sleep: serotonin to ordinary (slow wave) sleep and norepinephrine to paradoxical (rapid eye movement) sleep during which much dreaming occurs.

For the most part serotonergic and adrenergic pathways overlap and converge rostrally on the amygdala. Thus Cooper, Bloom and Roth (1978, p. 206) note that "most raphé neurons (the origin of the serotonergic systems) are more norepinephrine-like than dopamine-like in their topography. (One) group appears to furnish a very large component of the 5-HT innervation of the limbic system." This innervation reaches the amygdala via stria medularis and stria terminalis.

The regulation of sleep by the amygdala has not been quantitatively documented although sleep disturbances are commonplace immediately following amygdalectomy, the animals often falling into a torpor from which they are difficult to rouse for from several days to several weeks.

However, norepinephrine has been related to a behavioral function in which the amygdala systems are consistently implicated—the effects of reinforcing events (Stein, 1968). Norepinephrine has also been related to orienting and affective agonistic reactions. Once again a response to novelty—sensed against a background of familiarity—is norepinephrinergic, whereas "familiarity" in the guise of "territoriality" and "isolation" has been shown to some considerable extent to be dependent on a serotonergic mechanism (see reviews by Reis, 1974; Goldstein, 1974).

These data suggest that norepinephrine acts by regulating serotonergic substrate (which is determining one or another basic condition of the organism) to produce paradoxical sleep, reinforcement, orienting and perhaps other behaviorally relevant neural events that interrupt an ongoing state. In all likelihood there is a third level of modulation—the neuropeptides which also show some reciprocity in their activity. Thus substance P and the endorphines act reciprocally and both are found in abundance in the amygdala. More on this shortly.

Cholinergic-Dopaminergic Interactions

The most clear-cut evidence regarding neurochemical control systems is the now well established and dramatic findings of a dopaminergic nigrostriatal and mesolimbic (n. accumbens) mechanism that reaches the lateral frontal cortex (Fibiger, Phillips & Clouston, 1973; Ungerstedt, 1974; Goldman-Rakic & Schwartz, 1982). The evidence has been repeatedly reviewed to the effect that dopamine is involved in the maintenance of postural and targeted readiness (Matthysse, 1974; Snyder, Simantov & Pasternak, 1976).

In addition to the nigrostriatal and mesolimbic dopaminergic system, there is another that intimately involves the basal ganglia. This is the cholinergic system (reviewed extensively by Fuxe, 1977) which reaches the globus pallidus from which it innervates the cortex. It is also known that assertive agonistic behavior such as predatory aggression depends on the activation of cholinergic mechanism (see, *e.g.*, King & Hoebel, 1968). Thus it is likely that the dopaminergic process regulates a cholinergic substrate (see Fuxe, 1977) to determine the maintenance of targeted readiness of the organism.

Cholinergic-Aminergic Interactions

Cholinergic and aminergic (both serotonergic and norepinephrinergic) pathways converge on the septo-hippocampal system (Cooper *et al.*, 1978, p. 165, 206); a convergence which could account for the part this system plays in integrating the activity of the amygdala and basal ganglia systems. The regulation of septo-hippocampal cholinergic neurons by catecholamines has been delineated by Robinson, Cheney & Costa, (1981) and Butcher, Woolf, Albanese & Butcher, (1981). Oderfeld-Nowak and Aprison (1981) have presented evidence that those same cholinergic mechanisms are modulated by serotonergic indolamines. The interaction between cholinergic hippocampal neurons and adrenergic mechanisms on the one hand, and cholinergic hippocampal neurons and serotonergic mechanisms on the other, are, however, independent of one another (Ladinsky, Consolo, Tirelli, Forloni & Segal, 1981). We must therefore look at another "higher" level of neurochemical interaction for integration of these independently operating (perhaps opponent) processes. This higher level is reviewed in the next section.

Adrenocortical-Adrenocorticotrophic and Peptide Interactions

There is a matrix of steroid and peptide processes upon which and within which the cholinergic and aminergic mechanisms operate. For instance the amygdala systems are intimately interconnected with hypothalamic nuclei (supraoptic) which are rich in sex steroids and the nucleus of the amygdala is itself a site of concentration of such steroids. The hippocampal system is intimately involved in the pituitary-adrenocortical axis in the regulation of stress. Thus the receptors of adrenal cortical hormones can set the neural state which becomes regulated by ACTH. Bohus (1976) and McEwen (McEwen, Gerlach & Micco, 1976) showed that it is, in fact, the hippocampal formation that is the brain site most involved the selective uptake of adrenal cortical steroids. As McEwen states:

> It is only quite recently that we have come to appreciate the role of the entire limbic brain, and not just the hypothalamus, in these endocrine-brain interactions. Our own involvement in this revelation arose from studies of the fate of injected radioactive adrenal steroids, particularly corticosterone, when they entered the brain from the blood. These studies were begun, under the impetus of recent advances in molecular biology of steroid hormone action, to look for intracellular hormone receptors in brain tissue. We expected to find such putative receptors in the hypothalamus, where effects of adrenal steroids on ACTH secretion have been demonstrated (Davidson *et al.*, 1968; Grimm & Kendall, 1968). Much to our surprise, the brain region which binds the most corticosterone is not the hypothalamus but the hippocampus (McEwen *et al.*, 1976).

As the hippocampal circuit functions to coordinate familiarization with targeted readiness to make innovation possible, manipulations of any of the neurochemical mechanisms thus far described can be expected to produce a host of apparently conflicting results with very slight charges. An example is changing a one-way versus two-way conditioned avoidance task (see Pribram, Lim, Poppen & Bagshaw, 1966; van Wimersma, Greidanus & de Wied, 1976) which dramatically changes the results obtained under different drug conditions.

Effects on familiarization and readiness as well as on their coordination (effort) would be predicted. This expectation is borne out in the catalogue of results obtained with manipulations not only of ACTH but also of ACTH-related peptides: extinction of two-way but not one-way avoidance (de Wied, 1974); interference with passive avoidance (Levine & Jones, 1965); interference with learned taste avoidance (the Garcia-effect—Levine, Smotherman & Hennessay, 1977); interference with discrimination reversal (Sandman, George, Nolan & Kastin, 1976); facilitation of memory consolidation (van Wimersma *et al.*, 1976); and facilitation of exploratory behavior and conditioning (Endroczi, 1972).

Just as in the case of manipulation of hippocampal activity, *ongoing* behavioral activity (memory consolidation, exploratory behavior) is facilitated, while any change in behavior (two-way shuttle, passive avoidance, learned taste aversion, discrimination reversal) is interfered with. This appears initially as tilting the bias toward readiness. But as Pribram and Isaacson (1976) show for hippocampal function, and Sandman's group conclude (see Miller, Sandman & Kastin, 1977) such an interpretation is not valid. In the case of hippocampal research, the initial formulation states that after hippocampal resections, animals could not inhibit their responses (McCleary, 1961). This interpretation foundered when it was shown that such animals performed well in go/no-go alteration tasks (Pribram & Isaacson, 1976; Mahut, 1971) and that they could withhold behavioral responses despite an increase in reaction time when distractors were presented (Douglas & Pribram, 1969).

The most cogent analysis has been performed on discrimination reversals. Isaacson, Nonneman and Schualtz (1968) and Nonneman and Isaacson (1973) have shown that reversal learning encompasses three stages: extinction of the previously correct response, reversion to a position habit, and acquisition of the currently correct response. Pribram, Douglas and Pribram (1969) and Spevak and Pribram (1973) have shown that hippocampally lesioned monkeys are intact with regard to both the extinction and the new acquisition phases of the reversal training experience. However, such monkeys seem to become "stuck" in the 50% reinforcement phase or in the position response patterns. In short, the monkeys' behavior seems to be taken over by a relatively low variable interval schedule of reinforcement and they fail to "make the effort" to "pay attention" to the cues which would gain them a higher rate of reward. Champney, Sahley and Sandman (1976) have shown ACTH-related peptides to operate on just this aspect of the reversal experience—and, in fact, have shown interactions with sex differences.

Finally, ACTH and related peptides, the enkephalins, are endorphins—endogenous hormones that have morphine-like effects and, in fact, act as ligands on morphine receptors. These neuropeptides and the hippocampal circuit in which they are operative function therefore to modulate an effort-comfort dimension of experience and behavior.

Evidence such as this makes highly plausible the hypothesis that ACTH and ACTH-related peptides operate on the hippocampal circuit and therefore the "effort" process. Moreover, Strand, Cayer, Gonzalez and Stoboy (1976) present direct evidence that muscle fatigue is reduced by ACTH-related peptides and that this effect must be central. Before this study, the only evidence of metabolic shifts due to the effort of paying attention came from Berdina *et al.* (1972) (noted in the initial section of this review). It now appears that these peripheral anaerobic shifts affecting muscle tonicity may be a reflection of central processing modulated by ACTH and ACTH-related neuropeptides.

Test of the Model: Analysis of Event-Related Brain Electrical Potentials

The recording of brain electrical potential changes has added an all important dimension to the analysis of controls on attention. They have the advantage over other measures in that they are more immediate indicators of the brain activities that operate the relevant controls. They provide, therefore, an excellent opportunity to test, amend and add to the model of attention and para-attentional processes proposed in the previous sections.

To briefly summarize the nomenclature used in this section, event-related brain electrical potentials have been analyzed into the following process-related components: 1) The early components of event-related potentials which occur within approximately 50 milliseconds (depending on modality) reflect activity in the extrinsic systems. 2) The beginning of selection processing is heralded by a positivity occurring roughly at 60 msec to be followed by a processing negativity, occurring about 80–100 msec after the stimulus. This negativity is an indicator of sensory channel selection on the basis of sensory features. 3) Once again a new processing phase is reflected in a positive deflection followed by a negativity, which begins approximately 200 msec after the stimulus and may extend beyond the 400-msec range. This negativity has been shown to reflect within-channel selection. 4) Within-channel processing must be updated and the onset of this process is signalled by a positive component. 5) However, this positive component has two rather different sources; only one component, the P3b, reflects the initiation of the updating procedure. 6) The other, the P3a, which is usually found in a frontal location, reflects generalized orienting. 7) The P3b often, though not always, reflects a rebound from a prolonged negativity, the contingent negative variation (CNV). 8) But the CNV itself is not of unitary origin. This negativity also has a frontal component related to generalized orienting and a set of other components which are modality specific and include a motor readiness potential. Only a brief review of the evidence supporting this nomenclature is presented; more comprehensive reviews make up the remainder of this volume. Our aim here is to relate relevant findings to test and sharpen our model.

Positive Brain Electrical Potentials, Generalized and Targeted Orienting

There is an old observation made in the 1930s by Morison, Dempsey and Morison (1941) in which they reported that resections of the medial portion of the temporal lobe especially the amygdala, interfere with the production of secondary (*i.e.*, late components) responses evoked by sensory stimulation. In addition, Halgren *et al.* (1980) have recorded late (300-msec) components of event-related brain electrical activity (correlated with scalp recording) in the amygdala and hippocampus of human subjects during brain surgery.

An extensive set of studies has been performed in an attempt to determine the psychological process(es) coordinate with the occurrence of such positive deflections, especially those involving stimuli relevant to the organism. Hillyard and Squires thoroughly review this evidence (Hillyard, Squires, Baver & Lindsay, 1971) and conclude that these positive deflections reflect more than one process: a generalized orienting response and a more complex and active attentional process. Generalized orienting is reflected in a deflection which is early and maximal at frontal leads, while active attending produces later positive deflections that are maximal at posterior leads.

The positive deflection occurring around 300 msec after the stimulus, is made up of two subcomponents: a P3a and a P3b. The P3a component is related to generalized orienting and is largely frontal in distribution while the somewhat later P3b is influenced by a set of within-channel selection variables as is the prior processing negativity (Nd). What is of special interest is that this P3b component can be shown to occur—in reaction time experiments—*after* an overt response has already taken place. Thus the P3b cannot be a direct correlate of targeting but must reflect the initiation of a new phase of processing in which the sequelae, the consequences, of targeting are processed.

When the P3a component is prolonged, it is accompanied by desynchronization of the EEG (Grandstaff & Pribram, personal observation) and reflects the continuation of the response, usually in consummatory behavior (Clemente, Sterman & Wyrwicke, 1964). In such instances, the positivity is accompanied by a sharp increase in power both in the alpha (8–12H3) and in the theta (4–8H3) ranges (Grandstaff, 1969) recorded from the cortex of the cerebral convexity. (Conversely, negativity is accompanied by desynchronization; Pribram, 1971, p. 111.)

The P3b as recorded in the "odd-ball" task, signals the onset of an updating process in response to the unpredictable sequential structure of the task. Although updating has been ascribed to the P3 positivity [as attributed by Donchin (Donchin & Coles, 1988) to Pribram and McGuiness (1975)] a more likely interpretation is that updating is reflected in a late (400–600-msec) negativity. (See also the critique by Verlerger, 1988.)

The effects of generalized and targeted orienting are also reflected in the electrical activity recorded from the hippocampus. As a rule, however, synchronization (in the theta range) is recorded when desynchronization occurs in the cortical convexity and hippocampal desynchronization accompanies convexal synchronization. Lindsley (Macadar, Chalupa & Lindsley, 1974) in keeping with many other recent publications (*e.g.*, Fibiger *et al.*, 1973; Ungerstedt, 1974) has dissociated two systems of neurons that influence hippocampal synchronization and desynchronization. One system originates in the anterior portion of the median raphé and associated midline structures of the mesencephalon and courses through the medial portion of the hypothalamus. The other originates more laterally in the median forebrain bundle through the lateral hypothalamus. Electrical stimulations of the lateral mechanism produce hippocampal desynchronization and a momentary "locking on" to a specific aspect of the environment. Stimulations of the medial mechanisms result in a synchronized hippocampal theta rhythm (4–8 Hz), which is accompanied by isocortical desynchronization and in targeted orienting and exploration.

Theta frequencies were first recorded from the hippocampus by Jung and Kornmuller in 1938. Since this discovery theta has been implicated in generalized orienting (Green & Arduini, 1954; Grastyan, 1959; Grastyan, Lissak, Madarasz & Donoffer, 1959) and to intended movement, even when tested under curare (Dalton & Black, 1968; Black & Young, 1972; Black, Young & Batenchuck, 1970). Vanderwolf and his associates (Bland & Vanderwolf, 1972a; 1972b; Vanderwolf, 1969, 1971; Whishaw, Bland & Vanderwolf, 1972) noted that theta activity occurred almost exclusively when animals (rats) were making "voluntary" movements. Though synchronization in the form of a theta rhythm is not as obvious in records obtained in monkey and man, computer analysis has shown it to occur under similar circumstances in primates (Crowne, Konow, Drake & Pribram, 1972).

The results of the Lindsley studies (Lindsley & Wilson, 1976; in Isaacson & Pribram, *The Hippocampus*, Vol. II) as well as those of many others thus indicate that the hippocampal process can operate in at least two modes which regulate orienting: 1) Tonic inhibitory discharge of hippocampal neurones signified by theta rhythms leads to targeted exploration of more or less familiar territory during which the organism is presumably comfortable and updates his processing competence. 2) When generalized orienting occurs because something relevant (such as food) has been encountered, the inhibitory neurones are shut off, and hippocampal rhythms become desynchronized (while, as noted, those of the cortical convexity become synchronized), attention becomes focussed and, to a considerable extent, the organism is insulated from distracting explorations.

Negative Brain Electrical Potentials, the Selection of Sensory Input and the Targeting of Readiness

CNVs, TNVs, Generalized Arousal and Targeted Readiness. In the introduction we defined activation in terms of a readiness to respond, a readiness which allowed behavior to become or remain targeted by virtue of being resistant to generally destabilizing interruptions.

The simplest situation which demands that responses become or remain on target is one in which two successive input signals are separated by an interval. The first input signals the organism to become ready to make a response to the second, which determines the outcome. In this situation, a large body of data has been gathered regarding slow changes in brain electrical activity, *i.e.*, *contingent negative variations* (*CNVs*) (Walter *et al.*, 1964). In turn, these negativities have been related to the tonic slowing of heart rate (Lacey & Lacey, 1970) which was the psychophysiological basis of our definition of tonic activation.

The CNV was originally proposed to reflect an expectancy developed when a response was contingent on awaiting the second of two stimuli. This would suggest that the CNV reflects a central process activating the organism's neuronal model of this contingency. Other research indicated that the negtive shift in potential reflects intended motor activity (*e.g.*, Kornhuber & Deecke, 1965; Vaughan, Costa & Ritter, 1968). However, still another group of investigators (Weinberg, 1972; Donchin, Gerbrandt, Leifer & Tucher, 1972) demonstrated that a CNV occurs whether or not an overt motor or even a discriminative response is required, provided some set or expectancy is built into the situation. Such sets do, of course, demand postural motor readiness. Weinberg (1972), for instance, has shown that in man the CNV continues until feedback from the consequences of reinforcement of the response occurs. Similar evidence has been obtained in monkeys (Donchin, Otto, Gerbrandt & Pribram, 1971, 1973).

Teece, reviewing the literature on the CNV (1972) noted that, in humans, three types of negative potentials could interact depending upon demands of the experiment: (a) a CNV due to expectant attentional processes; (b) the motor readiness potential signaling intention to act; and (c) more or less "spontaneous" shifts. This classification was considerably sharpened by results obtained in a series of nonhuman primate studies (Donchin, Otto, Gerbrandt & Pribram, 1971, 1973) which specify more completely Tecce's last category. Bipolar (surface to depth) recordings were made from several cortical locations under a variety of conditions. These studies showed that sites which produced transcortical negative variations (TNVs) depended upon the type of task. Thus, far frontal TNVs were recorded sporadically early in the task and whenever the task was changed; precentral motor negative potentials were recorded only in anticipation of the necessity to make an overt response (release a depressed lever); while special sensory systems responded to their specific inputs—*e.g.*, parietal negativity occurred while the monkey was holding down the lever.

These data were paralleled by a study on humans (Gaillard, 1977) in which preparation was compared to expectancy in three tasks, one involving speed, another accuracy and the third, detection, but no response. The far frontal leads mirrored generalized expectancy in the no response condition in which no parietal CNV occurred. The other leads were affected by the task demands. The speed condition produced maximal CNV shifts in the parietal leads, the accuracy condition in the motor leads.

The evidence thus indicates that the CNV has a multiple composition: a frontal O (generalized orienting or arousal) wave which can peak as late as 500–800 msec and frequently occurs prior to a late parietal positivity; and a set of E (expectancy) waves which are modality specific and include as one of their manifestations the motor readiness potential.

Hillyard and the Squires (1971) identify the E waves with readiness on the basis of correlations with psychophysiological measures such as Lacy and Lacy's slowing of the heart rate, as do Pribram and McGuinness (1975). However, Hillyard and the Squires (1971) also identify the E waves of the CNV with effort. They show that the amplitude of these waves is a function of task difficulty. This constitutes a major disparity between the systematization attempted here and that which they provide in theirs.Their inference is based on the fact that when multiple tasks which are compatible are processed, the amplitude is additive. However, as they also note, when the tasks are incompatible (which to our view would increase the demand for effort) the amplitude is *reduced*: amplitude appeared in some instances at least to be inversely correlated with effort. It is thus more likely that readiness and effort reflect the operations of two separable neural systems, and that the E waves of the CNV reflect only the operation of the readiness system.

Event-Related Negative Potentials, Sensory Selectivity and the Targeting of Readiness. Analysis of event-related negative potentials has allowed a further processing distinction to be made. One process depends on "the rapid efficient selection of inputs by virtue of their physical attributes or features" (Hillyard *et al.*, 1971). This process corresponds to Broadbent's (1977) stimulus filtering process. A second slower, serial process (Naatanen, 1982, 1990) occurs whereby comparisons of input are made against "dictionary" units in memory prior to classification—Broadbent's pigeon-holing. This distinction has also been termed a between-channel vs a within-channel selection. Hillyard and collegues (1973) present data which relate the early component of the event-related potential to between-channel selection and the mid components to within-channel selection. It is the timing of these two processes—and some, dependent on matching the semantics of linguistic inputs may take as long or longer than 400 msec—that distinguishes the two. It appears that both stimulus filtering and pigeon-holing can proceed simultaneously but that the pigeon-holing process takes longer to complete.

Keys and Goldberg (unpublished manuscript) in an interesting study using microelectrodes have presented evidence regarding the nature of a variety of such parallel processes. Units in the primary sensory projection systems were found responsive to stimulus relevance (*i.e.*, reinforcing history) and "task difficulty independent of spatial location or task strategy." These results with unit recordings fit more general findings obtained in our laboratory from ensembles of units (Pribram, Spinelli & Kamback, 1967). In these studies stimulus features, response selection and reinforcing contingencies were all found to influence recordings from groups of neurons in the striate cortex of monkeys. Only the stimulus features (stimulus filtering), not response strategies, become encoded in primary visual cortex. Task difficulty determined by response strategy (pigeon-holing) is reflected in the electrical activity of the inferotemporal (posterior intrinsic) association cortex (Rothblatt & Pribram, 1972; Nuwer & Pribram, 1979; Pribram, Day & Johnston, 1976; Bolster & Pribram, in preparation).

In a set of beautifully executed studies on clinical patients, the Velascos (Velasco & Velasco, 1979; Velasco, Velasco, Machado & Olvera, 1973) confirm the distinction between the events initiated in sensory (lemniscal) systems and those which subsequently develop in (extralemniscal) systems whose connections are intrinsic, *i.e.*, restricted to brain stem and brain. Their evidence is in agreement with that obtained from scalp recordings that the early (under 60 msec) components of event-related potentials are related to the extrinsically connected sensory systems. In addition their results go one step further in confirming that indeed these potentials occur in, and only in, the extrinsically connected sensory (lemniscal) systems.

Late components of event-related potentials are shown by the Velascos to be due to processing in intrinsic systems. Lateness could be due to slower conduction times in collaterals from the lemniscal to extralemniscal pathways which is the classical view. Alternatively, generation of activity secondary to that evoked in the entire gamut of sensory connected structures could be responsible for the delayed processing. Timing of event-related activity as recorded from their implanted electrodes indicates that the classical view is in error, that in fact the late components originate in thalamocortical circuitry and only then involve the brain stem. Processing control is top-down.

Processing in the sensory systems is gated by a system of extralemniscal (brain stem tecto-tegmental) inputs to the reticular nucleus of the thalamus. Rose (1950) and Chow (1952, 1970) demonstrated a front-to-back arrangement of the projections from the reticular nucleus onto cortex. These projections have since been shown to be dependent on connections within the sensory projection thalamus. Furthermore, this nucleus receives an input from an equally exquisite arrangement of fibers from the mesencephalic tectum (n. cuneifornis) and possibly from the supradjacent deep tectum. These tectal inputs are multimodal and show marked spatial congruence: thus each tectal locus can be interpreted as coding for a point in the three-dimensional envelope surrounding the organism. Complimentary data on effector responses show the existence of a tegmental motor map closely matched to the sensory map.

Tecto-tegmental stimulation produces positive going slow waves and temporary *inhibition* of neuronal discharges in the thalamic reticular nucleus. An external stimulus or any prethalamic electrical stimulation of sensory pathways produces a similar inhibitory effect. By contrast, as shown by Skinner (1989), these thalamic reticular nucleus units are *driven* by stimulations of orbitofrontal cortex, inhibiting those of the sensory thalamus. Thus a reciprocal mechanism exists by virtue of the cells of the reticular nucleus of the thalamus: inhibition by tecto-tegmental inputs opens the "gates" for sensory processing: excitation by orbitofrontal activity closes those gates.

The orbitofrontal system is, of course, centered on the amygdala through the uncinate fasciculus (Pribram & MacLean, 1953). Skinner (1989) describes generalized arousal as characterized by "slow onset sustained potentials" elicited in frontal cortex by novel and other meaningful stimuli. Skinner also notes that generalized arousal involves visceroautonomic responses (Kimble, Bagshaw & Pribram, 1965; Grueninger & Grueninger, 1973) sustaining the process which would, when necessary close the thalamic gates to further sensory processing. Habituation occurs.

In the original model, targeted readiness was shown to be a function of the basal ganglia systems. In the analysis of the data obtained from studies using event-related electrical potentials, however, targeted readiness appears to depend on tecto-tegmental input to the reticular nucleus of the thalamus. Is there any evidence of a critical connection between these two sets of systems?

Recall the Velascos' finding that the latency of the responses evoked in the tecto-tegmental system (responses that correspond to the late components of simultaneously recorded scalp potentials) precluded an origin in the adjacent sensory systems. Rather, their data pointed to a top-down thalamic origin of the process.

Recall also the early experiments of Morison and Dempsey in which they showed the effects of amygdalectomy on the late components of the responses which could be evoked by stimulation of the midline and in intralaminar nuclei of the thalamus. The basal ganglia are intimately connected with these midline and intralaminar nuclei (*e.g.*, globus pallidus with the centromedian nucleus).

There is at present no direct evidence that the late components of scalp or tecto-tegmentally recorded event related potentials are the result of midline-intralaminar activity, nor is there any direct evidence of basal ganglia (and cortical) control of such activity. The indirect evidence just noted can only point to the locus of the initiation of inquiry that needs to be undertaken.

SUMMARY AND CONCLUSION

In 1972 when we began to analyze the vast amount of material from the laboratories of physiological psychologists, we had only a vague conceptualization of what a model of attention might look like. We began where everyone else had, with the view that everything had something to do with "arousal" but with Lacey's (1967) warning in mind that all of the dependent variables might not actually be measuring aspects of the same process.

With this warning in mind, we were forced by the data to organize them into a three-systems mode. Since the first publication of this model in 1975, we have found increasing amounts of evidence to support and extend it. This evidence is briefly reviewed in the present paper in terms of the techniques employed in various types of investigation.

Further, the current review of data has made it possible to specify the para-attentional substrate (the extrinsic lemniscal primary projection systems) upon which the three systems described in the earlier model operate. The earlier model was based on psychophysiological, neurobehavioral and neurochemical analyses while the current specification results from the results of recordings of event-related brain electrical responses. The conclusions derived from these results can be summarized as follows:

First. It has become possible to distinguish controlled attention from the para-attentional pre- and post-attentive automatic processes upon which controls operate.

Second. The pre- and post-attentive processes appear to be coordinate with activity in the extrinsic lemniscal primary sensory projection systems. Processing in these systems is reflected in the early components of event-related brain electrical potentials. These extrinsic systems are, however, not just throughputs for further processing. Rather, they are sensitive to the history of reinforcement which the subject has experienced. The concept of a limited channel capacity must, therefore, be modified to encompass this ability of organisms to improve, through practice, their competence to process a great deal of information in parallel. Competence, not capacity, limits central processing span.

Third. A set of intrinsic extralemniscal processing systems has been identified to operate via a tecto-tegmental pathway to the reticular nucleus of the thalamus. The later components (N_2,P_3, etc.) of event-related potentials have been shown to reflect processing in these systems and those that control them. Activity in these systems has been related to targeted conscious awareness.

Fourth. The late components of the event-related potentials recorded from the intrinsic extralemniscal systems are not due to activation of collaterals from the sensory systems but to top-down influences converging on them in the thalamus.

Fifth. According to our model these top-down influences are, on the one hand, the orbitofrontal-amygdala system responsible for familiarization and, on the other, the basal ganglia system responsible for targeted readiness. As yet, evidence for the latter relationship is only indirect.

Sixth. A third set of systems operates to enhance processing efficiency by modulating the functions of the orbitofronto-amygdala and nigrostriatal systems. This third set converges on the hippocampal system which exerts its influence on familiarization rostrally by way of frontocorticothalamic connections and on readiness posteriorly by way of brain stem connectivities.

Seventh. The components of the event-related electrical brain potentials, when carefully analyzed, differentially reflect the difference between automatic para-attentional and controlled attentional processes. However, little direct evidence regarding interconnections and operations of the systems involved in generating the late event-related components which reflect attentional processes is as yet available. Obtaining such evidence with depth recordings made in animals and in patients should be a high priority objective of future research.

REFERENCES

ABRAHAMS, V. C. & S. M. HILTON. 1958. Active muscle vasodilation and its relation to the "fight and flight reactions" in the conscious animal. J. Physiol. **140:** 16–17.

ABRAHAMS, V. C., S. M. HILTON & A. W. ZBROZYNA. 1964. The role of active muscle vasodilation in the alerting stage of the defense reaction. J. Physiol. **171:** 189–202.

ANAND, B. K. 1963. Influence of the internal environment on the nervous regulation of alimentary behavior. *In* Brain and Behavior. M. A. B. Brazier, Ed. Vol. II: 43–116. American Institute of Biological Sciences. Washington.

ANAND, B. K. & J. R. BROBECK. 1952. Food intake and spontaneous activity of rats with lesions in the amygdaloid nuclei. J. Neurophysiol. **15:** 421–430.

ANDERSON, N. S. & P. M. FITTS. 1958. Amount of information gained during brief exposures of numerals and colors. J. Exp. Psychol. **56:** 362–369.

BAGSHAW, M. H. & S. BENZIES. 1968. Multiple measures of the orienting reaction and their dissociation after amygdalectomy in monkeys. Exp. Neurol. **20:** 175–187.

BAGSHAW, M. H., D. P. KIMBLE & K. H. PRIBRAM. 1965. The GSR of monkeys during orienting and habituation and after ablation of the amygdala, hippocampus, and inferotemporal cortex. Neuropsychologia **3:** 111–119.

BAGSHAW, M. H. & J. D. PRIBRAM. 1968. Effect of amygdalectomy on stimulus threshold of the monkey. Exp. Neurol. **20:** 197–202.

BARCHAS, J. D., R. D. CIARANELLO, J. M. STOLK & D. A. HAMBURG. 1972. Biogenic amines and behavior. *In* Hormones and Behavior. S. Levine, Ed. 235–329. Academic Press. New York.

BATESON, P. G. R. 1972. Retardation of discrimination learning in monkeys and chicks previous exposed to both stimuli. Nature **237**(5351), 173–174.

BERDINA, N. A., O. L. KOLENKO, I. M. KOTZ, A. P. KUZETZOV, I. M. RODINOV, A. P. SAVTCHENCKO & V. I. THOREVSKY. 1972. Increase in skeletal muscle performance during emotional stress in man. Circ. Res. **6:** 642–650.

BLACK, A. H. & G. A. YOUNG. 1972. Electrical activity of the hippocampus and cortex in dogs operantly trained to move and to hold still. J. Comp. Physiol. Psychol. **79:** 128–141.

BLACK, A. H., G. A. YOUNG & C. BATENCHUCK. 1970. The avoidance training of hippocampal theta waves in flaxedized dogs and its relation to skeletal movement. J. Comp. Physiol. Psychol. **70:** 15–24.

BLAND, B. H. & C. H. VANDERWOLF. 1972a. Diencephalic and hippocampal mechanisms of motor activity in the rat: effect of posterior hypothalamic stimulation on behavior and hippocampal slow wave activity. Brain Res. **43:** 67–88.

BLAND, B. H. & C. H. VANDERWOLF. 1972b. Electrical stimulation of the hippocampal formation: behavioral and bioelectrical effects. Brain Res. **43:** 89–106.

BOHUS, B. 1976. The hippocampus and the pituitary adrenal system hormones. *In* The Hippocampus. R. L. Isaacson & K. H. Pribram, Eds. Vol. II: 323–353. Plenum Press. New York.

BOLSTER, R. B. & K. H. PRIBRAM. Cortical involvement in visual scan in the monkey. In preparation.

BOWEN, F. P. 1976. Behavioral alterations in patients with basal ganglia lesions. *In* The Basal Ganglia. M. D. Yahr, Ed. Raven Press. New York.

BROADBENT, D. E. 1977. The hidden preattentive process. Am. Psychol. **32**(2), 109–118.

BRUNER, J. S. 1957. On perceptual readiness. Psychol. Rev. **64:** 123–152.

BUTCHER, L. L., N. J. WOOLF, A. ALBANESE & S. H. BUTCHER. 1981. Cholinergic-monoaminergic interactions in selected regions of the brain: histochemical and pharmacologic analyses. *In* Cholinergic Mechanisms. G. Pepeu & H. Ladinsky, Eds. 723–738. Plenum Press. New York.

CHAMPNEY, T. F., T. L. SAHLEY & C. A. SANDMAN. 1976. Effects of neonatal cerebral ventricular injection of ACTH 4-9 and subsequent adult injections on learning in male and female albino rats. Pharmacol. Biochem. Behav. **5:** 3–10.

CHOW, K. L. 1952. Regional degeneration of the thalamic reticular nucleus following cortical ablations in monkeys. J. Comp. Neurol. **97**(1), 37–59.

CHOW, K. L. 1970. Integrative functions of the thalamocortical visual system of cat. *In* Biology of Memory. K. H. Pribram & D. Broadbent, Eds. 273–292. Academic Press. New York.

CLEMENTE, C. C., M. B. STERMAN & W. WYRWICKE. 1964. Post-reinforcement EEG synchronization during alimentary behavior. Electroencephalogr. Clin. Neurophysiol. **16:** 355–365.

COOPER, J. R., F. E. BLOOM & R. H. ROTH. 1978. The Biochemical Basis of Neuropharmacology. Oxford University Press. New York.

CROWNE, D. P., A. KONOW, K. J. DRAKE & K. H. PRIBRAM. 1972. Hippocampal electrical activity in the monkey during delayed alternation problems. Electroencephalogr. Clin. Neurophysiol. **33:** 567–577.

CROWNE, D. P. & W. I. RIDDELL. 1969. Hippocampal lesions and the cardiac component of the orienting response in the rat. J. Comp. Physiol. Psychol. **69:** 748–755.

DALTON, A. & A. H. BLACK. 1968. Hippocampal electrical activity during the operant conditioning of movement and refraining from movement. Commun. Behav. Biol. **2:** 267–273.

DAVIDSON, J. M., L. E. JONES & S. LEVINE. 1968. Feedback regulation of adrenocorticotropin secretion in "basal" and "stress" conditions: acute and chronic effects of intrahypothalamic corticoid implantation. Endocrinology **82:** 655–663.

DELONG, M. R. & P. L. STRICK. 1974. Relation of basal ganglia, cerebellum, and motor cortex to romp and ballistic limb movements. Brain Res. **71:** 327–335.

DENNY-BROWN, D. & N. YANAGISAWA. 1976. The role of the basal ganglia in the initiation of movement. *In* The Basal Ganglia. M. D. Yahr, Ed. Raven Press. New York.

DE WIED, D. 1974. Pituitary-adrenal system hormones and behavior. *In* The Neurosciences, Third Study Program. F. O. Schmitt & F. G. Worden, Eds. 653–666. MIT Press. Cambridge, MA.

DONCHIN, E. & G. H. COLES. 1988. Is the P300 component a manifestation of context updating? Behav. Brain Sci. **11:** 357–374.

DONCHIN, E., L. A. GERBRANDT, L. LEIFER & L. TUCKER. 1972. Is the contingent negative variation contingent on a motor response? Psychophysiology **9:** 178–188.

DONCHIN, E., D. OTTO, L. K. GERBRANDT & K. H. PRIBRAM. 1971. While a monkey waits: electrocortical events recorded during the foreperiod of a reaction time study. Electroencephalogr. Clin. Neurophysiol. **31:** 115–127.

DONCHIN, E., D. OTTO, L. K. GERBRANDT & K. H. PRIBRAM. 1973. While a monkey waits. *In* Psychophysiology of the Frontal Lobes. K. H. Pribram & A. R. Luria, Eds. 125–138. Academic Press. New York.

DOUGLAS, R. J. & K. H. PRIBRAM. 1966. Learning and limbic lesions. Neuropsychologia **4:** 197–220.

DOUGLAS, R. J. & K. H. PRIBRAM. 1969. Distraction and habituation in monkeys with limbic lesions. J. Comp. Physiol. Psychol. **69:** 473–480.

ELLIOTT, R., B. BANKART & T. LIGHT. 1970. Differences in the motivational significance of heart rate and palmar conductance: two tests of a hypothesis. J. Pers. Soc. Psychol. **14:** 166–172.

ENDROCZI, E. 1972. Pavlovian conditioning and adaptive hormones. *In* Hormones and Behavior. S. Levine, Ed. 173–207. Academic Press. New York.

ERDELYI, M. H. 1974. A new look at the new look: perceptual defense and vigilance. Psychol. Rev. **81**(1), 1–25.

FAIR, C. M. 1965. The Physical Foundations of the Psyche: a Neurophysiological Study. Wesleyan University Press. Middletown, CT.

FIBIGER, H. C., A. G. PHILLIPS & R. A. CLOUSTON. 1973. Regulatory deficits after unilateral electrolytic or 6-OHDA lesions of the substantia nigra. Am. J. Physiol. **225:** 1282–1287.

FULLER, J. L., H. E. ROSVOLD & K. H. PRIBRAM. 1957. The effect on affective and cognitive behavior in the dog of lesions of the pyriform-amygdala-hippocampal complex. J. Comp. Physiol. Psychol. **50:** 89–96.

FUXE, E. 1977. The dopaminergiz pathways. Proc. Am. Neuropathol. Assoc.

GAILLARD, A. W. K. 1977. The late CNV wave: preparation versus expectancy. Psychophysiology **14:** 563–568.
GARNER, W. R. 1962. Uncertainty and Structure as Psychological Concepts. Wiley New York.
GASTAUT, H. 1954. Interpretation of the symptoms of "psychomotor" epilepsy in relation to physiologic data on Rhinencephalic function. Epilepsia **3:** 84–88.
GERMANA, J. 1968. Response characteristics and the orienting reflex. J. Exp. Psychol. **78:** 610–616.
GERMANA, J. 1969. Autonomic-behavioral integration. Psycholphysiology **6:** 78–90.
GOLDSTEIN, M. 1974. Brain research and violent behavior. Arch. Neurol. **30:** 1–35.
GOLDMAN-RAKIC, P. S. & M. L. SCHWARTZ. 1982. Interdigitation of contralateral and ipsilateral columnar projections to frontal association cortex in primates. Science **216:** 755–757.
GRASTYAN, E. 1959. The hippocampus and higher nervous activity. *In* The Central Nervous System and Behavior. M. A. B. Brazier, Ed. Josiah Macy, Jr. Foundation. New York.
GRASTYAN, E., K. LISSAK, I. MADARASZ & H. DONOFFER. 1959. Hippocampal electrical activity during the development of conditioned reflexes. Electroencephalogr. Clin. Neurophysiol. **11:** 409–430.
GREEN, J. F. & A. ARDUINI. 1954. Hippocampal electrical activity in arousal. J. Neurophysiol. **17:** 533–557.
GRIMM, Y. & J. W. KENDALL. 1968. A study of feedback suppression of ACTH secretion utilizing glucocorticoid implants in the hypothalamus: the comparative effects of cortisol, cortisterone and their 20 acetates. Neuroendocrinology **3:** 55–63.
GROVES, P. M. & R. F. THOMPSON. 1970. Habituation: a dual-purpose theory. Psychol. Rev. **77:** 419–450.
GRUENINGER, W. E. & J. GRUENINGER. 1973. The primate frontal cortex and allassostasis. *In* Psychophysiology of the Frontal Lobes. K. H. Pribram & A. R. Luria, Eds. Academic Press. New York.
GRUZELIER, J. H. & P. H. VENABLES. 1972. Skin conductance orienting activity in a heterogeneous sample of schizophrenics. J. Nerv. Ment. Dis. **155:** 277–287.
HALGREN, E., N. K. SQUIRES, C. L. WILSON, J. W. ROHRBAUGH, T. L. BABB & P. H. CRANDALL. 1980. Endogenous potentials generated in the human hippocampal formation and amygdala by infrequent events. Science **210:** 803–806.
HEILMAN, K. M. & E. VALENSTEIN. 1972. Frontal lobe neglect. Neurology **28:** 229–232.
HEILMAN, K. M. & R. T. WATSON. 1977. Mechanisms underlying the unilateral neglect syndrome. *In* Advances in Neurology. Vol. 18. Hemi-Inattention and Hemisphere Specialization. E. A. Weinstein and R. P. Friedland, Eds. Raven Press. New York.
HERNANDEZ, L. & B. G. HOEBEL. 1978. Hypothalamic reward Vol. and aversion: a link between metabolism and behavior. *In* Current Studies of Hypothalamic Function. **2:** 72–92. Karger. Basel.
HILLYARD, S. A., R. F. HINK, U. L. SCHWENT & T. W. PICTOR. 1973. Electrical signs of selective attention in the human brain. Science. **182:** 177–180.
HILLYARD, S. A., K. C. SQUIRES, J. W. BAVER & P. H. LINDSAY. 1971. Evoked potential correlates of auditory signal detection. Science **172:** 1357–1360.
HIRST, W., E. S. SPELKE, C. C. REAVES, G. CAHARACK & U. NEISSER. 1980. Dividing attention without alternation or automaticity. J. Exp. Psychol. **109:** 98–117.
HOEBEL, B. G. 1974. Brain reward and aversion systems in the control of feeding and sexual behavior. J. K. Cole & T. B. Sanderegger, Eds. Nebr. Symp. Motiv. **22:** 49–112.
HOEBEL, B. G. 1976. Brain-stimulation reward and aversion in relation to behavior. *In* Brain-Stimulation Reward. A. Wauquier & E. T. Rolls, Eds. 335–372. North-Holland. Amsterdam.
ISAACSON, R. L., A. J. NONNEMAN & L. W. SCHUALTZ. 1968. Behavioral and aurtomical sequalae of the infant limbic system. *In* The Neuropsychology of Development: a Symposium. R. L. Isaacson, Ed. Wiley. New York.
JOUVET, M. 1974. Monoanesnergic regulation of the sleep-waking cycle in the cat. Neurosciences **3:** 499–508.

JUNG, R. & A. E. KORNMUELLER. 1938. Eine methodik der Abkitung lokalisierter Potentialschwankungen aus sobcorticalen Hirngebieten. Arch. Psychiatr. Nervenkr. **109:** 1–30.

KAADA, B. R., K. H. PRIBRAM & J. A. EPSTEIN. 1949. Respiratory and vascular responses in monkeys from temporal pole, insula, orbital surface and cingulate gyrus: a preliminary report. J. Neurophysiol. **12:** 347–356.

KEYS, W. & M. E. GOLDBERG. Unit potentials. Unpublished manuscript.

KIMBLE, D. P., M. H. BAGSHAW & K. H. PRIBRAM. 1965. The GSR of monkeys during orienting and habituation after selective partial ablations of the cingulate and frontal cortex. Neuropsychologia **3:** 121–128.

KING, M. B. & B. G. HOEBEL. 1968. Killing elicited by brain stimulation in rats. Commun. Behav. Biol. **2:** 173–177.

KONRAD, K. W. & M. H. BAGSHAW. 1970. Effect of novel stimuli on cats reared in a restricted environment. J. Comp. Physiol. Psychol. **70:** 157–164.

KORNHUBER, H. H. & L. DEECKE. 1965. Hirnpotentialaenderungen bei Wilkurbewegungen und passiven Bewegungen des Menschen: Bereitschaftspotential und reafferent Potentiale. Pflügers Arch. gesamte Physiol. Menschen Tiere **284:** 1–17.

LACEY, J. I. 1967. Somatic response patterning and stress: some revisios of activation theory. *In* Psychological Stress: Issues in Research. M. H. Appley & R. Trumball, Eds. Appleton-Century-Crofts. New York.

LACEY, J. I. & B. C. LACEY. 1970. Some autonomic central nervous system interrelationships. *In* Physiological Correlates of Emotion. P. Black, Ed. 205–227. Academic Press. New York.

LADINSKY, H., S. CONSOLO, A. S. TIRELLI, G. L. FORLONI & M. SEGAL. 1981. Regulation of cholinergic activity in the rat hippocampus: *in vivo* effects of oxotremorine and fenfluramine. *In* Cholinergic Mechanisms. G. Pepeu & H. Ladinsky, Eds. 781–793. Plenum Press. New York.

LEVINE, S. & L. E. JONES. 1965. Adrenocorticotropic hormone (ACTH) and passive avoidance learning. J. Comp. Physiol. Psychol. **59:** 357–360.

LEVINE, S., W. P. SMOTHERMAN & J. W. HENNESSY. 1977. Pituitary-adrenal hormones and learned taste aversion. *In* Neuropeptide Influences on the Brain and Behavior. L. H. Miller, C. A. Sandman & A. J. Kastin, Eds. Raven Press. New York.

LINDSLEY, D. B. 1961. The reticular activating system and perceptual integration. *In* Electrical stimulation of the brain. D. E. Sheer, Ed. Austin: University of Texas Press.

LINDSLEY, D. B. & C. L. WILSON. 1976. Brainstem-hypothalamic systems influencing hippocampal activity and behavior. *In* The Hippocampus. R. L. Isaacson & K. H. Pribram, Eds. Vol. **2:** 247–274. Plenum Press. New York.

LOGAN, G. D. 1979. On the use of a concurrent memory load to measure attention and automaticity. J. Exp. Psychol. **5:** 189–207.

LURIA, A. R., K. H. PRIBRAM & E. D. HOMSKAYA. 1964. An experimental analysis of the behavioral disturbance produced by a left frontal arachnoidal endothelloma (meningioma). Neuropsychologia **2:** 257–280.

MACADAR, A. W., L. M. CHALUPA & D. B. LINDSLEY. 1974. Differentiation of brain stem loci which affect hippocampal and neocortical electrical activity. Exp. Neurol. **43:** 499–514.

MAGOUN, H. W. 1958. The Waking Brain. Charles C. Thomas. Springfield, IL.

MAHUT, H. 1971. Spatial and object reversal learning in monkeys with partial temporal lobe ablations. Neuropsychologia **9:** 409–424.

MATTHYSSE, S. 1974. Schizophrenia: Relationship to Dopamine transmission, motor, control, and feature extraction. Neurosciences **3:** 733–737.

MCCLEARY, R. A. 1961. Response specificity in the behavioral effects of limbic system lesions in the cat. J. Comp. Physiol. Psychol. **54:** 605–613.

MCEWEN, B. S., J. L. GERLACH & D. J. MICCO. 1976. Putative glucocorticoid receptors in hippocampus and other regions of the rat brain. *In* The Hippocampus. R. L. Isaacson & K. H. Pribram, Eds. Vol. **2:** 285–322. Plenum Press. New York.

MCFARLAND, D. J. 1971. Feedback Mechanisms in Animal Behaviour. Academic Press. New York.

MEDNICK, S. A. & F.SCHULSINGER. 1968. Some premorbid characteristics related to breakdown in children with schizophrenic mothers. *In* The Transmission of Schizophrenia. D. Rosenthal & S. S. Kety, Eds. 267–291. Pergamon Press. New York.

MILLER, G. A. 1956. The magical number seven, plus or minus two, or, some limits on our capacity for processing information. Psychol. Rev. **63:** 81–97.

MILLER, G. A., E. H. GALANTER & K. H. PRIBRAM. 1960. Plans and the Structure of Behavior. Holt, Rinehart & Winston. New York.

MILLER, L. H., C. A. SANDMAN & A. J. KASTIN. 1977. Neuropeptide Influences on the Brain and Behavior. Raven Press. New York.

MILNER, B. 1958. Psychological defects produced by temporal lobe excision. *In* The Brain and Human Behavior. H. C. Solomon, S. Cobb & W. Penfield, Eds. Williams & Wilkins. Baltimore, MD.

MORISON, R. S., E. W. DEMPSEY & B. R. MORISON. 1941. Cortical responses from electrical stimulation of the brain stem. Am. J. Physiol. **131:** 732–743.

MORUZZI, G. & H. W. MAGOUN. 1949. Brain stem reticular formation and activation of the EEG. Electroencephalogra. Clin. Neurophysiol. **1:** 455–473.

NAATENEN, R. 1982. Processing negativity. Psychol. Bull. **92**(3), 605–640.

NAATENEN, R. 1990. The role of attention in auditory information processing as revealed by event-related potentials and other brain measures of cognitive function. Behav. Brain Sci. **13:** 201.288.

NONNEMAN, A. J. & R. L. ISAACSON. 1973. Task dependent recovery after early brain change. Behav. Biol. **8:** 143–172.

NUWER, M. R. & K. H. PRIBRAM. 1979. Role of the inferotemporal cortex in visual selective attention. J. Electroencephalogr. Clin. Neurophysiol. **46:** 389–400.

ODERFELD-NOWAK, B. & M. H. APRISON. 1981. On modulation of cerebral cholinergic mechanisms by endogenous indoleamines and their derivatives. *In* Cholinergic Mechanisms. G. Pepeu & H. Ladinsky, Eds. 739–761. Plenum Press. New York.

POSNER, M. I. 1973. Coordination of internal codes. *In* Visual Information Processing. W. G. Chase, Ed. 35–73. Academic Press. New York.

PRIBRAM, K. H. 1960a. The intrinsic systems of the forebrain. *In* Handbook of Physiology, Neurophysiology II. J. Field, H. W. Magoun & V. E. Hall, Eds. American Physiological Society. Washington, DC.

PRIBRAM, K. H. 1961. Limbic system. *In* Electrical Stimulation of the Brain. D. E. Sheer, Ed. 311–320. University of Texas Press. Austin.

PRIBRAM, K. H. 1962. Interrelations of psychology and neurological disciplines. *In* Psychology: a Study of a Science. Vol. 4. Biologically Oriented Fields: Their Place in Psychology and in Biological Sciences. S. Koch, Ed. 119–157. McGraw-Hill. New York.

PRIBRAM, K. H. 1963. Reinforcement revisited: a structural view. *In* Nebraska Symposium on Motivation. Vol. 11. M. Jones, Ed. 113–159. University of Nebraska Press. Lincoln.

PRIBRAM, K. H. 1969. Neural servosystems and the structure of personality. J. Nerv. Ment. Dis. **140:** 30–39.

PRIBRAM, K. H. 1971. Languages of the Brain: Experimental Paradoxes and Principles in Neuropsychology. Prentice-Hall. Englewood Cliffs, NJ.

PRIBRAM, K. H. 1974. How is it that sensing so much we can do so little? *In* Central Processing of Sensory Input. K. H. Pribram, Contrib. Ed. The Neurosciences Third Study Program. F. O. Schmitt & F. G. Worden, Eds. 249–261. MIT Press. Cambridge, MA.

PRIBRAM, K. H. 1976. Mind—it does matter. *In* Philosophical Dimensions of the Neuro-medical Sciences. S. F. Spicker & H. T. Englehardt, Jr., 97–111. D. Reidel. Dordrecht, Holland.

PRIBRAM, K. H. 1977. Peptides and protocritic processes. *In* Psychopathology and Brain Dysfunction. L. H. Miller, C. L. Sandman & A. J. Kastin, Eds. 77–95. Raven Press. New York.

PRIBRAM, K. H. 1987. Subdivisions of the frontal cortex revisited. *In* The Frontal Lobes Revisited. E. Brown and E. Perecman, Eds. 11–39. IRBN Press. New York.

PRIBRAM, K. H. & M. H. BAGSHAW. 1953. Further analysis of the temporal lobe syndrome utilizing frontotemporal ablations in monkeys. J. Comp. Neurol. **99:** 347–375.

PRIBRAM, K. H., R. U. DAY & V. S. JOHNSTON. 1976. Selective attention: distinctive brain electrical patterns produced by differential reinforcement in monkey and man. *In* Behavior Control and Modification of Physiological Activity. D. I. Mostofsky, Ed. 89–114. Prentice-Hall. Englewood Cliffs, NJ.

PRIBRAM, K. H., R. J. DOUGLAS & B. J. PRIBRAM. 1969. The nature of nonlimbic learning. J. Comp. Physiol. Psychol. **69:** 765–772.
PRIBRAM, K. H. & R. L. ISAACSON. 1976. The Hippocampus, Vol. 2: Neurophysiology and Behavior. Plenum. New York.
PRIBRAM, K. H. & L. KRUGER. 1954. Functions of the "olfactory brain". Ann. N. Y. Acad. Sci. **58:** 109–138.
PRIBRAM, K. H., H. LIM, R. POPPEN & M. H. BAGSHAW. 1966. Limbic lesions and the temporal structure of redundancy. J. Comp. Physiol. Psychol. **61:** 368–373.
PRIBRAM, K. H. & P. D. MACLEAN. 1953. Neuronographic analysis of medial and basal cerebral cortex. II. J. Neurophysiol. **16:** 324–340.
PRIBRAM, K. H. & D. MCGUINNESS. 1975. Arousal, activation, and effort: separate neural systems. *In* Brain Work: the Coupling of Function, Metabolism and Blood Flow in the Brain. D. H. Ingvar & N. A. Lassen, Eds. 428–451. Alfred Benzon Foundation. Copenhagen.
PRIBRAM, K. H., S. REITZ, M. MCNEIL & A. A. SPEVACK. 1979. The effect of amygdalectomy on orienting and classical conditioning in monkeys. Pavlovian. J. **14**(4), 203–217.
PRIBRAM, K. H., D. N. SPINELLI & M. C. KAMBACK. 1967. Electrocortical correlates of stimulus response and reinforcement. Science **157:** 94–96.
RAPHELSON, A. C., R. L. ISAACSON & R. J. DOUGLAS. 1965. The effect of distracting stimuli on the runway performance of limbic damaged rats. Psychon. Sci. **3:** 483–484.
REIS, D. J. 1974. The chemical coding of aggression in brain. *In* Neurohumeral Coding of Brain Function. R. D. Myers & R. R. Drucker-Colin, Eds. 125–150.
RIDDELL, W. L., L. A. ROTHBLAT & W. A. WILSON, JR. 1969. Auditory and visual distraction in hippocampectomized rats. J. Comp. Physiol. Psychol. **67:** 216–219.
ROBINSON, S. E., D. L. CHENEY & E. COSTA. 1981. Regulation of septal-hippocampal cholinergic neurons by catecholamines. *In* Cholinergic Mechanisms. G. Pepeu & H. Ladinsky, Eds. 705–713. Plenum Press. New York.
ROSE, J. 1950. The cortical connections of the reticular complex of the thalamus. Patterns of Organization in the Central Nervous System **30:** 454–479.
ROSVOLD, H. E., A. F. MIRSKY & K. H. PRIBRAM. 1954. Influence of amygdalectomy on social interaction in a monkey group. J. Comp. Physiol. Psychol. **47:** 173–178.
ROTHBLAT, L. & K. H. PRIBRAM. 1972. Selective attention: input filter or response selection? Brain Res. **39:** 427–436.
RUSSELL, R. W., G. SINGER, F. FLANAGAN, M. STONE & J. W. RUSSELL. 1968. Quantitative relations in amygdala modulation of drinking. Physiol. Behav. **3:** 871–875.
SANDMAN, C. A., J. GEORGE, J. D. NOLAN & A. J. KASTIN. 1975. Enhancement of attention in man with ACTH/MSH 4-10. Physiol. Behav. **15:** 427–431.
SAUERLAND, E. K. & C. D. CLEMENTE. 1973. The role of the brain stem in orbital cortex induced inhibition of somatic reflexes. *In* Psychophysiology of the Frontal Lobes. K. H. Pribram & A. R. Luria, Eds. 167–184. Academic Press. New York.
SCHWARTZBAUM, J. S., W. A. WILSON, JR. & J. R. MORRISSETTE. 1961. The effects of amygdalectomy on locomotor activity in monkeys. J. Comp. Physiol. Psychol. **54:** 334–336.
SEMMES, J., S. WEINSTEIN, L. GHENT & H. L. TEUBER. 1963. Correlates of impaired orientation in personal and extrapersonal space. Brain **86:** 747–772.
SHARPLESS, S. & H. JASPER. 1956. Habituation of the arousal reaction. Brain **79:** 655–680.
SHERRINGTON, C. 1955. Man on His Nature. Doubleday. Garden City, NY.
SIMON, H. A. 1974. How big is a chunk? Science **183:** 482–488.
SKINNER, B. F. 1989. The origins of cognitive thought. Am. Psychol. **44**(1), 13–18.
SKINNER, B. F. & D. B. LINDSLEY. 1973. The nonspecific medio-thalamic-fronto-cortical system: its influence on electrocortical activity and behavior. *In* Psycholphysiology of the Frontal Lobes. K. H. Pribram & A. R. Luria, Eds. 185–234. Academic Press. New York.
SNYDER, S. H., R. SIMANTOV & G. W. PASTERNAK. 1976. The brain's own morphine, "enkelphalin:" a peptide neurotransmitter? Soc. Neurosci. **1:**
SOKOLOV, E. N. 1960. Neuronal models and the orienting reflex. *In* The Central Nervous System and Behavior. M. A. B. Brazier, Ed. Josiah Macy, Jr. Foundation. New York.
SOKOLOV, E. N. 1963. Perception and the Conditioned Reflex. MacMillan. New York.

SOLOMON, R. L. 1980. The opponent-process theory of acquired motivation: the cost of pleasure and the benefits of pain. Am. Psychol. **35:** 691–712.

SPEVACK, A. & K. H. PRIBRAM. 1973. A decisional analysis of the effects of limbic lesions in monkeys. J. Comp. Physiol. Psychol. **82:** 211–226.

STEIN, L. 1968. Chemistry of reward and punishment. *In* Psychopharmacology. a Review of Progress, 1957–1967. D. H. Effron, Ed. 105–135. U.S. Government Printing Office. Washington, DC.

STRAND, F. L., A. CAYER, E. GONZALES & H. STOBOY. 1976. Peptide enhancement of neuromuscular function: animal and clinical studies. Pharmacol. Biochem. Behav. **5:** 179–188.

TEECE, J. J. 1972. Contingent negative variation (CNV) and psychological processes in man. Psychol. Rev. **69:** 74–90.

TEITELBAUM, P. 1955. Sensory control of hypothalamic hyperphagia. J. Comp. Physiol. Psychol. **48:** 156–163.

TEITELBAUM, P. & P. MILNER. 1963. Activity changes following partial hippocampal lesions in rats. J. Comp. Physiol. Psychol. **56:** 284–289.

TEITELBAUM, P. & A. N. EPSTEIN. 1962. The lateral hypothalamus syndrome: recovery of feeding and drinking after lateral hypothalamic lesions. Psychol. Rev. **69:** 74–90.

UNGERSTEDT, U. 1974. Brain dopamine neurons and behavior. *In* The Neurosciences Third Study Program. F. O. Schmitt & F. G. Worden, Eds. 695–704. MIT Press. Cambridge, MA.

URSIN, H. & B. R. KAADA. 1960. Functional localization within the amygdala complex within the cat. Electroencephalogr. Clin. Neurophysiol. **12:** 120.

VAN WIMERSMA GREIDANUS, TJ. B. & D. DE WIED. 1976. The dorsal hippocampus: a site of action of neuropeptides on avoidance behavior? Pharmacol. Biochem. Behav. **5:** 29–34.

VANDERWOLF, C. H. 1969. Hippocampal electrical activity and voluntary movement in the rat. Electroencephalogr. Clin. Neurophysiol. **26:** 407–418.

VANDERWOLF, C. H. 1971. Limbic-diencephalic mechanisms of voluntary movement. Psychol. Rev. **78:** 83–113.

VAUGHAN, H. G., L. D. COSTA & W. RITTER. 1968. Topography of the human motor potential. Electroencephalogr. Clin. Neurophysiol. **25:** 1–10.

VELASCO, F. & N. VELASCO. 1979. A reticulo-thalamic system mediating propriceptive attention and tremor in man. Neurosurgery **4:** 30–36.

VELASCO, N., F. VELASCO, J. MACHADO & A. OLVERA. 1973. Effects of novelty, habituation, attention, and distraction on the amplitudes of the various components of the somatic evoked responses. Int. J. Neurosci. **5:** 30–36.

VERLEGER, R. 1988. Event-related potentials and cognition: a critique of the context updating hypothesis and an alternative interpretation of P3. Behav. Brain Sci. **11:** 343–427.

WALL, P. D. & G. D. DAVIS. 1951. Three cerebral cortical systems affecting autonomic function. J. Neurophysiol. **14:** 507–517.

WALTER, W. G., R. COOPER, V. J. ALDRIDGE, W. C. MCCALLUM & A. L. WINTER. 1964. Contingent negative variation: an electric sign of sensorimotor association and expectancy in the human brain. Nature **23:** 380–384.

WEINBERG, H. 1972. The contingent negative variation: its relation to feedback and expectant attention. Neuropsychologia **10:** 299–306.

WHISHAW, I. Q., B. H. BLAND & C. H. VANDERWOLF. 1972. Hippocampal activity, behavior, self-stimulation, and heart rate during electrical stimulation of the lateral hypothalamus. J. Comp. Physiol. Psychol. **79:** 115–127.

WICKLEGREN, W. O. & R. L. ISAACSON. 1963. Effect of the introduction of an irrelevant stimulus on runway performance of the hippocampectomized rat. Nature **200:** 48–50.

WILCOTT, R. C. & C. E. HOEL. 1973. Arousal response to electrical stimulation of the cerebral cortex in cats. J. Comp. Physiol. Psychol. **85**(2), 413–420.

WRIGHT, J. J. 1979. Changed cortical activation and the lateral hypothalamic syndrome: a study in the split-brain cat. Brain Res. **151:** 632–636.

WRIGHT, J. J. 1980a. Intracranial self-stimulation, cortical arousal, and the sensorimotor neglect syndrome. Exp. Neurol. **64**.

WRIGHT, J. J. 1980b. Visual evoked response in lateral hypothalamic neglect. Exp. Neurol. **65**.

Selected Problems of Analysis and Interpretation of the Effects of Sleep Deprivation on Temperature and Performance Rhythms

HARVEY BABKOFF,[a,b] MARIO MIKULINCER,[a]
TAMIR CASPY,[a] AND HELEN C. SING[c]

[a]*Department of Psychology*
Bar-Ilan University
Ramat-Gan, Israel

[b]*Naval Health Research Center*
San Diego, California 92138-9174

[c]*Department of Behavioral Biology*
Walter Reed Army Institute of Research
Washington, DC 20307

INTRODUCTION

Performance Rhythms

Although diurnal variations in performance were reported as early as the end of the 19th century (Colquhoun, 1982), it is only in the past 50 years that behavioral scientists began to study rhythmicities systematically. The turning point in the study of periodicity and performance was the series of extensive studies by Kleitman, starting in the early 1940s. Kleitman (1963) (for additional review, see Lavie, 1980) measured body temperature together with performance and concluded that generally there is a positive correspondence between fluctuations of performance and body temperature over a 24-hour (circadian) period even with acute shifts in routine.

Since the early studies of Kleitman, numerous studies of periodicities in performance and their relation to physiological circadian fluctuations have been performed. These studies have been diverse and have employed a large variety of tasks (Colquhoun, 1981; Folkard & Monk, 1982; Minors & Waterhouse, 1981; Naitoh, 1982; Wever, 1982; Wilkinson, 1982). Evidence from the accumulated findings indicate a marked rhythmicity in human performance, in normal as well as in unusual environmental and social conditions. The predominant observed rhythms have primarily a period of approximately 24 hours, with peak and trough times varying according to the task measured (Colquhoun, 1981; Folkard & Monk, 1982, 1984, 1985; Minors & Waterhouse, 1981). The circadian periodicities in performance have been shown to be quite stable over time (Colquhoun, 1972, 1982; Colquhoun, Blake & Edwards, 1968; Kronauer, 1984; Monk, Fookson, Kream, Moline, Pollak & Weitzman, 1985). It has been argued that the very persistence of circadian rhythms during periods of extended sleep deprivation proves their endogenous nature (Minors & Waterhouse, 1985).

There are two major ways of demonstrating performance circadian rhythms.

One procedure minimizes the sleep loss inherent in a continuous 24-hour testing paradigm by utilizing a shift work or service watch system in which task performance is maintained over the 24-hour period by subjects working different parts of the day and night. Sections of the 24-hour curve are assembled either from different work periods for the same subject spread over several days or weeks, or from different people where there is no shift rotation over the testing period (Minors & Waterhouse, 1981; Wilkinson, 1982). There is clearly a disadvantage in this procedure, since it disrupts testing continuity in the same subject or depends upon a between-subject design to synthesize performance rhythms. However, the advantage of this paradigm is that it minimizes, although it may not totally eliminate, the possible interaction of sleep loss with performance rhythms. The alternative procedure is to test performance periodicially around the clock for durations extending 24 hours and beyond. The advantage of this latter method is that performance is measured continually in the same individual, eliminating inter-individual variance as well as intra-individual variance due to large gaps in testing. The disadvantage of this procedure is the presence of sleep loss, its effect on performance and possible interaction with performance rhythms. This paper attempts to address some of the methodological and theoretical problems associated with assessing performance rhythms frequently and continually in the same individuals for periods of 24 hours and longer.

Sleep Deprivation and Performance Rhythms

A wide variety of behaviors become unstable under conditions of continuous work and/or sleep loss. Research over the past decades has demonstrated the effects of sleep loss on biochemical, physiological, psychological and performance variables (Dinges, 1988; Horne, 1978, 1988; Johnson, 1982; Johnson & Naitoh, 1974; Kjellberg, 1977; Medis, 1982). The effects of sleep deprivation have been tested in a very broad and diverse range of behaviors including cognitive performance, psychomotor tasks, subjective ratings of sleepiness, mood, and activation. Most of these behaviors have been shown to be sensitive to sleep loss (Dinges, 1988; Horne, 1978, 1985, 1988; Johnson, 1982; Johnson & Naitoh, 1974; Kjellberg, 1977; Minors & Waterhouse, 1981; Naitoh, 1976). It seems, therefore, well established that cognitive functions are subject to inherent cyclicities and sensitive to the effects of sleep loss.

Rhythmic components in any data set are uniquely identified by the duration (or period) of the cycle, its amplitude and phase, any or all of which may be affected by sleep deprivation. Sleep loss is necessarily incurred when rhythms are observed in the same subjects by continual long-term testing lasting 24 hours or longer. One of the theoretical and empirical concerns regarding the performance rhythms measured in this manner relates to their stability. Does fatigue or sleep loss alter the characteristics of performance rhythms? Although few studies of sleep deprivation address this issue directly, some reports suggest possible changes in the rhythmic components of performance over periods of long-term sleep deprivation (Babkoff, Mikulincer, Caspy, Kempinski & Sing, 1988; Bugge, Opstad & Magnus, 1979).

Performance rhythm stability can be examined by testing for changes in the characteristics of the rhythmic components over time (Babkoff *et al.*, 1988) and by comparing the data set to well-known "stable" physiological rhythms, such as body temperature (Minors & Waterhouse, 1981). Two questions addressed by the comparison of performance to physiological data are: To what extent is performance rhythmicity coupled to body temperature in normal and abnormal conditions? What are the phase relations between performance rhythms and the temperature rhythm and are these phase relations altered by disrupting the sleep-wake cycle?

The classical statistical analyses, *e.g.*, ANOVA and regression, are not designed to provide estimates of the characteristics of the rhythmic components of oscillating data (Monk *et al.*, 1985). In order to investigate the issues raised above, it is necessary to isolate the monotonic from the rhythmic components in data sets and analyze them separately. This requires the use of time-series analytic techniques which can detect and identify the individual rhythmic components; quantify their parameters; isolate them from the monotonic trend; and identify the proportion of the variance accounted for by the trend and by each rhythmic component. There are a number of different time series analysis techniques. How to choose an appropriate technique for analysis of a given performance rhythm has been examined elsewhere and will not be discussed here (Babkoff, Mikulincer, Caspy & Sing, 1991; Monk *et al.*, 1985; Sing, Thorne, Hegge & Babkoff, 1985). This paper illustrates the application of complex demodulation (CD), a stochastic time series technique (Box & Jenkins, 1976) to the analysis and interpretation of performance data whose cyclical characteristics may be changing under conditions of sleep deprivation (Babkoff, Genser, Sing, Thorne & Hegge, 1985a; Babkoff, Thorne, Sing, Genser, Taube & Hegge, 1985b; Orr & Hoffman, 1974; Orr, Hoffman & Hegge, 1976; Redmond, Sing & Hegge, 1982; Sing, Hegge & Redmond, 1984; Sing *et al.*, 1985).

Two examples of changing cyclical characteristics are discussed, peak-to-trough amplitude and phase, as measured by the time of occurrence of the peak/trough of performance rhythms generated during 72 hours of sleep deprivation.

Comparison of Physiological and Performance Data Generated in Sleep Deprivation Experiments

Performance and Body Temperature

Comparing sets of data generated under the same conditions, either indirectly by curve comparisons or directly by correlation, has been a time-honored means of testing for the interrelation of underlying mechanisms. Such techniques have been used to compare performance data to a variety of physiological and biochemical measures generated during periods of extended sleep loss (Folkard & Monk, 1982, 1983, 1984, 1985; Froberg, 1985; Wilkinson, 1982). One of the earlier and still popular comparisons has been that of performance with body temperature, which serves as a standard measure of circadian rhythmicity. As noted above, Kleitman had reported in-phase coupling between body temperature and a variety of performance tasks even concluding some causal relationships between the variables (Kleitman, 1963).

Following Kleitman, body temperature has been used quite extensively as a standard against which to compare performance. The rationale for the central role of temperature as a standard in the study of performance rhythms stems from the fact that the temperature rhythm is well insulated from short-term external disturbances (Froberg, 1985). The stability reflects the strong endogenous component possessed by the generator of the temperature rhythm which, it has been argued, is related to its anatomical proximity to the hypothesized central oscillator responsible for circadian rhythmicity (Minors & Waterhouse, 1981). Traditionally, therefore, the data analysis has included the correlation of performance with body temperature curves so as to identify phasic relations, such as coincident peaks and troughs of the respective rhythms (Colquhoun, 1971; Folkard & Monk, 1979, 1982, 1983, 1984, 1985; Froberg, 1985; Wilkinson, 1982). A graphic example of phasic coupling of a short-term memory performance task (digit span) with body temperature during sleep deprivation was shown by Froberg (Froberg, 1985).

The meaning of the relationship between body temperature and performance rhythms (Blake 1967; Kleitman, 1963; Wever, 1979) has been debated quite often over the past quarter century (Blake, 1967, 1971; Colquhoun, 1971; Minors & Waterhouse, 1981; Rutenfranz, Aschoff & Mann, 1972; Wilkinson, 1982). Evidence for and against a direct causal link has been assessed by several reviewers (Minors & Waterhouse, 1981; Wilkinson, 1982) who concluded that the two may be correlated under certain conditions, but little evidence exists for concluding a causal relationship. Wilkinson believes that the major evidence for "linkage" between temperature and performance comes from studies of post transmeridian flight adaptations (Wilkinson, 1982). Minors and Waterhouse argue that ". . . it is likely that temperature is a poor predictor of performance in part because other factors can influence brain function . . ." (Pp. 131–132) (Minors & Waterhouse, 1981).

Multioscillator Rhythms

The results of a number of studies of biological rhythms in humans over the past two decades have led several researchers to conclude that there are at least two master clocks or internal oscillating mechanisms which control physiological functions and behavior (Aschoff, 1979; Aschoff & Wever, 1976; Folkard & Monk, 1982, 1983, 1984, 1985; Minors & Waterhouse, 1982; Wever, 1979). Free run experiments have provided the main evidence for the existence of two independent, although interacting, oscillating systems. The evidence suggests two master clocks which are coupled under normal conditions; however, when exogenous "Zeitgebers" (synchronizers) are removed and subjects can "free run," the two oscillators may decouple. Body temperature, a marker for one of the endogenous systems (Group I), is quite stable and remains close to a 24-hour cycle (25 ± 0.5 hrs (Wever, 1982)) even under free-run conditions. The sleep-wake cycle is, apparently, a marker for another group of oscillators (Group II), whose period is less stable, and may range widely in free-run conditions, (Minors & Waterhouse, 1981; Wever, 1979, 1982) reaching 30–32.5 hrs (Wever, 1982).

The two oscillating systems differ in strength, not only in terms of their stability, *i.e.*, maintenance of fairly fixed cycle duration, but also in terms of their ability to control one another and of superimposing the controller's cycle duration on the organism as a whole. The major difference in period between the temperature and the sleep-wakefulness cycle during free run occurs when the cycles are out of phase. When the two cycles are almost in phase, they couple, *i.e.*, they "phase-link" and their periods become similar temporarily. The "coupled" period is much closer to that of the more stable temperature cycle. This results in more irregular shifts in cycle periods for sleep-wakefulness than for temperature. These findings have served as evidence for the argument that the Group I oscillator, of which temperature is a marker, is stronger (by an estimated factor of 15) than the Group II oscillator, of which the sleep-wakefulness cycle is a marker (Wever, 1979). It is quite natural, therefore, that the temperature cycle has been used as a rhythmic marker of Group I oscillators, which are mainly endogenously controlled, in studies of performance rhythms (Folkard & Monk, 1982, 1983, 1984, 1985; Froberg, 1985; Minors & Waterhouse, 1981; Wever, 1975, 1979). Data from several laboratories have been interpreted to indicate that certain cognitive functions are coupled to the Group I oscillators. Others are coupled to the Group II set of oscillators, which are more sensitive to exogenous influences. These latter may even drift and change their period when not under the influence of an exogenous circadian clock, *e.g.*, the 24-hour light-dark cycle (Wever, 1979).

The persistence of some performance rhythms even in the absence of a sleep-wake cycle such as during sleep deprivation suggests that they are, at least, partly controlled by an endogenous oscillator (Folkard & Monk, 1985). Examples of such performance rhythms continuing even over 72 hours of sleep deprivation have been reported previously (Babkoff *et al.*, 1985a; Babkoff *et al.*, 1988; Babkoff, Caspy, Mikulincer, Carrasso & Sing, 1989; Froberg, 1985; Mikulincer, Babkoff, Caspy & Sing, 1989). However, the rate of adjustment of performance rhythms after transmeridian flight (jet lag) and after changes of work shift to night work have been reported to differ according to the cognitive function being tested (Folkard & Monk, 1983, 1985). Folkard & Monk have argued that these findings suggest either the presence of more than one oscillating system controlling performance rhythms or alternatively that performance rhythms differ with respect to the extent to which they are masked by the sleep-wake cycle (Folkard & Monk, 1985). The circadian performance rhythms of temporally isolated and desynchronized subjects sometimes followed that of the body temperature and sometimes that of the sleep-wake cycle (Wever, 1979). These results provided additional support for the view that some types of performance are influenced by the group of oscillators controlling body temperature (I) and others by the group of oscillators controlling the sleep-wake cycle (II) (Folkard & Monk, 1985).

The Methodology and Interpretation of Temperature-Performance Comparisons

Long-term sleep deprivation with frequent testing, which provides multisampling of circadian rhythms could, theoretically, serve as an important means for testing a variety of performance curves and determining to which of the "internal clocks" they are coupled. However, the foregoing argument raises the problem addressed in this paper. Can a direct test be made of the extent to which the oscillatory system controlling body temperature is related to the oscillatory system controlling performance by correlating performance with temperature data?

A basic methodological question must be addressed regarding the correlation of two or more sets of data having monotonic and rhythmic components. The question may be stated as follows: Are we to correlate the raw data sets or are we to correlate the best fitting models generated by the respective data sets? Correlating the raw data is not equivalent to correlating the best fitting models generated by the respective data sets. The best technique to use depends on the question addressed by the experimental paradigm. If the question is mainly empirical, *i.e.*, do two (or more) data sets interrelate, and can one predict changes in one from changes in the other, then the global solution of correlating the raw data seems to be more appropriate. This focuses on the dependent variables and tests the extent of their relationship. The mechanisms causing these changes are of lesser interest; rather the immediate or predictive value is the focus of analysis. If, however, the experimental question relates to the identification of the underlying oscillating mechanisms of the respective data sets and the extent of their phasic coupling, correlating the raw data may not be the more appropriate technique. Since data generated in sleep deprivation studies are subject to various monotonic and rhythmic effects (Babkoff *et al.*, 1991), the raw data include the influence of multiple sources of variance, not all of which are, even theoretically, coupled to one another. Consequently, correlating the raw performance and temperature data may provide a distorted picture of the extent of coupling between the rhythmic components of the respective data sets. The alternative is to first isolate the rhythmic components and to correlate the transformed data sets of the best fitting rhythmic models. This is a stepwise procedure that begins with the analysis of each of the data sets for each subject by a time series technique, such as complex demodulation, which provides the best fitting model of the desired frequency after the removal of the monotonic component (remodulate) (Babkoff *et al.*, 1991; Sing, Redmond & Hegge, 1980; Sing *et al.*, 1984, 1985). The remodulates are treated as transformed data, which are then correlated.

Examples of temperature-performance relationships during long-term sleep deprivation are presented to illustrate the use of both analytic methods. As discussed below, the problem of a choice of analysis has theoretical, as well as methodological, connotations. Both methods are illustrated using two performance tasks and body temperature. These data are from a study of the effects of 72 hours of sleep deprivation on performance.

The experimental protocol required the presence, under supervision, of subjects for almost five full days, including the two nights they slept in the laboratory (Babkoff *et al.*, 1988).

In a typical experimental run, subjects arrived at the laboratory Saturday evening, and slept from approximately 2200–2300 to 0600 Sunday morning. Sunday was spent training and gathering baseline data (6 sessions) following the exact procedure to be used during the experiment. Training began at 0800 and ended at 2000, followed by 2 free hours and bedtime between 2200 and 2300. The experiment began at 0800 on Monday morning and continued until 0800 on Thursday morning at which time subjects were debriefed, allowed to nap if they wished, and accompanied home. Each subject was constantly accompanied by a laboratory technician who was responsible for testing and for keeping the subject awake. Laboratory technicians worked eight-hour shifts.

The protocol included physiological, psychological and performance testing every two hours, 36 test sessions for each of the tasks over the 72 hours of sleep deprivation. The testing segment of each two-hour block of time ranged from approximately 55 to 70 minutes providing a work-to-rest ratio of 0.45 to 0.60 (Babkoff *et al.*, 1985b).

Body Temperature: Body temperature was measured orally twice every two hours by a Diatek Model 500E digital thermometer, and the average recorded. Subjects did not eat, drink nor smoke for approximately 30 minutes prior to the measurement of body temperature (Babkoff *et al.*, 1988). The average group temperature (N = 12) is shown in FIGURE 1 as a function of hours of sleep deprivation scaled as time of day. The data show a small downward trend marked by very strong rhythmic oscillations. The results of the two-way repeated ANOVA indicate that there are significant changes across hours of each day ($F(11, 132) = 16.25$, $p < 0.0001$) and across the three days of sleep deprivation ($F(2, 24) = 4.03$, $p < 0.01$). Body temperature varies between the peak of 36.6°C at around 1800–2000 and 35.98°C at around 0400–0600, a range of 0.62°C. The average temperature on the third day (36.25°C) is significantly lower than on the first day (36.39°C).

The body temperature curve of each subject (N = 12) was analyzed by complex

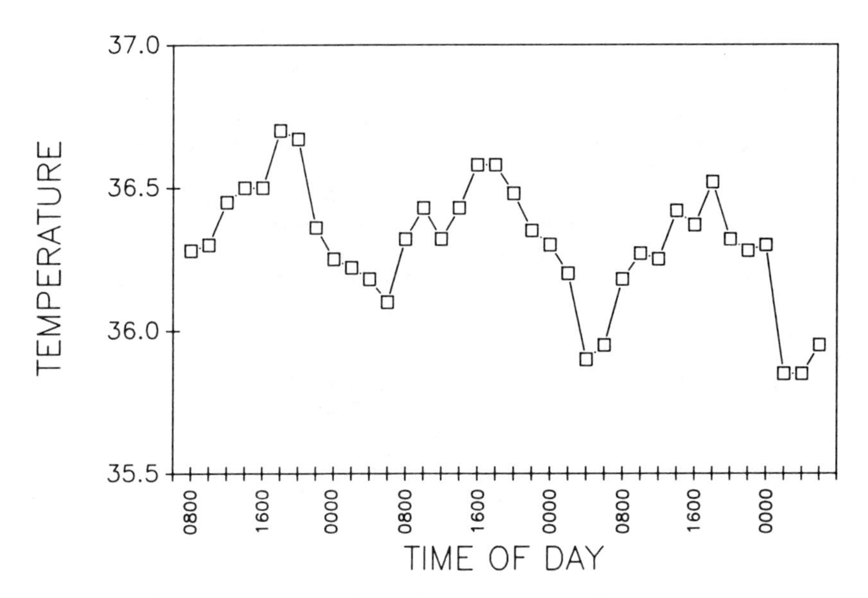

FIGURE 1. Body Temperature in °C: Group data (N = 12) plotted as a function of 72 hours of sleep deprivation.

demodulation (Redmond *et al.*, 1982; Sing *et al.*, 1984; Sing *et al.*, 1985) after removal of the linear component by regression analysis. The circadian remodulates were generated from the residuals. The amount of variance accounted for by the linear and circadian components are shown in TABLE 1. The linear component accounts for approximately 10.86% (± 9.45%) of the variance on the average; while the circadian component accounts of approximately 34.87% (± 13.84%) of the variance, a total of 45.73% of the variance. The finding that the rhythmic component accounts for much more of the variance than the linear component provides quantitative support for the visual impression of strong rhythmic oscillations with a relatively smaller overall decrease in temperature during sleep deprivation (FIG. 1).

Performance: The logical reasoning task was derived from Baddeley (1968) and has been used by several researchers (Angus & Heslegrave, 1985; Englund, Ry-

TABLE 1. Percent Variance Accounted for by Linear and Rhythmic Components[a]

	Components	
	Linear	Circadian
Temperature	10.86 ± 9.45	34.87 ± 13.84
Logical reasoning	14.92 ± 10.60	27.52 ± 11.49
Digit-symbol substitution	17.62 ± 14.26	27.02 ± 11.15

[a] The mean amount of variance explained by all components in this table is significant ($0.006 \leq p \leq 0.0001$).

man, Naitoh & Hodgdon, 1985; Haslam, 1982, 1985). The version of the task used in the present experiment consisted of 20 statements in Hebrew regarding the relative position of the Hebrew letters equivalent to A and B placed adjacent to each other. One pair of letters with a randomly determined order was placed to the left of each statement. The subject had 1 minute to complete as many of the problems as he could.

The logical reasoning task provided accuracy as well as general performance data (defined as the percent of the 20 possible problems attempted, whether the response is correct or not). Only general performance (or speed of performance) is examined since, in the authors' opinion, these data are best suited to illustrate the temperature-performance comparison. Furthermore, logical reasoning accuracy (computed in terms of errors of commission) is relatively insensitive to sleep loss using the Baddeley paradigm (Baddeley, 1968) when the relative position of two letters are judged on each trial (Angus & Heslegrave, 1985; Englund *et al.*, 1985; Heslegrave & Angus, 1985). The issues of the choice of the response class (*i.e.*, dependent variable) which best represent sleep loss data and the problem of the comparison of dependent variables across different studies of sleep deprivation have been raised and discussed by the present authors and by others (Angus & Heslegrave, 1985; Babkoff *et al.*, 1985a; Babkoff *et al.*, 1985b; Babkoff *et al.*, 1988; Carskadon & Dement, 1979) but will not be presented in detail in this paper.

The digit-symbol substitution was based on a task used by Haslam (Haslam, 1982, 1985) and Angus and Heselgrave (Angus & Heslegrave, 1985). The subject was presented with a sheet, at the top right of which were symbols (code) associated with numbers from 0 to 9 (stimuli to be encoded). The symbols associated with a given number differed from session to session. Below the codes and associated numbers were 10 double rows. The upper row consisted of 15 numbers, again differing from session to session, the lower row consisted of 15 boxes for encoding. Task duration was 1-½ minutes. The general performance data are examined in detail. Accuracy ranges between 98% and 100% with almost no variations, and no significant reduction over the three days of sleep deprivation. Although the hour of the day is significant ($F(11, 132) = 1.98$, $p \leq 0.035$), the Duncan Multiple Range Test ($p \leq 0.05$) indicates that this is due to the decrease in accuracy at 0600 from 99.3% to 98% as compared to the other 22 hours of the day. Although significant, a decrease of 1.3% in accuracy is hardly impressive as an indication of deficit, given the much larger decrease in general encoding performance (28%).

General performance on the two tasks is shown in FIGURE 2 as a function of hours of sleep deprivation scaled as time of day. Both curves show large overall decreases marked by oscillations. The results of two-way repeated ANOVAs indicate significant differences over days of sleep deprivation ($F(2, 24) = 19.53$; $p \leq 0.0001$; ($F(2.24) = 11.8$; $p \leq 0.0003$ for logical reasoning and digit symbol substitution respectively) and across hours of the day ($F(11, 132) = 6.46$; $p \leq 0.0001$; $F(11, 132) = 9.98$; $p \leq 0.0001$) in determining performance. The percent of trials attempted on the second and third days were significantly lower than on the first day for both performance tasks (a difference of 9.45% for logical reasoning and 7.45% for digit symbol substitution; Duncan Multiple Range test; $p \leq 0.05$). The percent of trials attempted between 0200 and 0600 was significantly lower than between 1600 and 2000 (14.9% and 18% for logical reasoning and digit symbol substitution respectively, Duncan Multiple Range test; $p \leq 0.05$). The difference between the performance during peak (1600–2000) and trough (0200–0600) hours was greater on the second and third days of sleep deprivation (19.7% on the average for logical reasoning, 22% for digit symbol substitution) than on the first day (8%

for logical reasoning, 12.5% for digit symbol substitution; $F(22, 259) = 2.56; p \leq 0.0002$; $F(22, 259) = 1.95; p \leq 0.008$).

The curves of the individual subjects were analyzed by complex demodulation as described above for body temperature. The amount of variance accounted for by the linear and circadian components is 14.9% and 27.52% for logical reasoning and 17.6% and 27.02% for digit symbol substitution.

Temperature and performance: Cross-Correlations and Curve Topology

Cross-correlations of the three (raw) data sets were performed for each of the subjects for each of the three days of sleep deprivation and are shown in TABLE

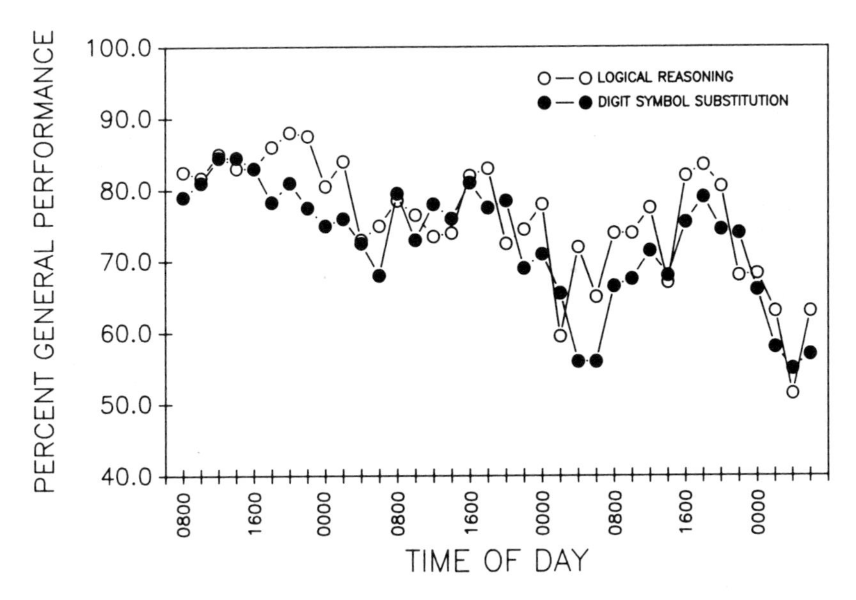

FIGURE 2. General performance is plotted separately for the logical reasoning and symbol substitution tasks as a function of hours of sleep deprivation.

2. The temperature-digit symbol substitution correlations are significant but low and fairly constant across the three days of sleep deprivation. The temperature-logical reasoning correlations are not significant. The logical reasoning-digit symbol substitution correlations are significant and increase over the three days of sleep deprivation. This finding, along with the curves shown in FIGURE 2, support the conclusion that the general decrease in performance of logical reasoning and digit symbol substitution coupled with the strong increasing oscillations during sleep deprivation represent not only a group trend, but also the trend of the individual subjects. On the other hand, the lack of systematic correlations between performance and temperature could be interpreted to mean that these dependent variables are not affected by the same circadian oscillators.

As noted previously, since each set is composed of monotonic and rhythmic components, cross-correlation of raw data reflects the extent to which all the factors within the data sets interrelate, but not the extent to which any one of the individual components of the different data sets correlates with another. The alternative analysis correlates the transformed data sets for each subject obtained by a time series technique such as complex demodulation. The complex demodulation analysis generates remodulates, which are smoothed functions of the rhythmic components extracted from the data and retransformed back into the time domain (Redmond *et al.*, 1982; Sing *et al.*, 1985). The remodulate is the best fitting model of a given frequency generated from the data set after the linear and general trends have been removed. The characteristics of the remodulate of each subject's

TABLE 2. Cross-Correlation of Raw Data and Circadian Remodulates of Temperature and General Performance on Logical Reasoning and Digit-Symbol Substitution Tasks

	Logical Reasoning	Digit-Symbol Substitution
A. Raw Data		
Day 1		
Temperature	−0.05	0.27*
Logical reasoning		0.42*
Day 2		
Temperature	0.00	0.21*
Logial reasoning		0.53*
Day 3		
Temperature	0.08	0.25*
Logical reasoning		0.59*
B. Circadian remodulate		
Day 1		
Temperature	0.495*	0.49*
Logical reasoning		0.42*
Day 2		
Temperature	0.66*	0.68*
Logical reasoning		0.52*
Day 3		
Temperature	0.67*	0.72*
Logical reasoning		0.58*

* $0.02 < p \leq 0.0001$.

performance curve, as determined by complex demodulation, are treated as transformed data which can be correlated and subjected to ANOVA (Babkoff *et al.*, 1991). Since the circadian remodulate represents the best fitting 24-hour rhythm, cross-correlating the remodulates generated by the three data sets should yield the size and degree of the interphase relations of this component.

The average circadian remodulates of body temperature and performance are plotted in FIGURE 3. Several features of these curves should be noted. First, there is only a moderate alignment of the three curves on the first day; by the second and third day, the performance and temperature curves nearly overlap and reach their peaks (maximums) and troughs (minimums) approximately coincidentally.

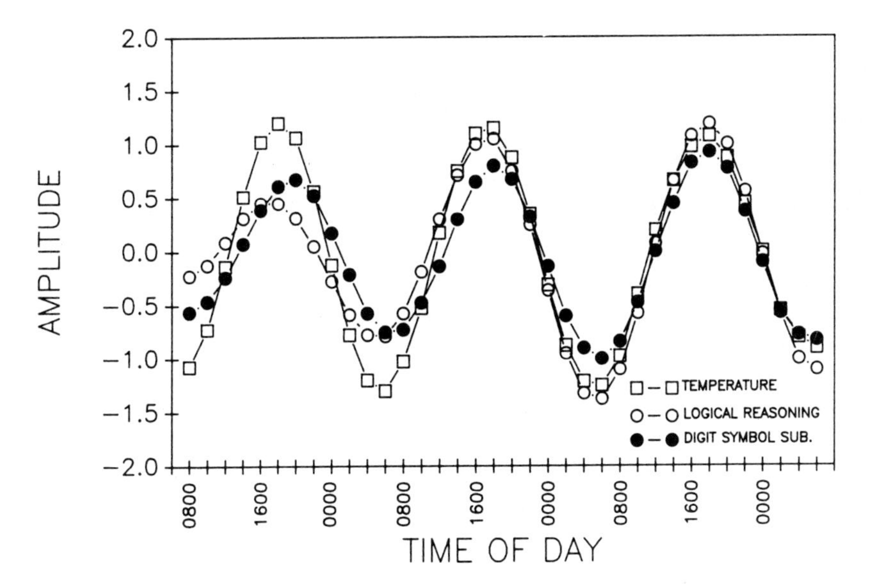

FIGURE 3. Circadian remodulates of body temperature, Logical reasoning and digit symbol substitution are plotted as a function of hours of sleep deprivation.

Second, there is an increase in the peak-to-trough (maximum to minimum) amplitudes of the performance circadian remodulates from the first to the third day.

Verification of the increasing phasic coincidence of the circadian remodulates over the three days is obtained by cross-correlating the circadian remodulates over the three days (TABLE 2). All of the correlations of the circadian remodulates are significant. Several clear trends are also evident. The correlation of temperature and logical reasoning increases from 0.50 on the first day to 0.66 and 0.67 on the second and third day. The temperature-digit symbol substitution correlations show a similar trend increasing from 0.49 to 0.72 across the three days. The logical reasoning-digit symbol substitution correlations of the circadian remodulate are very similar to those of the raw data.

A complementary analysis of the time domain distribution of the maxima and minima of the circadian remodulates of body temperature and performance was

TABLE 3. Maxima and Minima Times for Circadian Remodulates of Temperature and General Performance[a]

	Temperature		Logical Reasoning		Digit-Symbol Substitution	
	Max	Min	Max	Min	Max	Min
Day 1	1822(1.5)	0516(1.8)	1600(5.87)	1033(5.94)	1455(5.95)	0548(2.1)
Day 2	1655(1.64)	0538(2.16)	1433(5.73)	0538(4.1)	1749(2.27)	0455(1.86)
Day 3	1727(2.01)	0611(2.28)	1749(2.9)	0538(2.5)	1716(3.26)	0505(3.44)

[a] Average in 24-h clock time; standard deviations (in parentheses) in hours.

made. This analysis provides an estimate of intersubject variability and the extent to which phase locking occurs between temperature and performance.

The average times of occurrence of the maxima and minima of the circadian remodulates (±SD) over the three days of sleep loss are shown in TABLE 3. Body temperature yields the most stable circadian maxima and minima (smallest standard deviations). The three curves show increasing coincidence of maxima and minima over the 3 days of sleep deprivation, so that by day 3, the average maxima for the three variables range between 1716 and 1749, and the average minima range between 0505 and 0611.

The complex demodulation and cross-correlation of the characteristics of the rhythmic components of the performance curves with temperature confirm the conclusion of an increasing correspondence of the peaks and troughs of the circadian components of the three curves over time.

Curve Topology: Amplitude

Both performance curves share a common feature with body temperature: progressive phase locking and temporal correspondence of maxima and minima as a function of cumulative sleep loss. On the other hand, both performance curves have a feature which they do not share with body temperature: amplified circadian component as a function of accumulating sleep loss.

The interrelation of sleep loss and inherent rhythmicities may be determined by examining the stability of the circadian component over the three days of sleep deprivation. Is this relationship orthogonal or interactive? In an orthogonal relationship a basic rhythmicity may ride on an underlying monotonic curve whose removal leaves a rhythmic curve without change in period, amplitude or phase. An interactive relationship, on the other hand, may take several forms (Babkoff *et al.*, 1991). For example, the rhythm may either be damped (*i.e.*, the peak-to-trough amplitude may decrease over time) or enhanced (*i.e.*, the peak-to-trough amplitude may increase over time).

In several recent studies the sleep deprivation protocol was designed as a multiple of 24-hour periods (*e.g.*, 48 or 72 hrs) and performance testing was frequent and at fixed intervals. Such data can be analyzed by repeated measures ANOVA with days of sleep deprivation as one variable and hours of the day as another variable (Babkoff *et al.*, 1985a, 1988). The expectation is for decreased performance over the 2 or 3 days of sleep loss and for changes over the hours of the day within each day (diurnal effect). If both variables are significant, but the interaction is not, this implies that the amount of prior wakefulness (days of sleep deprivation) and the hour of the day effect are probably orthogonal to one another. If the interaction is significant, this implies that the change over the hours of the day differs from day 1 to days 2 and 3, of from days 1 and 2 to day 3. In the performance data sets under examination in this paper, the ANOVA yielded a day × hour interaction reflecting the exaggerated troughs in the second and third days of sleep loss (Colquhoun, 1981).This suggests an interactive relation between prior wakefulness and the circadian rhythm. However, the ANOVA is not capable of identifying and separating out the circadian from other rhythms in the data nor of defining the relationship between this component and a variable which changes monotonically (*e.g.*, days of sleep loss) (Monk *et al.*, 1985). Consequently, although the ANOVA can point to the existence of an orthogonal or interactive relationship between variables, it cannot clarify its nature. A time series technique, such as complex demodulation (CD) extracts constituent rhythms from the data and provides the best fitting remodulates. The rhythmic characteristics of the remodulates may then be tested statistically for changes over time. Complex demodulation can thus aid in determining whether and how sleep loss changes the nature of the circadian component of performance data. After removal of the linear component, changes in the peak-to-trough amplitude of the circadian component, changes in the peak-to-trough amplitude of the circadian component over days of sleep deprivation were tested by ANOVA. The results confirm the visual impressions of an increase in maxima to minima amplitude of both logical reasoning and digit symbol substitution over the three days of sleep loss ($F(2, 24) = 3.84$; $p \leq 0.036$; $F(2, 24) = 14.87$; $p \leq 0.0001$, respectively). No similar increase was found for temperature ($F(2, 24) = 0.66$). The digit symbol substitution and logical reasoning performance curves are examples of an interactive relationship between sleep loss and circadian rhythm. Even after the removal of the linear component, the amplitude of the circadian component increases significantly over the three days of sleep loss. On the other hand, the amplitude of the circadian component of body temperature does not change significantly over time, hence, body temperature conforms to the orthogonal model.

Interactive relationships between sleep deprivation and circadian rhythmicity in performance curves have been reported often in the recent literature. Alluisi and Chiles (1967) suggested that circadian oscillations become more prominent when subjects are sleep deprived. Bugge, Opstad and Magnus (1979) reported data from three cognitive tasks, generated during a four-day field study in which participants slept a total of approximately 1–2 hours. They reported a general decrease in performance from the first to the fourth day as well as an increase in the variations around the mean which led them to suggest that sleep loss potentiates the circadian rhythm. Several other recent studies have shown performance curves whose oscillation amplitude changes over time (Babkoff *et al.*, 1988, 1989, 1991), or whose deviations around a fixed-amplitude fitted cosinor function increase over time (Naitoh, Englund & Ryman, 1985).

Evidence is accumulating that although performance rhythms persist during sleep deprivation, thus supporting the argument for their endogenous nature (Minors & Waterhouse, 1981), the characteristics of the major component (circadian) may change over time. However, the change does not reflect weak rhythmicity in the face of accumulated sleep debt, as would be the case, were the amplitudes of the circadian rhythm to decrease, but rather, strong rhythmicity, since the amplitude of the rhythm is potentiated.

The theoretical implications of the choice of analytic procedures used to compare performance with temperature become clearer when comparing the correlations obtained by the two methods with the predictions of the hypothesis linking performance with temperature. In addition to potentiating circadian amplitudes, accumulating sleep loss also serves to phase lock the circadian cycle of performance with temperature (see FIG. 3 and TABLE 2). The version of logical reasoning used in this study is an example of an "immediate processing" task which makes very little, if any, demands on memory (Folkard & Monk, 1982, 1985). This point is equally valid for digit symbol substitution, which presents all of the information necessary for solution to the subject while he is performing the task. Low memory loading tasks are thought to be highly correlated with temperature, and are, hypothetically, generated by the stable, fairly strong Group I oscillators (Folkard & Monk, 1979, 1982, 1984, 1985; Wever, 1975). Consequently, the two performance tasks should correlate significantly with body temperature.

The prediction of linkage between temperature and performance can be tested directly by examining the correlations in TABLE 2. The results of the cross-correlation of the raw temperature and performance data (TABLE 2A) do not support this hypothesis, as neither the logical reasoning nor the digit symbol substitution task correlate highly with temperature although they do correlate with each other. However, the cross-correlations of the circadian remodulates (TABLE 2B) are highly significant and indicate increased phase locking between temperature and logical reasoning as sleep loss progresses. This latter analysis supports the argument that the circadian rhythms of all three variables share a source of variance and are likely generated by the same group of oscillators.

Two major changes in the circadian component of the performance data as a function of sleep deprivation have been observed, increased peak-to-trough amplitude and increased phase locking with body temperature. Both changes imply more manifest rhythmicity and stronger coupling to body temperature due to sleep loss. As noted, the performance tasks are relatively simple. In some instances, circadian rhythmicity has not been reported for simple performance tasks under normal sleep-wake conditions. This has been interpreted to mean either that such a rhythm does not exist for those tasks (Minors & Waterhouse, 1981) or that subjects may avoid the variation in performance due to the diurnal effect by expending extra effort if the task is simple and the subject not fatigued (Alluisi & Chiles, 1967; Hockey & Colquhoun, 1972). If the former interpretation is correct, then altering stimulus or response characteristics of the task should not change the form of the performance curve and produce oscillations. If the latter interpretation is correct, a diurnal rhythm may appear if the subject becomes fatigued by prolonged work or accumulating sleep loss, or if intratask characteristics are altered (*e.g.*, increasing stimulus discrimination difficulty or complicating the motor response) so as to make corrective compensation by the subject more difficult (Babkoff *et al.*, 1988, 1989). "Unmasking" of an inherent rhythm under the influence of accumulating fatigue due to sleep loss might explain the increased amplitude of the circadian component from the first to the third day of sleep deprivation. Rutenfranz and Colquhoun (1979) noted that performance rhythms appear to be related to the daily cycle of sleep "need" and the largest decrements are noted when the need is greater.

However, unmasking is not sufficient to explain the phase shifting of the two performance circadian rhythms during sleep deprivation. If sleep loss only serves to reduce the ability of the subjects to correct their performance so as to maintain their high level and, consequently, the inherent circadian rhythmicity is revealed, then the phase of that rhythmicity should remain intact. The progressive phase locking of the two performance circadian rhythms with the circadian temperature rhythm during accumulating sleep deprivation implies that sleep loss can also serve as an entrainer. In an isolated environment which promotes "free running," sleep can serve as an entrainer (anchor sleep) of body temperature (Mills, Minors & Waterhouse, 1974; Minors & Waterhouse, 1980, 1981), perhaps fatigue or sleepiness associated with lack of sleep can also lead to entrainment of performance. As sleep loss has a linearly progressive rather than a cyclically varying character it is unlikely that it could serve, by itself, as an entrainer. However, sleep loss might act as a facilitator of entrainment, making the Group I related oscillators which control performance more sensitive to entrainment by the oscillator controlling the temperature rhythm. This could explain the phase locking of the three variables as well as amplification of the peak-to-trough amplitudes of the performance circadian remodulates.

In summary, there is evidence for the inherent coupling of the circadian rhythms of logical reasoning and digit symbol substitution with the group of oscillators generating body temperature and for the role of sleep loss as an entrainer or facilitator of their entrainment by these oscillators. The important methodological point is that correlating the raw data sets of performance and temperature yielded low or insiginficant correlations; while correlating the circadian remodulates generated by the complex demodulation yielded highly significant correlations, indicating strong underlying phase locking between performance and temperature.

SUMMARY

One of the major methodological-analytic problems encountered by researchers in sleep deprivation involves the examination and analysis of the relationship between sleep loss and rhythmic influences on performance. The comparison of performance rhythms with physiological rhythms, *e.g.*, body temperature, generated under the same conditions of sleep deprivation, has become an important means of testing for an endogenous source of the rhythmicity in the data and for clarifying the nature of the proposed oscillator system. Should the data sets be correlated before or after their separation into monotonic and rhythmic parts? Correlating the raw data without separating them into their components can yield negative results, while, in reality, some of the major underlying rhythms may be highly related. The example used in this chapter showed strong cross correlations of the circadian components of temperature and two performance tasks. Sleep deprivation is thus seen to interact with performance rhythms. This interaction is only revealed after the data are analyzed and broken into their component parts. This procedure leads to the conclusion that certain performance rhythms and temperature may share the same generating oscillators.

ACKNOWLEDGMENT

The authors would like to thank Dr. Tamsin L. Kelly, Naval Health Research Center, San Diego, CA, for her reading and editorial corrections.

REFERENCES

ALLUISI, E. A. & W. D. CHILES. 1967. Sustained performance, work rest scheduling and diurnal rhythms in man. Acta Psychol. **27:** 436–442.

ANGUS, R. G. & R. J. HESLEGRAVE. 1985. Effects of sleep loss on sustained cognitive performance during a command and control simulation. Behav. Res. Methods Instrum. Comput. **17:** 55–67.

ASCHOFF, J. & R. WEVER. 1976. Human circadian rhythms: a multioscillatory system. Fed. Proc. **35:** 2326–2332.

ASCHOFF, J. 1979. Circadian rhythms. General features and endocrinological aspects. *In* Endocrine Rhythms. D. T. Krieger, Ed. 1–61. Raven Press. New York.

BADDELEY, A. D. 1968. A 3 min reasoning test based on grammatical transformation. Psychon. Sci. **10:** 341–342.

BABKOFF, H., T. CASPY, M. MIKULINCER, R. CARRASSO & H. C. SING. 1989. The implications of sleep loss for circadian performance accuracy. Work Stress **3:** 3–14.

BABKOFF, H., T. CASPY, M. MIKULINCER & H. C. SING. 1991. Monotonic and rhythmic influences: a challenge for sleep deprivation research. Psychol. Bull. **109:** 411–428.

BABKOFF, H., S. G. GENSER, H. C. SING, D. R. THORNE & F. W. HEGGE. 1985a. The effects of progressive sleep loss on a lexical decision task: response lapses and response accuracy. Behav. Res. Methods Instrum. Comput. **17:** 614–622.

BABKOFF, H., M. MIKULINCER, T. CASPY, D. KEMPINSKI & H. C. SING. 1988. The topology of performance curves during 72 hours of sleep loss: a memory and search task. Quart. J. Exp. Psychol. **40:** 737–756.

BABKOFF, H., D. R. THORNE, H. C. SING, S. G. GENSER, S. L. TAUBE & F. W. HEGGE. 1985b. Dynamic changes in work/rest duty cycles in a study of sleep deprivation. Behav. Res. Methods Instrum. Comput. **17:** 604–613.

BLAKE, M. J. 1967. Time of day effects on performance in a range of tasks. Psychon. Sci. **9:** 349–350.

BLAKE, M. J. 1971. Temperament and time of day. *In* Rhythms and Human Performance. W. P. Colquhoun, Ed. 109–148. Academic Press. London.

BOX, G. E. P. & G. M. JENKINS. 1976. Time Series Analysis: Forecasting and Control. Holden Day. San Francisco.

BUGGE, J. F., P. K. OPSTAD & P. M. MAGNUS. 1979. Changes in the circadian rhythm of performance and mood in healthy young men exposed to prolonged, heavy physical work, sleep deprivation, and caloric deficit. Aviat. Space Environ. Med. **50:** 663–668.

CARSKADON, M. A. & W. C. DEMENT 1979. Effects of total sleep loss on sleep tendency. Percept. Mot. Skills **48:** 495–506.

COLQUHOUN, W. P., M. J. BLAKE & R. S. EDWARDS. 1968. Experimental studies of shift work I: a comparison of 'rotating' and 'stabilized' 4-hour shift systems. Ergonomics **11:** 437–454.

COLQUHOUN W. P. 1971. Circadian variations in mental efficiency. *In* Biological Rhythms and Human Performance. W. P. Colquhoun, Ed. 39–107. Academic Press. London.

COLQUHOUN, W. P., ed. 1972. Aspects of Human Efficiency: Diurnal Rhythm and Loss of Sleep. English Universal Press. London.

COLQUHOUN, W. P. 1981. Rhythms in performance. *In* Handbook of Behavioral Neurobiology. J. Aschoff, Ed. Vol. 4: 333–348. Plenum Press. New York.

COLQUHOUN, W. P. 1982. Biological rhythms and performance. *In* Biological Rhythms, Sleep, and Performance. W. B. Webb, Ed. 59–86. Wiley, New York.

DINGES, D. F. 1988. The nature of sleepiness: causes, context and consequences. *In* Perspective in Behavioral Medicine. A. Baum & A. S. Stunkard, Eds. Erlbaum. Hillsdale, NJ.

ENGLUND, C. E., D. H. RYMAN, P. NAITOH & J. A. HODGDON. 1985. Cognitive performance during successive sustained physical work episodes. Behav. Res. Methods Instrum. Comput. **17:** 75–85.

FOLKARD, S. & T. H. MONK. 1979. Time of day and processing strategy in free recall. Quart. J. Exp. Psychol. **31:** 461–475.

FOLKARD, S. & T. H. MONK. 1982. Circadian rhythms in performance—one or more oscillations? *In* Psychophysiology 1980—Memory, Motivation and Event-Related Potentials in Mental Operations. R. Sinez & M. R. Rosenzweig, Eds. Elsevier. Amsterdam.

FOLKARD, S. & T. H. MONK. 1983. Circadian rhythms and human performance. *In* Psychological Correlates of Human Behavior: Attention and Performance. A. Gale & J. A. Edwards, Eds. Vol. 2. Academic Press. London.

FOLKARD, S. & T. H. MONK. 1984. Do we need a multioscillator model of circadian rhythms in human performance? *In* Chronobiology 1982–1983. E. Haus & H. F. Kabat, Eds. 185–191. Karger. Basle.

FOLKARD, S. & T. H. MONK. 1985. Circadian performance rhythms. *In* Hours of Work. S. Folkard & T. H. Monk, Eds. 37–52. John Wiley & Sons. New York.

FROBERG, J. E. 1985. Sleep deprivation and prolonged working hours. *In* Hours of Work. S. Folkard & T. H. Monk, Eds. John Wiley & Sons. New York.

HASLAM, D. R. 1982. Sleep loss, recovery, sleep, and military performance. Ergonomics **25:** 163–178.

HASLAM, D. R. 1985. Sleep deprivation and naps. Behav. Res. Methods Instrum. Comput. **17:** 46–54.

HESLEGRAVE, R. G. & R. G. ANGUS. 1985. The effects of task duration and work-session location on performance degradation induced by sleep loss and sustained cognitive work. Behav. Res. Methods Instrum. Comput. **17:** 592–603.

HOCKEY, G. R. J. & W. P. COLQUHOUN. 1972. Diurnal variation in human performance: a review. *In* Aspects of Human Efficiency: Diurnal Rhythm and Loss of Sleep. W. P. Colquhoun, Ed. English Universities Press. London.

HORNE, J. A. 1978. A review of the biological effects of total sleep deprivation in man. Biol. Psychol. **7:** 55–102.

HORNE, J. A. 1985. Sleep function, with particular reference to sleep deprivation. Ann. Clin. Res. **17:** 199–208.

HORNE, J. A. 1988. Why We Sleep: the Functions of Sleep in Humans and Other Mammals. Oxford University Press. Oxford.

JOHNSON, L. C. & P. NAITOH. 1974. The operational consequences of sleep deprivation and sleep deficit (AGARD Report No. AG-193). 1–43. North Atlantic Treaty Organization. London.

JOHNSON, L. C. 1982. Sleep deprivation and performance. *In* Biological Rhythms, Sleep and Performance. W. B. Webb, Ed. 111–141. Chichester, England. Wiley.

KJELLBERG, A. 1977. Sleep deprivation and some aspects of performance: II. Lapses and other attentional effects. Waking Sleeping **1:** 145–148.

KLEITMAN, N. 1963. Sleep and Wakefulness. 2nd Edit. University of Chicago Press. Chicago.

KRONAUER, R. E. 1984. Modeling principles for human circadian rhythms. *In* Mathematical Models of the Circadian Sleep Wake Cycle. M. C. Moore-Ede & C. A. Czeisler, Eds. Raven Press. New York.

LAVIE, P. 1980. The search for cycles in mental performance from Lombard to Kleitman, Chronobiologia **7:** 247–256.

MEDDIS, R. 1982. Cognitive dysfunction following loss of sleep. *In* The Pathology and Psychology of Cognition. E. Burton, Ed. 225–252. Methuen. London.

MIKULINCER, M., H. BABKOFF, T. CASPY & H. C. SING. 1989. The effects of 72 hours of sleep loss on psychological variables. Br. J. Psychol. **80:** 145–162.

MILLS, J. N., D. S. MINORS & J. M. WATERHOUSE. 1974. The circadian rhythms of human subjects without timepieces or indication of the alternation of day and night. J. Physiol. **240:** 567–594.

MINORS, D. S. & J. M. WATERHOUSE. 1980. Does 'anchor sleep' entrain the internal oscillator that controls circadian rhythms? J. Physiol. **308:** 92P–93P.

MINORS, D. S. & J. M. WATERHOUSE. 1981. Circadian Rhythms and the Human. Chapters 2, 6, and 12. Wright. Bristol.

MINORS, D. S. & J. M. WATERHOUSE. 1985. Introduction to circadian rhythms. *In* Hours of Work. S. Folkard & T. H. Monk, Eds. John Wiley & Sons. New York.

MONK, T. H., F. E. FOOKSON, J. KREAM, M. L. MOLINE, C. P. POLLAK & M. B. WEITZMAN. 1985. Circadian factors during sustained performance: background and methodology. Behav. Res. Methods Instrum. Comput. **17:** 19–26.

NAITOH, P. 1976. Sleep deprivation in human subjects: a reappraisal. Waking Sleeping **1:** 53–60.

NAITOH, P. 1982. Chronophysiological approach for optimizing human performance. *In* Rhythmic Aspects of Behavior. F. M. Brown & R. A. Graeber, Eds. 41–103. Erlbaum. Hillsdale, NJ.

NAITOH, P., C. E. ENGLUND & D. H. RYMAN. 1985. Circadian rhythms determined by cosine curve fitting: analysis of continuous work and sleeploss data. Behav. Res. Methods Instrum. Comput. **17:** 630–641.

ORR, W. C. & H. J. HOFFMAN. 1974. A 90-minute cardiac biorhythm: methodology and data analysis using modified periodiograms and complex demodulation. IEEE Trans. Biomed. Eng. **21:** 130–143.

ORR, E. C., H. J. HOFFMAN & F. W. HEGGE. 1976. Ultradian rhythms in external performance. Aerosp. Med. **45:** 995–1000.

REDMOND, D. P., H. C. SING & F. W. HEGGE. 1982. Biological time series analysis using complex demodulation. *In* Rhythmic Aspects of Behavior. F. W. Brown & R. C. Graeber, Eds. 429–457. Erlbaum. Hillsdale, NJ.

RUTENFRANZ, J., J. ASCHOFF & H. MANN. 1972. The effects of a cumulative sleep deficit, duration of preceding sleep period and body-temperature on multiple choice reaction time task. *In* Aspects of Human Efficiency. W. P. Colquhoun, Ed. 217–229. English University Press. London.

RUTENFRANZ, J. & W. P. COLQUHOUN. 1979. Circadian rhythms in human performance. Scand. J. Work Environ. Health **5:** 167–177.

SING, H. C., D. P. REDMOND & F. W. HEGGE. 1980. Multiple complex demodulation: a method for rhythmic analysis of physiological and biological data. Proceedings of the 4th Annual Symposium on Computer Application in Medical Care. 151–158. Institute of Electrical and Electronic Engineers. New York.

SING, H. C., F. W. HEGGE & D. P. REDMOND. 1984. Complex demodulation technique and application. *In* Chronobiology 1982–1983. E. Haus & H. F. Kabat, Eds. Karger. Basel.

SING, H. C., D. R. THORNE, F. W. HEGGE & H. BABKOFF. 1985. Trend and rhythm analysis of time-series data using complex demodulation. Behav. Res. Methods Instrum. Comput. **17:** 623–629.

WEVER, R. A. 1975. The circadian multi-oscillator system of man. Int. J. Chronobiol. **3:** 19–55.

WEVER, R. A. 1979. The Circadian System in Man: Results of Experiments Under Temporal Isolation. Springer Verlag. New York.

WEVER, R. A. 1982. Behavioral aspects of circadian rhythmicity. *In* Rhythmic Aspects of Behavior. F. M. Brown & G. C. Graeber, Eds. 105–171. Erlbaum. Hillside, NJ.

WILKINSON, R. T. 1982. The relationship between body temperature and performance across circadian phase shifts. *In* Rhythmic Aspects of Behavior. F. M. Brown & G. C. Graeber, Eds. 213–240. Erlbaum. Hillside, NJ.

Hippocampal Function and Schizophrenia

Experimental Psychological Evidence[a]

PETER H. VENABLES

Department of Psychology
University of York
Heslington York YO1 5DD, England

INTRODUCTION

Some brief explanation is required about the decision to present a paper on hippocampal function in schizophrenia in a Festschrift for Sam Sutton.

My first reaction on meeting with Sam over a quarter of a century ago was that here was a man who was essentially an experimental psychologist (in spite of his origins as an auditory physiologist). Although he was working in the fields of neurophysiology and psychopathology, his approach was the same as a worker from mainstream experimental psychology. His astute manipulation of experimental variables was masterly, indeed on some occasions it took all one's time to follow the intricacies of his designs. It was this approach which engendered my admiration, and consequently I wanted, on this occasion, to present a discussion of work which has its foundations in experimental psychology but which I feel his implications for the study of psychopathology, an enterprise which I hope he would have approved.

However, while the essence of the approach of experimental psychology is the hypotheticodeductive method, the application of this method in the field of psychopathology is often not easy or straightforward. Hypotheses do not come easily, rather a productive approach is that of trying to see how a jigsaw of often disparate findings fits together. Sam himself (Sutton, 1973) advocated the process of constructive validation "the use of converging operations—the process of making and testing additional inferences in order to interpret the finding properly" (p. 210). While the preferred method is to use these converging operations in one's own experiments, it is a good second best to attempt to examine how data in the literature may be used in the same way. This is the aim of the present paper. Almost inevitably, the selection of experiments to be discussed will be biassed. In this context the bias is towards studies which in particular have their origins in general experimental psychology and thus permit the development of ideas of schizophrenic dysfunction from a sound footing.

The starting point of the work to be reviewed is the contention that one of the fundamental dysfunctions of schizophrenia is to be found in the area of attention. This was a view being put forward (Venables, 1964) on the occasion of my first

[a] This paper was written while the author was in receipt of an Emeritus Fellowship from the Leverhulme Trust.

meeting with Sam and the emphasis did not change markedly in chapters in books of which he was co-editor (Venables, 1973, 1975). The other aspect of work in this area which appears to be fundamental is the recognition that we cannot talk about 'schizophrenia' as a unitary phenomenon. At least, it is important to introduce dichotomies, *e.g.*, chronic/acute, poor/good premorbid, process/reactive, non-paranoid/paranoid or in more current terminology, negative/positive symptom schizophrenia. These are workable dichotomies; however, it should not be assumed that they are immediately equivalent as contrasting categories, even though in some instances one end of the dichotomy matches that in another category. Nor should the use of dichotomies lull us into accepting a categorical view of schizophrenia. Much more appropriate is a dimensional approach. This is inherent in a 'stress/diathesis' view of the etiology of the disorder(s) and also in the recognition that experimental work on schizophrenia should not confine itself solely to work on patients but should also employ studies on normal subjects having some characteristics of the patient. The idea of a dimensional, rather than a categorical view of schizophrenia probably received its greatest initial emphasis in an article by Meehl (1962), but has been put into a practical perspective by the work of the Chapmans (*e.g.*, Chapman & Chapman, 1985). Here again, there is strong evidence that schizotypy is not unidimensional but rather there is a parallel to the concept of positive, paranoid, acute schizophrenia in aspects of schizotypy measured by, for instance, the Chapmans' perceptual aberration and magical ideation scales, while the negative, process, type of schizophrenia has reflection in anhedonia scales (see Venables, Wilkins, Mitchell, Raine & Bailes, 1990 for a recent review).

In relation to attentional dysfunction, one starting point, chosen from a wide selection of others which make roughly the same point (*e.g.*, McGhie, 1970; Neale & Cromwell, 1970), might be from a statement (Venables, 1964, p. 41):

> Chronic schizophrenic patients—and possibly included in this category are process patients—tend to be characterised by a state of restriction of the attentional field. . . . In contrast to the chronic patient, the acute (and possibly the reactive and the paranoid) patient is characterised by an inability to restrict the range of his attention so that he is flooded by sensory impressions from all quarters.

This quotation may serve as a point of departure, but only as such, for there have since been many advances in the field. Among the most important have been the integration of theories of attention from experimental psychology, such as those of Broadbent (1958, 1971), on early and late processing, and Schneider and Shiffrin (1977), as well as Shiffrin and Schneider (1977), on automatic and controlled processing, into hypotheses of attentional disturbance in schizophrenia (*e.g.*, Hemsley, 1975, 1987). Early views of the attentional dysfunction in schizophrenia, emphasising the inability of patients to select relevant from irrelevant aspects of the environment, suggested that this was due to a breakdown in an early 'filtering' process. However, more considered approaches (Hemsley, 1987; Venables, 1984) suggest that the disturbance is likely not to be found in an early process but more likely in a later information processing (or 'pigeonholing') mechanism where the selection of the relevant from the irrelevant is not made on the simple or physical characteristics of the incoming stimuli but rather on the processed assessment of their importance. This proposal, as it stands, is also probably too simple, insofar as it can be suggested that adequate full processing leads to the creation of models which enable 'automatic' processing to take place (Schneider & Shiffrin, 1977). There is thus the likelihood of an appearance of inadequacy of both early and late processing but due initially to a failure in one of the processes.

Assuming that there is a failure of selective attention in schizophrenia, the review to be undertaken has the *limited* aim of attempting to examine how far this failure is compatible with the idea that its physiological substrate is in the hippocampus. It would be foolish to suggest that other mechanisms are not involved, bearing in mind the centrality of the hippocampus in the totality of limbic and corticolimbic systems. Furthermore, the review makes no attempt to take into account the neurotransmitters involved, again a limiting viewpoint taking into account our knowledge in particular of dopamine involvement. Nevertheless, it is the aim of the review to concentrate specifically on experimental psychological studies which are based on animal studies of the effect of hippocampal lesions in the hope that by doing so some ideas may be brought together to stimulate future studies of schizophrenia.

BACKGROUND

Before reviewing up to date literature which reinforces the idea of hippocampal involvement as a possible source of disorder in schizophrenia, a brief review of the background material which first raised the hypothesis is warranted.

There are probably two origins to be considered. One is the first report in 1970 by Mednick from the Copenhagen High-Risk study that those of the 'at-risk' subjects who at that time had become psychiatrically unwell, had experienced more perinatal birth complications than those who remained well or who were in the control group. Mednick considered brain sites of selective vulnerability, and where damage was particularly due to anoxic factors. After a review of the literature, he suggested that the hippocampus was a strong candidate. *Inter alia* he quoted Friede (1966) who stated "the hippocampus represents one of the most striking examples of selective vulnerability in the brain." Later, Mednick and Schulsinger (1973) expanded the review of the relevant literature on this topic to include consideration of the relevant animal literature on the behavioral effects of hippocampal lesions and also work on the orienting response in schizophrenic patients. In this review, Mednick and Schulsinger stated that "it may well be that hippocampal dysfunction is an important contributing predispositional factor in only some types of schizophrenia, perhaps the more typical process, chronic or poor premorbid types." Part of the function of the present review will be to see how far the direction of this proposed relationship between hippocampal dysfunction and schizophrenic subdiagnosis may be substantiated.

The second, parallel line of work going on at roughly the same time, was that which resulted in 1972 in a paper on the orienting response in schizophrenia by Gruzelier and Venables. The work concerned the attempt to resolve the controversy which was presented by the then available literature; some of which suggested that schizophrenic patients exhibited less autonomic orienting than normals and other reports which suggested that schizophrenic patients were characterized by increased orienting responses which habituated more slowly. The outcome of the Gruzelier and Venables study was that there appeared to be a dichotomy in an unselected schizophrenic population with some 40% of the patients showing no orientation (the nonresponders), while an equal sized group appeared to show exaggerated orientation and less than normal habituation (the responders). It was suggested, in seeking for possible explanation of these findings that 'nonresponding' and 'hyperresponding' might respectively involve amygdala[1] and hippocampal dysfunction. In a review of the then available literature, Venables (1973) cites, in particular, the work of Bagshaw, Kimble and Pribram (1965) which was (and still is) one of the few studies where autonomic aspects of the orienting response had been studied in animals with lesions of the amygdala and hippocampus. The former lesion tended to produce animals who showed hyporesponsivity, whereas the latter lesion produced animals whose ORs did not habituate.

This is not the place to start yet another review of the literature on the orienting response in schizophrenia. However, it is perhaps worthwhile summarizing what appears to be the present position. Bernstein, Frith, Gruzelier, Patterson, Straube, Venables and Zahn (1982) reviewed the wide range of studies that they had severally carried out and reported that "one universal dysfunction emerged. Consistently, schizophrenics displayed an abnormally high incidence of non-responsiveness, involving nearly 50% of the schizophrenic sample on average." However, the Gruzelier and Venables (1972) finding of more or less equal numbers of nonresponders and hyperresponders was not replicated universally; "evidence for a dysfunction simultaneously involving SCOR hypo- and hyperresponsiveness within schizophrenia was obtained, but in a minority of studies." Ohman (1981), in an extensive review, reached a similar conclusion, namely, that there are two distinct forms of autonomic responsiveness in schizophrenia. He concluded that the two groups differ from each other in clinical picture and that the nonresponding pattern "may be secondary to a clinical picture of withdrawal and confusion, whereas the responder pattern may index vulnerability to schizophrenic episodes." Two important outcomes result from these studies. Firstly, that the group of patients labelled schizophrenics contains at least two subgroups and that, if as generally argued (*e.g.*, Ohman, 1981), the orienting response is best considered as involving attentional/information processing systems, then it would appear that the two different subgroups of schizophrenics might employ different processes and systems. It is the hyperresponding subgroup, who may have higher vulnerability to schizophrenic episodes (Ohman 1981), who exhibit largely 'positive' symptoms and little improvement during the course of treatment (Frith, Stevens, Johnstone & Crow, 1979; Zahn, Carpenter & McGlashan, 1981; Bartfai, Levander, Edman, Schalling & Sedvall, 1983). If this group, as suggested above, are characterised by hippocampal dysfunction, then this clearly warrants a further examination of the evidence from other sources, and this is what this paper will now undertake.

Some Animal Studies on Hippocampal Function

The proposal, that consideration of nonhuman work is important, arises from animal studies which suggest that the role of the hippocampus in the control of attention has direct relevance to theories of the nature of the disturbance of attention in schizophrenics.

Douglas and Pribram (1966) and Douglas (1967) put forward the notion that it was the function of the hippocampus "to exclude stimulus patterns from attention through a process of efferent control known as gating" and further distinguished

[1] It should be noted that a recent report states that it is possible to elicit skin conductance responses in humans even after bilateral amygdala damage (Tranel & Damasio, 1989).

two types of gating: nonspecific gating, which "has been postulated to have the function of protecting memory traces from interference during consolidation" and specific gating, which "acts to inhibit reception of specific stimuli which have been associated with nonreinforcement." The position was summed up epigrammatically as follows: "The amygdaloid system makes stimuli more figural, while the hippocampal system converts figure into ground" (Douglas, 1967, pp. 435, 436).

Later, Black, Nadel and O'Keefe (1977) proposed that the main function of the hippocampus is to process information about place, in other words to provide spatial maps. Insofar as the dysfunctions found in schizophrenia are not confined to experience of disorientation in space (although such disorientation is, of course, sometimes found, for instance, as a deficit in auditory localization in space (Balogh, Schuck & Leventhal, 1979), it would appear that hippocampal dysfunction is not involved in schizophrenia if we adhere to the limited view of hippocampal function espoused by Black *et al.* (1977).

Moore (1979) and Solomon (1979), however, suggested that the view of Black *et al.* (1977) did not account for all the then available data. They put forward a view that the hippocampus is "involved in the processing of temporal as well as spatial information" (Solomon, 1979, p. 1272) and that "the hippocampus enables the animal to ignore (tune out) irrelevant stimuli and events" (Moore, 1979, p. 224). In doing so, to reconcile his position with that of Black *et al.* (1977), Moore suggested that "the essential point is that although hippocampal damage can disrupt performance in spatial memory tasks, the psychological processes underlying such disruption may have a significant temporal dimension" (Moore, 1979, p. 227). In summary, Moore puts forward a position (not too different from the first element of the Douglas-Pribram 1966 position) which appears to be particularly relevant to the forms of disturbance seen in schizophrenia namely that: "the hippocampus is not solely concerned with spatial learning, but that the selective character of behavioral deficits can be interpreted in terms of a general information processing function of the hippocampus. Instead of regarding the hippocampus as the repository of spatial engrams or cognitive maps, it may be more useful to regard the function of the hippocampus as one of tuning out irrelevant stimuli or events" (Moore, 1979, p. 231).

Solomon and Moore's tuning-out hypothesis is based to a large extent on studies which show that hippocampal animals exhibit deficits in latent inhibition (Lubow & Moore, 1959; Lubow, 1973), and it thus is particularly relevant initially to examine studies with schizophrenic patients which involve this paradigm.[2]

In a latent inhibition experiment, the subject is given repeated preexposure to a stimulus that is to be the conditioned stimulus (CS) in a classical conditioning paradigm. When this stimulus is later paired with the unconditioned stimulus, the normal subject shows impaired conditioning in comparison to the subject not previously exposed to the CS. This phenomenon is latent inhibition. Animals with hippocampal damage, on the other hand, do not show signs of latent inhibition; their performance is as though they had not previously been exposed to the CS

[2] Although the emphasis in this paper is on the role of the hippocampus in selective attention, it should be noted that many of the studies on latent inhibition which are to be reviewed later take as their starting point the finding that latent inhibition is disrupted by amphetamine (ingestion of which causes the release of dopamine) and restored by neuroleptics. Thus if it can be shown that all or some schizophrenic patients exhibited impaired latent inhibition this provides support for the dopamine hypothesis of schizophrenic disorder (*e.g.*, Crow, 1980).

(Kaye & Pearce, 1987).[3] The interpretation is that the intact animal has been able to 'tune out' the preexposed CS as it signalled nothing that was relevant, whereas the hippocampal animal does not 'tune out' the irrelevant CS and therefore behaves as though it had not previously been exposed to it, in other words it had not become habituated to it. This view is in line with that put forward earlier by Kimble (1968), *viz.*,

> The behavior of the rats with hippocampal lesions . . . can be interpreted as an example of behaviour of organisms relatively devoid of the neural machinery necessary for the production of stimulus-induced internal inhibition and the normal decrement in response strength which presumably results from internal inhibition (p. 291).

Studies on Latent Inhibition, and Negative Priming on Schizophrenic Patients and Schizotypic Normals

Two sets of studies will be reviewed; in one set the authors have deliberately set out to use paradigms which are adaptations of those used in the experiments on animals reviewed above, and in the other set, the direct parallel with animal experiments is not drawn but the experimental paradigm used has a structure which makes it somewhat similar to that in the first group. The main body of experiments in this second group are those on 'negative priming' (Tipper, 1985).

The first attempt to employ the latent inhibition (LI) paradigm with schizophrenic patients was by Lubow, Weiner, Schlossberg and Baruch (1987). These workers used medicated schizophrenic patients who on the basis of their mean total hospitalization and numbers of admissions would be classed as chronic. They were divided into paranoid and nonparanoid groups and their performance was compared to that of normal undergraduates. There was a clear difference in age between the normal and patient subjects, which impaired the outcome of the study. Additionally, if as stated earlier, it would be expected that it would only be the acute patients with positive symptoms who would be characterized by hippocampal dysfunction and would therefore exhibit attenuation of latent inhibition, it would seem unlikely that this attenuation would be shown in the chronic patients in this study. In fact normal LI was shown in both paranoid and nonparanoid groups of chronic schizophrenics. If, as suggested by Venables (1964), the chronic patient is characterised by narrowed attention, it is likely that he might in any case have the capacity to ignore irrelevant stimuli and thus have the capacity to show latent inhibition.

What is important to note, however, is the paradigm which was used in this experiment, as it is used in later studies. The material consisted of recordings of two lists of the same 40 nonsense syllables repeated 5 times. On one of the recordings the target stimulus, a series of white noise bursts, was recorded. The experimental groups were exposed initially to the nonsense syllables plus noise and told that they should count the number of occurrences of a particular nonsense syllable. This direction of attention to the nonsense syllables was intended to act as a mask to the presentation of the noises. The control groups were exposed only to the nonsense syllables, which they also had to count. In the critical second part of the study, both control and experimental subjects were exposed to the syllable

[3] It is also relevant to note that Kaye and Pearce also showed that in animals with hippocampal lesions there was an impairment of habituation of the orienting response.

plus noise tape. In this instance, they were shown how scores were increased on a scoring device and asked to say as soon as they knew the rule for the score being increased. The score was, in fact, augmented every time a noise was presented.

The finding of the study was that both schizophrenic groups and the normal group who had been preexposed to the noise stimuli took longer to guess the rule than the not-preexposed groups, thus demonstrating the LI effect.

As stated above, it would not be expected that the schizophrenic patients in the experiment of Lubow *et al.* would show the LI effect. A second study, Baruch, Hemsley and Gray (1988a), however, remedied this position by using both acute and chronic schizophrenics as subjects. The procedure of the study was essentially the same as that of the first study by Lubow *et al.* The first group of subjects were acute schizophrenics who were described as being either in the first stage of a psychotic breakdown or in the acute phase of a chronic disorder. The second group were remitted or day patients, chronic schizophrenics who were currently free from hallucinations, delusions or other major symptoms. The normal subjects were more closely age matched to the patients than in the previous study. The findings were that both the chronic schizophrenics and the normal subjects showed evidence for latent inhibition while this was not exhibited by the acute patients. In commenting on the results the authors state that the finding with the acute patients

> cannot be attributed simply to a nonspecific loss of efficient functioning, as might be expected in the acute phase of a schizophrenic illness: it points rather to a specific loss of ability to ignore, or learn to ignore irrelevant stimuli.

(Note the close similarity in wording to that put forward as the function of the hippocampus as being to tune out irrelevant stimuli.)

The third study in this series (Baruch, Hemsley & Gray, 1988b) employed the same experimental procedure as the two earlier studies but used as its subjects the normal group of the previous experiment on whom measures of psychoticism, P (Eysenck & Eysenck, 1975); schizotypal personality, STA scale (Claridge & Broks, 1984) and hallucination proneness, LSHS scale (Launay & Slade, 1981) were available. The subject group was dichotomised separately for each scale into low and high ‘psychotic proneness’ subgroups.

The results raise some rather important questions in relation to the status of dimensions of ‘psychosis proneness.’ What is expected is that there would be an interaction term in the analyses of variance carried out between the subgroup division on ‘psychosis proneness’ and preexposure to the noise stimulus. The results showed that such was the case for groups subdivided on P ($p < 0.02$), barely so for STA ($p < 0.08$), and not at all for LSHS.

Venables, Wilkins, Mitchell, Raine and Bailes (1990) review currently available data examining the relationship between scales of psychosis proneness. As a general statement, what emerges is that a set of scales tapping cognitive/perceptual aspects of schizotypy, such as STA, LSHS, ‘magical ideation’ (Eckblad & Chapman, 1983), body image (perceptual) aberration (Chapman, Chapman & Raulin, 1978), schizophrenism (Neilsen & Petersen, 1976) and schizophrenism (Venables *et al.*, 1990) tend to correlate together to make up measures of what might be called ‘positive’ aspects of schizotypy. Largely unrelated to this group of scales are those which measure anhedonia (Chapman, Chapman & Raulin, 1976; Venables *et al.*, 1990). These scales might be thought of as tapping ‘negative’ aspects of schizotypy. The relationship of P to these two groups of scales varies from study to study. Chapman, Chapman & Miller (1982) report correlations of around 0.3 between the positive scales and P, while Claridge and Broks (1984) and Muntaner, Garcia-Sevilla, Fernandez and Torrubia (1988) indicate that P does not correlate with

STA but does correlate with STB, the subscale, which with STA forms the STQ (Claridge & Brocks, 1984). The STB scale was designed to be a questionnaire measure of DSM III borderline personality disorder. It is also worthy of note that Chapman, Chapman, Numbers, Edell, Carpenter and Beckfield (1984) report that their 'impulsive nonconformity scale' "constructed to measure impulsive antisocial behavior of the sort often reported in the premorbid adjustment of some psychotics" correlated 0.68 with P.

Bearing these findings in mind, it is therefore perhaps somewhat surprising that the study of Baruch *et al.* (1988b) shows that it is high scorers on the P scale and not on the STA or LSHS scales who might be thought to parallel more the state of the acute schizophrenic, who shows attenuation of the LI effect. If the P scale does indeed measure general 'psychosis proneness,' including proneness to antisocial behavior, then it might be expected that LI attenuation might be shown in (for instance) depressed or delinquent groups. No data appear to be available to examine this possibility. Equally, no theoretical position appears to have been advanced, to suggest hippocampal involvement in depression or antisocial behavior.

The next set of studies to be reviewed are those which do not nominally involve latent inhibition but are concerned with the concept of 'negative priming.' However, because of the apparent parallel nature of the two concepts the comparison of data on negative priming with that which has just been reviewed is important.

The paradigm is of particular relevance in that it is concerned, in the general context of experimental psychology, with elucidation of the mechanisms underlying the phenomenon and hence with mechanisms of selective attention in general. Although the concept was discussed initially by Dalrymple-Alford and Budayr (1966) in relation to the Stroop (1935) test, at the present time it is probably more useful to introduce it on the basis of a paper by Tipper (1985). The simplest version of the negative priming experiment is as follows: a priming visual display is presented containing two objects, one of which is to be named and the other to be ignored as a distractor; the objects are identified by colour as target and distractor. If then, the ignored item is the one to be named in a subsequent probe display, again containing two objects, naming latencies are impaired. The parallel with the latent inhibition paradigm is close. In this, as outlined above, the first preexposure session contains target and distractor, respectively nonsense syllables and noise bursts, and the second test phase again contains nonsense syllables and noise, but in this second session the noise stimuli, previously designated as distractors and to be ignored are now the 'targets.'

The parallel may also be considered on a theoretical level. In work on negative priming the discussion is in part centred around whether the increased latency of response to a previously ignored target is due to inhibition built up to that stimulus at the time it is ignored, or whether other processes are involved, such as decay or increased cost of alteration of processing strategy at the time of probe presentation (Allport, Tipper & Chmiel, 1985). The second point of issue is the one which arises in other contexts, namely, whether the selectivity involved in the process of separating target from distractor occurs early or late in processing. (This issue is, of course, of particular relevance in the context of work on schizophrenia insofar as it returns us to the point of whether the schizophrenic's inability to select relevant from irrelevant stimuli is at the 'filtering' or at a later information processing stage.) Tipper and colleagues present a series of studies addressed to elucidating the position (Tipper & Cranston, 1985; Tipper & Driver, 1988; Tipper, Macqueen & Brehaut, 1988), and the position, which also takes into account the work to be discussed in the next section on individual differences in the negative priming phenomenon, is best summed up in a quotation from the last paper:

> . . . We prefer to interpret negative priming as reflecting an inhibitory mechanism of attention. . . . Inhibition is not located in the perceptual representations that encode the physical properties of ignored objects. Rather, inhibition must be at some more central abstract semantic internal representation or must occur at the response stages (Tipper *et al.*, 1988, p. 51).

If Tipper is correct in espousing the notion of inhibition as an appropriate mechanism to explain the phenomenon of negative priming, then the closeness of the concept to latent inhibition is made more plausible. Additionally, if he is again correct in placing the site of action of the inhibition at a central locus, then this provides material with which to support or refute the idea that the locus of disorder in selective attention in schizophrenia is at a late information processing and not at an early filtering stage. Where the parallel between the two types of paradigm possibly breaks down is that in the latent inhibition experiment the stimuli which are to be ignored in the preexposure stage (the noise bursts) are presented repeatedly, thus allowing habituation to occur (and if habituation may be thought of as an active inhibitory process, then a situation in which inhibition to the presentation of the noise bursts increases). On the other hand, in the negative priming experiment, only a single presentation of the relevant and irrelevant stimuli is made on each trial and the inhibition generated by the ignored aspect of the priming stimulus is concluded on a single trial. As will be seen later, the fine grain of the duration of the prime stimulus is critical in the negative priming experiment, whereas in the latent inhibition studies the duration of the noise bursts is long (mean 1.25 sec) and variable (0.5–2.0 sec).

The use of negative priming to examine the performance of schizophrenics and schizotypics is made in three studies by Beech and colleagues. In the first (Beech & Claridge, 1987), the performance of normal subjects divided into low and high schizotypic scorers on the STA (Claridge & Broks, 1984) scale was examined using Stroop-like stimuli in a priming task. In the priming situation a colour bar to be overtly named was presented simultaneously with a distractor. This could be either a neutral stimulus, (+++++), or an experimental distractor, a colour name written in white. The stimuli were presented on the colour monitor of a computer, the bar was shown in the centre of the screen and the word or neutral distractor above or below it. In the probe situation, three seconds after the prime, a Stroop word was presented in the centre of the screen and the subject was instructed to name the hue of the word rather than the colour name. In half the trials the target to be named had been preceded by the prime with the colour word distractor, and in half by the prime with the neutral distractor. Negative priming was measured by subtracting the naming RT for neutral trials from the RT for experimental trials. Also obtained from the priming part of the experiment was a measure of 'interference' calculated as the difference in naming RT between the experimental and neutral trials.

The results showed that the low and high STA groups did not behave differently on the probe trial preceded by the neutral distractor prime. However, with the experimental prime, the low STA subjects showed a marked and significant negative priming effect, whereas the high STA subjects showed a small (but not significant) facilitatory effect. Interestingly there was a strong overall 'stroop interference effect' but there was no difference between the groups in this effect. Also in this study, the subjects completed the Eysenck and Eysenck (1975) EPQ scale. Correlations were run between the negative priming effect and EPQ variables. In relation to the Baruch (1988b) study outlined earlier, it is noteworthy that in this instance P was not significantly related to negative priming.

A paper by Beech, Baylis, Smithson and Claridge (1989) describes an extension of the earlier study. A set of studies were conducted using an extended range of prime conditions; these were: (1) priming distractor (PD), which was like the experimental distractor condition in the earlier experiment; (2) neutral distractor (ND), as in the previous study; (3) unrelated distractor (UD), where the colour word and the hue colour of the later probe word were unrelated; and (4) repeated distractor (RD), where the distractor colour word was the same throughout all trials. Negative priming was calculated as the difference in probe RT in UD and PD conditions, while interference was measured as the difference in UD and ND RTs. When the conditions of timing of stimuli were the same in the 1989 as in the 1987 experiment, namely, that the prime stimuli were presented for 100 msec, then the 1989 study replicated the earlier one; however, no difference in negative priming between high and low STA groups was found when prime presentation times were 500 msec.

A third study (Beech, Powell, McWilliam & Claridge, 1989) employed the same experimental paradigm as the second and used a 100-msec prime exposure time but employed as its subjects acute schizophrenic patients all of whom had one or more Schneiderian first rank symptoms and were thus to be considered as 'positive symptom' patients. Controls were psychiatric patients without major psychotic illness. The results showed that the acute schizophrenics showed significantly less (but some) negative priming than the control subjects, in accord with expectations. An explanation for the existence of some negative priming in the patients is made in terms of the fact that they were in relative remission or were on medication which might be expected to normalize their condition.

The comparison of latent inhibition and negative priming experiments, both of which involve the ability to ignore irrelevant stimuli, presents several inconsistencies. On the diagnostic front, although both sets of studies show that the performance of acute schizophrenic patients is impaired, the latent inhibition studies show that impaired performance is characteristic of normal subjects with high P scores. On the other hand, normal subjects with high STA scores and not high P scores are those who show impairment on the negative priming task.

Possibly of more importance, is the finding of the critical nature of the duration of the prime stimuli in the negative priming experiment, whereas the timing of the noise bursts in the latent inhibition experiments appears not to be important. The two sets of experiments do, of course, differ in other major critical aspects. In the LI studies, the stimuli are all auditory and successive, while in the negative priming studies the stimuli are visually presented and 'to be attended to' and distractor stimuli are presented simultaneously.

The importance of successive versus simultaneous input parameter differences may be examined by consideration of other studies. Two commonly used experimental paradigms, concerned to test the ability of subjects to ignore irrelevant stimuli, are the 'Continuous Performance Test' (CPT) (Orzack & Kornetsky, 1966) and the Span of Apprehension Test (SOA) (Neale, McIntyre, Fox & Cromwell, 1969; Asarnow, Steffy, MacCrimmon & Cleghorn, 1977). Both are usually presented visually, but the CPT involves the ability to select target stimuli from a series of irrelevant stimuli which are presented *successively*, while the SOA involves the selection of target stimuli from irrelevant stimuli presented *simultaneously*. The ability of the tests to show differences between schizophrenic (or schizophrenic-related) and normal groups depends on the difficulty of the versions of the tasks used (thus suggesting that schizophrenics are unable to cope with considerable processing loads), but in general both are able to distinguish pathological from nonpathological groups. Neuchterlein and Dawson (1984) and Erlenmeyer-Kimling (1987) have extensively reviewed the data (*inter alia*) on the CPT and in particular have shown that it is capable of distinguishing children at risk for Schizophrenia from those not at risk. In relation to the subdiagnostic findings of the latent inhibition and negative priming studies reviewed earlier, Venables (1990) has shown that performance on both the SOA and the CPT is impaired in subjects scoring highly on 'schizophrenism' (related conceptually to the STA, see above) but not on anhedonia (related to negative aspects of schizotypy). This finding is a virtual replication of that by Asarnow, Nuechterlein and Marder (1983), which showed impaired SOA performance in normal subjects with high scores on Perceptual Aberration, and Magical Ideation (Chapman, Chapman & Raulin, 1978; Eckblad & Chapman, 1983). Performance on the SOA was not related to status on questionnaire scores of anhedonia. It thus seems unlikely that the parameter difference of simultaneous versus successive presentation need necessarily suggest that LI and negative priming studies are inherently different.

Whether the modality of presentation is important in experiments on selective attention is specifically addressed in a study by Harris, Ayers and Leek (1985), who showed that the findings of the usual visual version of the SOA were substantially replicated in an auditory analogue.

If as suggested above by Tipper *et al.* (1988), it is at a central processing level that the inhibitory effect is manifest in a negative priming experiment and not in "the perceptual representations that encode the physical properties of ignored objects" then, as suggested above, the simultaneous or successive presentation and the modality of that presentation are possibly irrelevant in these studies. Where the importance of the timing of the presentation of the prime stimulus in the negative priming experiment arises is possibly at the level of perceptual mechanisms. It should perhaps be seen as a limitation of the negative priming experiments that they involve possible deficits at both peripheral and central levels which need to be disentangled.[4] The selection of the relevant from the irrelevant in the environment would seem most clearly to be largely (in the first instance at least) a central process and likely therefore one in which the hippocampus plays a major role.

Further analysis of the statement above leads to the issue of the distinction between concepts, relevant and irrelevant. In certain instances the aspect of the environment which could possibly be thought of as irrelevant and therefore given the status of 'background' and to be generally ignored, becomes important. This background may be referred to as 'context' and as such becomes, in many instances, far from irrelevant. There are two particular instances in which the consideration of the background becomes important. One arises in the difficulty which proved theoretically seminal in reconciling the fact that while hippocampal lesions in man lead to substantial memory disturbances, no such memory deficits were found in animals with similar lesions. The suggestion, put forward by Iversen (1973) to reconcile the contradiction was that tests of memory used in

[4] There is, of course, a major area of research which is concerned more with deficits in peripheral than in central mechanisms in schizophrenia (*e.g.*, Saccuzzo & Braff, 1981; Balogh & Merritt, 1987). It is also important to note the reports which suggest that increased durations of early stimulus processing are associated with negative rather than positive symptomatology (Braff, 1989; Green & Walker, 1986).

animal experiments were virtually interference free and thus dissimilar from those used in human experiments, where in particular large possibilities exist for interference from linguistic stimuli. The failure of those subjects with hippocampal lesions to suppress the interfering stimuli was what brought about the memory defects in man (Weiskrantz & Warrington, 1975), and it was not until studies in animals using contexts with the high possibility of interference were used that memory deficits in hippocampal animals were also shown.

Another view, put forward for instance by Hirsh (1974), was that the hippocampus was essential for the process of the retrieval of the context in which the original memorization took place. In consequence the loss of memory is the result of a loss of ability in hippocampal animals to reconstruct the representation of the original context serving as cue for accurate retrieval. Winocur and Olds (1978) conducted a study to investigate whether context manipulation did indeed impair the performance of hippocampal rats. When rats formed a discrimination habit and were subsequently tested for retention, both normal and hippocampal animals responded at an equally high level when the context remained constant, but the hippocampal animals were impaired when the context at recall was not the same as that at original learning. In a later study, Winocur, Rawlins and Gray (1987) used a Pavlovian conditioning paradigm in which a CS predicted a US, footshock, always, never, or half the time. Conditioning always took place in a black painted box and subsequent assessment of extent of conditioning to context was made by testing the length of time the subjects spent in a black box in preference to an adjacent white one. In normal rats there was graded contextual conditioning, that is when the footshock was well predicted by the CS, there was no fear of the (black) context when no CS was presented, as was always the case on testing trials. However, hippocampal animals showed no differential avoidance of the black compartment as a function of the original CS-shock relationship. Winocur, Rawlins and Gray interpret their result as "underscoring the importance of the hippocampus with respect to stimulus selection and the process of distguishing relevant and irrelevant information" (p. 625). The usefulness of these studies on the importance of context in the present instance is that they suggest again that the selection process which is moderated by the hippocampus occurs at a high level of complexity.

An early study (Venables, 1963), although originally conceived in a different theoretical context, does, to some extent, permit the examination of the more recent views on the role of context in performance. The task employed was that of card sorting; the subjects had to sort cards into two piles depending on whether they had a B or a Z on them. These letters appeared in any one of nine possible positions on the cards. In one pack the other eight positions were filled with a set of eight other letters and in a second pack the positions were filled with a another different set of eight letters. The subject made four sortings with one pack and then without warning the fifth sorting was made with the second, alternative, pack. At no time was reference made to the irrelevant contextual letters. The measure was the time taken to sort the pack. Sorting times for the second, third and fourth sorts formed a reasonably linear trend and individual regression lines were used to predict the time of the fifth sort, had there been no change in context letters. The difference between estimated and observed times was used as the measure to which the irrelevant context had been noted.

The subjects in this experiment were all chronic schizophrenics, half were on medication and half were not. Their behavior was rated by nurses on scales to measure withdrawal (Venables, 1957) and paranoid tendencies (Venables &

O'Connor, 1959). Patients were dichotomised on the second scale into nonparanoids and paranoids. Two similar studies were carried out with patients in different hospitals. In both instances there was no relation between the card sorting measure and withdrawal in the paranoid patients.[5] However, in the nonparanoid patients there were correlations of 0.47 ($p < 0.02$) and 0.56 ($p < 0.02$) between withdrawal and the extent to which the change of letter context had *not* disrupted card sorting. Thus, if the measure of withdrawal then used can be considered as being equivalent to a measure of negative symptom schizophrenia, then it is the positive symptom patients who are unable to ignore the irrelevant letters in the task and who are disrupted by change of context. This result, although clearly a late interpretation of much earlier data does appear to fit with the results which suggest that it is positive-symptom patients who are unable to ignore irrelevant stimuli to the detriment of task performance. It is perhaps noteworthy too that this result does not appear to be the result of subject deterioration, insofar as it is the most withdrawn subjects who perform the fifth sorting best. On the other hand, in contrast to the experiments on latent inhibition described earlier, these results were achieved in a population designated as chronic on account of their stay in hospital being greater than two years. Additionally, any normalizing effects due to the use of medication appears not to have brought about any reduction of the effect.

The idea was discussed earlier that memory deficits due to hippocampal dysfunction may be expected in situations where the possibility of interference from irrelevant stimuli is maximum. This can be examined by the comparison of subjects' performance on recall and recognition memory tasks. In the case of recall tasks the subject has to select the correct item to recall from ostensibly a very large set of possible interfering items. On the other hand, in a recognition task the selection is only from a limited set of items presented. It would thus be expected that subjects with hippocampal dysfunction would do worse on the recall task. However, as recognition is inherently easier than recall, any deficits shown by schizophrenics may be due to task difficulty rather than to a particular dysfunction. This is the issue raised by Chapman and Chapman (1978). Their recommendation that tasks should be matched for difficulty in order to overcome this possible bias was incorporated in the methodology of a study by Calev (1984). He very carefully matched the levels of difficulty of verbal recall and recognition items prior to using them with a group of chronic schizophrenic patients. His findings were that performance on the recall task was significantly inferior to that on the recognition task in schizophrenic but not in normal subjects. Where the finding is possibly out of line with the position taken at the beginning of this paper is the fact that the subjects used in this study were chronic patients on medication (Calev does indeed address the drug issue and finds that there were no differences in recognition or recall between those on and off drugs in his sample). Although he espouses the notion of a chunking deficit at encoding to explain his findings, Calev does entertain the possibility of hippocampal dysfunction as a possible mechanism. Insofar as this finding does generally fit with the other material reviewed above it is possible to say that it provides some additional support for the hippocampal dysfunction hypothesis.

[5] In the context of other studies carried out at about the same time it is likely that the patients rated as paranoid were relatively intact and monosymptomatic and unlikely to be classed as positive symptom schizophrenics in the current sense.

CONCLUSION

The studies which have been reviewed are compatible (but no more than that) with the idea that the selective attention deficit in some schizophrenics and some schizotypics is one of inability to adequately distinguish, and where necessary 'tune out,' irrelevant from relevant stimuli; a function which appears to be (as a majority view) assigned to the hippocampus.

Because more than simple selection on the basis of easily identifiable features is usually required to distinguish relevant from irrelevant aspects of the environment, the position of the mechanism whose failure is responsible for the phenomena described must be late in the information processing chain.

As expressed here, the hypothesis of the schizophrenics' failure to distinguish relevant from irrelevant signals is possibly more basic than, but not incompatible with, other hypotheses (*e.g.*, Frith, 1979; Frith & Done, 1988; Hemsley, 1987). The first of these emphasises the difficulty of the schizophrenic in suppressing awareness of processes that are normally carried out automatically, while the last is concerned with the failure of the schizophrenic to integrate stored memories in the control of current perception.

Where there is major need for further study is in the extent to which the attentional disturbance which has been discussed is a feature solely of schizophrenics in their acute phase or when characterised by positive symptoms at whatever stage of temporal chronicity.

The neuroanatomical evidence of volume reduction of the hippocampus (amongst other structures) in postmortem studies of those who had exhibited both negative and positive symptoms (Bogerts, Meertz & Schonfeldt-Bausch, 1985) would suggest that hippocampal deficits may not be confined to any one type of schizophrenic state. Bogerts (1989) suggests that limbic pathology may already be present in early childhood, and Suddath, Christison, Torrey, Casanova and Weinberger (1990) provide evidence of smaller hippocampi in monozygotic twins discordant for schizophrenia. They suggest (p. 793) "some subtle failure of development" may explain the findings. Thus it is possible that hippocampal disorder may be present as a general precursor, with the possibility, following the data reviewed in this paper, that it may appear as attentional dysfunction in those presently normal subjects having high scores on 'psychosis proneness' questionnaires.

REFERENCES

Allport, D. A., S. P. Tipper & N. Chmiel. 1985. Perceptual integration and post categorical filtering. *In* Attention and Performance XI. M. Posner & O. S. M. Marin, Eds. 107–132. Erlbaum. Hillside, NJ.

Asarnow, R. F., K. H. Nuechterlein & S. R. Marder. 1983. Span of apprehension performance, neuropsychological functioning, and indices of psychosis proneness. J. Nerv. Ment. Dis. **171:** 662–669.

Asarnow, R. F., R. A. Steffy, D. J. MacCrimmon & J. M. Cleghorn. 1977. An attentional assessment of foster children at risk for schizophrenia. J. Abnorm. Psychol. **86:** 267–275.

Bagshaw, M. H., D. P. Kimble & K. H. Pribram. 1965. The GSR of monkeys during orienting and habituation and after ablation of the amygdala, hippocampus and inferotemporal cortex. Neuropsychologia **3:** 111–119.

Balogh, D. B. & R. D. Merritt. 1987. Visual masking and the schizophrenia spectrum: interfacing clinical and experimental methods. Schizophr. Bull. **13:** 679–698.

BALOGH, D. W., J. R. SCHUCK & D. B. LEVENTHAL. 1979. A study of schizophrenics' ability to localize the source of sound. J. Nerv. Ment. Dis. **167:** 484–487.

BARTFAI, A., S. LEVANDER, G. EDMAN, D. SCHALLING & G. SEDVALL. 1983. Skin conductance responses in unmedicated recently admitted schizophrenic patients. Psychophysiology **20:** 180–187.

BARUCH, I., D. R. HEMSLEY & J. GRAY. 1988a. Differential performance of acute and chronic schizophrenics in a latent inhibition task. J. Nerv. Ment. Dis. **176:** 598–606.

BARUCH, I., D. R. HEMSLEY & J. A. GRAY. 1988b. Latent inhibition and "psychotic proneness' in normal subjects. Pers. Individ. Differ. **9:** 777–783.

BEECH, A., G. C. BAYLIS, P. SMITHSON & G. CLARIDGE. 1989. Individual differences in schizotypy as reflected in measures of cognitive inhibition. Br. J. Clin. Psychol. **28:** 117–129.

BEECH, A. & G. CLARIDGE. 1987. Individual differences in negative priming: relations with schizotypal personality traits. Br. J. Psychol. **78:** 349–356.

BEECH, A., T. POWELL, J. MCWILLIAM & G. CLARIDGE. 1989. Evidence of reduced 'cognitive inhibition' in schizophrenia. Br. J. Clin. Psychol. **28:** 109–116.

BERNSTEIN, A. S., C. D. FRITH, J. H. GRUZELIER, T. PATTERSON, E. STRAUBE, P. H. VENABLES & T. P. ZAHN. 1982. An analysis of the skin conductance orienting response in samples of American, British and German schizophrenics. Biol. Psychol. **14:** 155–211.

BLACK, A. H., L. NADEL & J. O'KEEFE. 1977. Hippocampal function in avoidance learning and punishment. Psychol. Bull. **84:** 1107–1129.

BOGERTS, B. 1989. The role of limbic and paralimbic pathology in the aetiology of schizophrenia. Psychiatry Res. **29:** 255–256.

BOGERTS, B., E. MEERTZ & R. SCHONFELDT-BAUSCH. 1985. Basal ganglia and limbic system pathology in schizophrenia. Arch. Gen. Psychiatry **42:** 784–791.

BRAFF, D. L. 1989. Sensory input deficits and negative symptoms in schizophrenic patients. Am. J. Psychiatry **146:** 1006–1011.

BROADBENT, D. E. 1958. Perception and Communication. Pergamon Press. London.

BROADBENT, D. E. 1971. Decision and Stress. Academic Press. London.

CALEV, A. 1984. Recall and recognition in chronic nondemented schizophrenics: use of matched tasks. J. Abnorm. Psychol. **93:** 172–177.

CHAPMAN, L. J. & J. P. CHAPMAN. 1978. The measurement of differential deficits. J. Psychiatr. Res. **14:** 303–311.

CHAPMAN, L. J. & J. P. CHAPMAN. 1985. Psychosis proneness. *In* Controversies in Schizophrenia. M. Alpert, Ed. 157–173. Guilford. New York.

CHAPMAN, L. J., J. P. CHAPMAN & E. N. MILLER. 1982. Reliabilities and interrelations of eight measures of proneness to psychosis. J. Consult. Clin. Psychol. **50:** 187–195.

CHAPMAN, L. J., J. P. CHAPMAN, J. S. NUMBERS, W. S. EDELL, B. N. CARPENTER & D. BECKFIELD. 1984. Impulsive nonconformity as a trait contributing to the prediction of psychotic-like and schizotypal symptoms. J. Nerv. Ment. Dis. **172:** 681–691.

CHAPMAN, L. J., J. P. CHAPMAN & M. L. RAULIN. 1976. Scales for physical and social anhedonia. J. Abnorm. Soc. Psychol. **85:** 374–382.

CHAPMAN, L. J., J. P. CHAPMAN & M. L. RAULIN. 1978. Body-image aberration in schizophrenia. J. Abnorm. Soc. Psychol. **87:** 399–407.

CLARIDGE, G. & P. BROKS. 1984. Schizotypy and hemisphere function—I. Theoretical considerations and the measurement of schizotypy. Pers. Individ. Differ. **5:** 615–632.

CROW, T. J. 1980. Positive and negative symptoms and the role of dopamine. Br. J. Psychiatry **137:** 383–386.

DALRYMPLE-ALFORD, E. C. & B. BUDAYR. 1966. Examination of some aspects of the Stroop colour-word test. Percept. Mot. Skills **23:** 1211–1214.

DOUGLAS, R. J. 1967. The hippocampus and behavior. Psychol. Bull. **67:** 416–442.

DOUGLAS, R. J. & K. PRIBRAM. 1966. Learning and limbic lesions. Neuropsychologia **4:** 283–284.

ECKBLAD, M. & L. J. CHAPMAN. 1983. Magical ideation as an indicator of schizotypy. J. Consult. Clin. Psychol. **51:** 215–225.

ERLENMEYER-KIMLING, L. 1987. Biological markers for the liability to schizophrenia. *In* Biological Perspectives of Schizophrenia. H. Helmchen & F. A. Henn, Eds. 33–56. John Wiley. Chichester.

EYSENCK, H. J. & S. G. B. EYSENCK. 1975. Psychoticism as a Dimension of Personality. Hodder and Stoughton. London.
FRIEDE, R. 1966. The histochemical architecture of Ammons Horn as related to its selective vulnerability. Acta Neuropathol. **6:** 1–13.
FRITH, C. D. 1979. Consciousness, information processing and schizophrenia. Br. J. Psychiatry **134:** 225–235.
FRITH, C. D. & D. J. DONE. 1988. Towards a neuropsychology of schizophrenia. Br. J. Psychiatry **153:** 437–443.
FRITH, C. D., M. STEVENS, E. C. JOHNSTONE & T. J. CROW. 1979. Skin conductance responsivity during acute episodes of schizophrenia as a predictor of symptomatic improvement. Psychol. Med. **9:** 101–106.
GREEN, M. & E. WALKER. 1986. Symptom correlates of vulnerability to backward masking in schizophrenia. Am. J. Psychiatry **143:** 181–186.
GRUZELIER, J. H. & P. H. VENABLES. 1972. Skin conductance orienting activity in a heterogeneous sample of schizophrenics. J. Nerv. Ment. Dis. **155:** 277–287.
HARRIS, A., T. AYERS & M. R. LEEK. 1985. Auditory span of apprehension deficits in schizophrenia. J. Nerv. Ment. Dis. **173:** 650–657.
HEMSLEY, D. R. 1975. A two-stage model of attendion in schizophrenia research. Br. J. Soc. Clin. Psychol. **14:** 81–89.
HEMSLEY, D. R. 1987. An experimental psychological model for schizophrenia. *In* Search for the Causes of Schizophrenia. H. Hafner, W. F. Gattaz & W. Janzarik, Eds. 179–188. Springer-Verlag. Berlin.
HIRSH, R. 1974. The hippocampus and contextual retrieval of information from memory: a theory. Behav. Biol. **12:** 421–444.
IVERSEN, S. D. 1973. Brain lesions and memory in animals. *In* The Psychological Basis of Memory. J. A. Deutsch, Ed. 204–232. Academic Press. New York.
KAYE, H. & J. M. PEARCE. 1987. Hippocampal lesions attenuate latent inhibition and the decline of the orienting response in rats. Quart. J. Exp. Psychol. **39B:** 107–126.
KIMBLE, D. P. 1968. Hippocampus and internal inhibition. Psychol. Bull. **70:** 285–295.
LAUNAY, G. & P. SLADE. 1981. The measurement of hallucinatory predisposition in male and female prisoners. Pers. Individ. Differ. **2:** 221–234.
LUBOW, R. E. 1973. Latent inhibition. Psychol. Bull. **79:** 398–407.
LUBOW, R. E. & A. U. MOORE. 1959. Latent inhibition: the effect of nonreinforced preexposure to the conditioned stimulus. J. Comp. Physiol. Psychol. **52:** 416–419.
LUBOW, R. E., I. WEINER, A. SCHLOSSBERG & I. BARUCH. 1987. Latent inhibition and schizophrenia. Bull. Psychon. Soc. **25:** 464–467.
MCGHIE, A. 1970. Attention and perception in schizophrenia. *In* Progress in Experimental Personality Research. B. A. Maher, Ed. Vol. 5: 1–66. Academic Press. London.
MEDNICK, S. A. 1970. Breakdown in individuals at high risk for schizophrenia: possible predispositional perinatal factors. Ment. Hyg. **54:** 50–63.
MEDNICK, S. A. & F. SCHULSINGER. 1973. A learning theory of schizophrenia: thirteen years later. *In* Psychopathology: Contributions from the Social, Behavioral and Biological Sciences. M. Hammer, K. Salzinger & S. Sutton, Eds. 343–360. Wiley. New York.
MEEHL, P. E. 1962. Schizotaxia, schizotypy and schizophrenia. Am. Psychol. **17:** 827–838.
MOORE, J. W. 1979. Information processing in space-time by the hippocampus. Physiol. Psychol. **7:** 224–232.
MUNTANER, C., L. GARCIA-SEVILLA, A. FERNANDEZ & R. TORRUBIA. 1988. Personality dimensions, schizotypal and borderline personality traits and psychosis proneness. Pers. Individ. Differ. **9:** 257–268.
NEALE, J. M. & R. L. CROMWELL. 1977. Attention and schizophrenia. *In* Contributions to the Psychopathology of Schizophrenia. B. A. Maher, Ed. 99–134. Academic Press. London.
NEALE, J. M., C. W. MCINTYRE, R. FOX & R. L. CROMWELL. 1969. Span of apprehension in acute schizophrenics. J. Abnorm. Psychol. **74:** 593–596.
NEILSEN, T. C. & K. E. PETERSEN. 1976. Electrodermal correlates of extraversion, trait anxiety and schizophrenism. Scand. J. Psychol. **17:** 73–80.
NUECHTERLEIN, K. H. & M. E. DAWSON. 1984. Information processing and attentional

functioning in the developmental course of schizophrenic disorders. Schizophr. Bull. **10:** 160–203.

OHMAN, A. 1981. Electrodermal activity and vulnerability to schizophrenia: a review. Biol. Psychol. **12:** 87–145.

ORZACK, M. H. & C. KORNETSKY. 1966. Attention dysfunction in chronic schizophrenia. Arch. Gen. Psychiatry **14:** 323–326.

SACCUZZO, D. P. & D. L. BRAFF. 1981. Early information processing deficit in schizophrenia. Arch. Gen. Psychiatry **38:** 175–179.

SCHNEIDER, W. & R. M. SHIFFRIN. 1977. Controlled and automatic human information processing. I. Detection, search and attention. Psychol. Rev. **84:** 1–66.

SHIFFRIN, R. M. & W. SCHNEIDER. 1977. Controlled and automatic human information processing. II. Perceptual learning, automatic attending and a general theory. Psychol. Rev. **84:** 127–188.

SOLOMON, P. R. 1979. Temporal versus spatial information processing theories of hippocampal function. Psychol. Bull. **86:** 1272–1279.

STROOP, J. R. 1935. Studies of interference in serial verbal reactions. J. exp. Psychol. **18:** 643–662.

SUDDATH, R. L., G. W. CHRISTISON, E. F. TORREY, M. F. CASANOVA & D. R. WEINBERGER. 1990. Anatomical abnormalities in the brains of monozygotic twins discordant for schizophrenia. N. Engl. J. Med. **322:** 789–794.

SUTTON, S. 1973. Fact and artifact in the psychology of schizophrenia. *In* Psychopathology: Contributions from the Social, Behavioral and Biological Sciences. M. Hammer, K. Salzinger & S. Sutton, Eds. 197–213. Wiley. New York.

TIPPER, S. P. 1985. The negative priming effect: inhibitory priming by ignored objects. Quart. J. Exp. Psychol. **37A:** 571–590.

TIPPER, S. P. & G. C. BAYLIS. 1987. Individual differences in selective attention: the relation of priming and interference to cognitive failure. Pers. Individ. Differ. **8:** 667–676.

TIPPER, S. P. & M. CRANSTON. 1985. Selective attention and priming: inhibitory and facilitatory effect of ignored primes. Quart. J. Exp. Psychol. **37A:** 591–611.

TIPPER, S. P. & J. DRIVER. 1988. Negative priming between pictures and words in a selective attention task: evidence for semantic processing of ignored stimuli. Mem. Cognit. **16:** 64–70.

TIPPER, S. P., G. M. MACQUEEN & J. C. BREHAUT. 1988. Negative priming between response modalities: evidence for the central locus of inhibition in selective attention. Percept. Psychophys. **43:** 45–52.

VENABLES, P. H. 1957. A short scale for rating "activity-withdrawal" in schizophrenics. J. Ment. Sci. **103:** 197–199.

Cognitive Alterations as Markers of Vulnerability to Schizophrenia

BONNIE SPRING

Department of Psychology
University of Health Sciences
The Chicago Medical School
and
Veterans Affairs Medical Center
3333 Green Bay Road
North Chicago, Illinois 60064

> "Your daughter has schizophrenia," I told the woman.
> "Oh, my god, anything but that," she replied. "Why couldn't she have leukemia or some other disease instead?"
> "But if she had leukemia she might die," I pointed out. "Schizophrenia is a much more treatable disease."
> The woman looked sadly at me, then down at the floor. She spoke softly. "I would still prefer that my daughter had leukemia."
> (E. Fuller Torrey, *Surviving Schizophrenia*, 1983, p. xv)

So begins a book entitled *Surviving Schizophrenia*, by E. Fuller Torrey. In most respects the woman is fortunate to receive news of her daughter's illness in an era when antipsychotic drugs are available to help treat the symptoms of schizophrenia. Had she learned her daughter's diagnosis in the 1950s, she might also have been treated to an explanation of the illness according to the prevailing psychoanalytic theories of the day. She might, for example, have been told that schizophrenia is likely to be produced by a cold and rejecting "schizophrenogenic" mother (sometimes called an "empty refrigerator mother"). Or she might have learned of Gregory Bateson's proposal (Bateson, Jackson, Haley & Weakland, 1956) that parents create schizophrenia-producing home environments by engaging in double-bind communications: mixed messages which a child can neither acknowledge nor ignore. Finally, she might have been told that marital dyads are abnormal among parents of schizophrenics, involving much overt or covert conflict (Lidz, Cornelison, Fleck & Terry, 1957). Alternatively, she might have been informed that schizophrenia is likely to develop when a thin veneer of "pseudomutuality" covers a fractious and toxic home environment (Wynne, Ryckoff, Day & Hirsh, 1958).

But times have changed since the 1950s, and many confusing data have muddied the waters of pristine theories that laid parental behavior at the wellspring of schizophrenia. Indeed there *is* often something odd about the thought processes, communication and behavioral interaction of families that include schizophrenic offspring (Cook, Kenny & Goldstein, 1991; Miklowitz, Strachan, Goldstein, *et al.*, 1986). Clinicians will not have escaped noticing that the woman with whom Dr. Torrey spoke produced an odd rhyming association when she summoned up leukemia upon being told that her daughter had schizophrenia. Nor will they have missed finding it unusual that a parent could wish for a fatal disease to be visited upon her child, in preference to a merely unfortunate one.

What has changed since the 1950s is our lack of certainty about whether psychopathologic features found among parents of schizophrenics represent a causal factor in the etiology of the offspring's disorder, a response to the stress of interacting with a peculiar youngster (Liem, 1974), or an epiphenomenal sign of a more potent causal agent: the presence of a genetic predisposition toward schizophrenia in both parent and offspring.

The other thing that has changed is our reigning model of the etiology of schizophrenia. Contemporary theories are biological in emphasis, giving greatest weight to genetic factors (Gershon, Merril, Goldin, DeLisi, Berrettini & Nurnberger, 1987; Pardes, Kaufmann, Pincus & West, 1989), biochemical alterations (Davis, Kahn, Ko & Davison, 1991) and neuropathological changes (Berquer & Ashton, 1991) in the onset and course of schizophrenia. Current approaches are liberating in the respect that they remove the tremendous burden of blame that was once laid squarely on the shoulders of parents who "produced" schizophrenic offspring. But biological theories also drag in tow some of their own metacommunications and arbitrary assumptions.

Patients and their families are one group who may be influenced by these tacit belief systems. A study by Fisher and Farina (1979) illustrates some of the implicit assumptions that can stem from biological explanations about the causes of psychological discomfort. Fisher and Farina told one group of college students that psychological troubles arise primarily as a result of genetic and somatic causes. They told another group that mental disturbances result primarily from learning maladaptive social behaviors. By the end of the semester, the two groups had formulated different opinions about how to manage their own psychological troubles. Unlike the psychologically oriented group, the biologically oriented students felt that there was little they could personally do to control their problems. They found small value in thinking about the causes or solutions to their life difficulties, and made greater use of alcohol or drugs to alleviate distress.

Inevitably, the impact of theories of psychopathology is felt in social and human terms. Reigning theories influence the burden of responsibility or the sense of control that patients and their families assume for a mental disorder. But the primary yardstick by which we must evaluate such theories is the criterion of scientific merit and truth.

The Disease Model of Schizophrenia versus the Vulnerability Model

Kraepelin's (1919) disease model of schizophrenia is without doubt the biological paradigm that has exerted greatest influence on the study of schizophrenia. Kraepelin regarded schizophrenia as a brain disease that is characterized by an early onset, and a continuous irreversible, downhill dementing course, fundamentally different from the more episodic manic-depressive illness. From this, it was a short step to prevailing early 20th century beliefs that social and psychological factors play no important role in the pathogenesis of schizophrenia. The neglect of psychological variables was consistent with the dominant model in medicine at that time, as exemplified by Virchow's and Pasteur's theories of the "specific cause of disease." Scientific investigation aimed to identify a discrete pathogen that endogenously gives rise to schizophrenia and to localize the brain abnormalities that constitute the disease's underpinnings.

After a temporary sojourn into the psychoanalytic theories of the 1950s, an important discovery reignited interest in biological explanations of schizophrenia. In 1952, the French physicians Henri Laborit and Jean Delay observed that thorazine was effective in reducing the hallucinations, delusions, violent behavior, and communication disturbances that trouble schizophrenic patients and those around them. The discovery of antipsychotic medications suffused a new energy into biological research on schizophrenia. It seemed only logical that a disorder that yielded to pharmacological intervention must have a primarily biological etiology.

The ensuing years have delivered an increasingly sophisticated outpouring of genetic and biochemical theories of schizophrenia (Faraone & Tsuang, 1985; Moises, Gelernter, Giuffra, Zarcone *et al.*, 1991). One might also speculate that the growing complexity of biological models has been most directly stimulated by schizophrenia's refusal to comply with expectations about how this disease entity ought to behave. Schizophrenia's obstreperousness has occurred on two primary fronts. First, even Kraepelin noted that not all schizophrenics would lie down and progress towards deterioration. Approximately 13% of Kraepelin's patients actually recovered. A generation later, based on an extensive longitudinal follow-up of 208 patients, Manfred Bleuler (1978) painted a surprising picture of the course and outcomes of schizophrenia in a comparable cohort of patients. He found that the majority of cases of schizophrenia were episodic and few progressed towards the type of dire outcome that Kraepelin described.

> If the long-term recovery periods, the recoveries that are interrupted only by brief psychotic episodes, and the mild chronic "end states" are regarded as characteristic of benign schizophrenia, and the moderate and severe "end states" as characteristic of malignant ones, the following general rule applies: Some two-thirds to three-fourths of schizophrenia are benign, and only about one-third or less are malignant. . . . About one-fourth to one-third of the "end states" are long-term recoveries, and about one-tenth to one-fifth are the severe chronic psychoses. . . .
>
> Most frequently schizophrenia runs in acute phases with intermittent recoveries (a good third of all cases). About a quarter of all schizophrenias run in phases, and end in mild or moderate chronic psychoses. Third in order of frequency are the schizophrenias with chronic beginnings and outcomes in moderate or mild chronic states (about one-fifth). Then follow the schizophrenias with chronic courses to severe chronic psychoses. Each of the remaining course types occurs only in a very low percentage of all cases.
>
> (M. Bleuler, 1978, p. 414)

The second respect in which schizophrenia has refused to comply with a straightforward medical model is that drugs do not completely cure the patient. The positive or florid symptoms (*e.g.*, delusions, hallucinations, formal thought disorder) are most responsive to antipsychotic medication. On the other hand, negative or deficit symptoms (*e.g.*, work impairment, social isolation, anhedonia, flat affect, and lack of motivation) are less drug-responsive than positive symptoms. In fact, they can sometimes even be triggered by high-dose antipsychotics.

Biologically and psychologically oriented theorists have each inspected these same discrepancies, but have taken their interpretations in different directions. Observing the great variability in the course of schizophrenia, biologically oriented psychopathologists have suggested that schizophrenia is a heterogeneous collection of diseases (Helmchen, 1988). The heterogeneity argument stipulates that there are many different schizophrenias, each characterized by a specific etiology and pathogenesis:

> The symptoms of a disorder depend largely upon the system or systems affected, whereas the course tends to reflect the etiology and nature of the pathological process. If that process is fairly abrupt in onset, relatively short-lived, and reversible, the course of illness will have the same characteristics. If it persists and produces effects repetitively, the course may be intermittent with periods of remission and exacerba-

> tion. If the disease process persists and progresses, the course of illness will have the same chronicity and progression. It is reasonable to assume that a system in the brain may be affected by different etiological agents that produce the same symptoms, but over a different course—acute, intermittent, or chronic and progressive—depending upon the nature of the agent and its interactions with the brain."
>
> (Henn & Kety, 1982, p. 9)

For example, viral, other infectious or autoimmune processes could produce acute schizophrenia, intermittent schizophrenia or chronic schizophrenia as a function of whether they only caused acute brain inflammation, or produced recurrent flare-ups of inflammation, or left permanent scarring of brain tissue.

Psychologically oriented theorists have examined the same variability in the course and outcome of schizophrenia, but have turned to another locus to explain the vicissitudes of the patient's illness. They have looked to the patient's life. According to the vulnerability or diathesis-stress model of schizophrenia (Zubin & Spring, 1977; Spring & Coons, 1982), the varying illness histories of schizophrenic patients can be understood by looking at the stressful life events, the family circumstances and the social supports that impinge upon or cushion the patient. Psychotic episodes come and go as life stressors surpass and recede below the patient's threshold of tolerance (Ventura, Nuechterlein, Lukoff & Hardesty, 1989). This threshold is, in turn, established by the patient's vulnerability, on the one hand and his or her stress-buffering resources on the other.

A similar divergence of perspectives has occurred in response to the datum that not all of the schizophrenic individual's difficulties are cured by antipsychotic medications. Biologically oriented theorists have again replied by proposing that there is more than one syndrome of schizophrenia. According to Crow (Crow, Frith, Johnstone & Owens, 1980; Crow, 1985) only patients with an active Type I process, who exhibit many positive symptoms and a hyperdopaminergic disease, respond well to antipsychotic medications. Since Type I schizophrenia may be caused by a proliferation of dopamine receptors, these patients can be helped by pharmacologic agents that block dopamine receptors. Schizophrenics affected by a Type II process, in contrast, are presumed to have a different disease (although they may have progressed through a Type I phase early in the course of illness). Type II patients suffer primarily from deficit symptoms. Crow and his associates propose that these negative symptoms may arise because of the enlarged brain ventricles that are visible upon CAT scanning. At present, controversy continues over whether positive and negative symptoms define two subtypes of schizophrenia, two stages of the disease or two independent symptom dimensions (Andreason, Flaum, Swayze, Tyrrell & Arndt, 1990; Kay, 1990; Lenzenweger, Dworkin & Wethington, 1989). Greater consensus exists that a new and different form of somatic treatment may be needed to treat deficit symptoms (Lewine, 1990).

Psychologically oriented researchers have again made different interpretations of these same data. According to the vulnerability model, antipsychotic drugs treat psychotic symptoms. But episodes of florid psychosis are only one part of the patient's difficulty. The other part is the enduring vulnerability that was present before the psychosis and that lingers on afterward. The patient carries and manifests this vulnerability at many levels, in the form of cognitive dysfunctions, impairments in social competence and sometimes a withdrawn and isolative personality. These disturbances may not be simply and perhaps not even complexly explicable in biological terms. Instead they may be better understood with reference to behavioral and cognitive constructs.

Even if enormous scientific advances were ultimately to make it possible to reduce vulnerability to biological explanatory constructs, two scientific tasks would remain. One is the task of understanding the path whereby the biological process germinates into the psychological manifestations of vulnerability: the cognitive disturbances, the unusual personality and the awkward social behavior. Just as we cannot now fully understand the gestalt of memory functioning in terms of the activity of hippocampal neurons, it is unlikely that further progress in biology will eliminate the need for greater psychological understanding of vulnerability.

The second task that will remain is that of finding ways to reduce vulnerability. Although pharmacologic interventions would be a welcome part of this solution, the need for psychological treatments is likely to remain. Liberman, Wallace, Vaughn, Snyder and Rust (1980) speak very strongly to this point:

> Despite their widespread and accepted usage, neuroleptic drugs leave much to be desired as a satisfactory treatment for schizophrenia. Drug therapy has delayed but not solved the problem of recurrent relapse . . . : has led to a decrease in census, but an increase in readmissions to mental hospitals; has been accompanied by a host of unpleasant and sometimes irreversible side effects; has had its full efficacy limited by a staggering noncompliance rate; and does not enable the schizophrenic person to learn those social and life skills necessary for survival and satisfactory functioning in the community. While medication undoubtedly reduces symptomatology and facilitates interpersonal contact, more is needed for a comprehensive treatment of the person with schizophrenia.
>
> (Liberman *et al.*, 1980, p. 50)

In sum, within a psychological framework, it is not necessary that negative or deficit symptoms be seen as signs of a second disease process. Nor does the relative nonresponsivity of these phenomena to antipsychotic agents necessarily call for the development of a new somatic treatment. Rather, from a psychological perspective, certain of the personality, cognitive and social skill deficits may be construed as part of the patient's premorbid personality, social competence, and chronic vulnerability. Not only the social skills deficits (Halford & Hayes, 1991; Hogarty, Anderson, Reiss *et al.*, 1991) but even the persistent cognitive impairments that accompany them may be more responsive to psychological interventions than has usually been assumed (Brenner, Kraemer, Hermanutz & Hodel, 1990; Spring & Ravdin, 1992).

The next section of this paper considers some of the cognitive attributes that may be associated with vulnerability to schizophrenia. The research to be discussed relies mostly on group comparisons at a fixed point in time. Our knowledge base is very sparse concerning the state changes in cognition that may occur as patients pass from relative remission into episodes of psychosis and back. Nonetheless, it remains imperative to understand how biological and information processing events interdigitate to produce the cognitive appraisals that occur as patients pass into and out of episodes.

Strategies for Identifying Vulnerability Markers

A generally accepted strategy for determining whether a cognitive characteristic is a marker of vulnerability to schizophrenia relies on the dovetailing of three types of evidence (Spring & Zubin, 1978; Cromwell & Spaulding, 1978; Garver, 1987).

(1) The characteristic is found among first-degree biological relatives of schizophrenics to a greater extent than among normal controls.

(2) Deviations in performance persist among schizophrenic individuals who are no longer highly psychotic.
(3) The anomaly is present to a greater than normal degree among individuals judged to be at risk for schizophrenia because of schizotypal personality features or biological characteristics associated with schizophrenia.

Biological Relatives of Schizophrenics

Studies involving relatives of schizophrenics are plagued by several uncertainties. One is that, since first-degree relatives share on average only 50% of their genes with the proband, not all are expected to possess the same genetic liabilities. Consequently, mean comparisons of scores of relatives versus controls represent an extremely stringent test of a potential vulnerability marker. Mean comparison procedures are conservative because the average score for the group of relatives pools the data from genetically predisposed and nonpredisposed individuals. This problem of heterogeneity in design and data analysis has been handled innovatively by various researchers (cf. Nuechterlein, 1983; Erlenmeyer-Kimling, 1987; Erlenmeyer-Kimling, Golden & Cornblatt, 1989).

The study of biological relatives is also complicated by the possibility that some types of schizophrenia might be nongenetic in origin. If so, then vulnerability markers would not be expected to be prevalent among relatives of patients with a nongenetically derived risk, unless of course the proband and relatives shared other sources of vulnerability. The existence of both genetically and nongenetically transmitted forms of schizophrenia is one possible explanation for the data that the risk for schizophrenia among monozygotic cotwins of schizophrenia, who share 100% of genetic material, is only 50% and that 90% of schizophrenics do not have a schizophrenic parent. For example, it may be the case that perinatal and intrauterine complications are sources of risk for a nongenetically based vulnerability to schizophrenia, and that different behavioral characteristics mark genetically and nongenetically based forms of vulnerability (Cannon, Mednick & Parnas, 1990; Reveley, Reveley & Murray, 1984).

A fundamental question persists about whether cognitive alterations should be construed as markers of the predisposition to schizophrenia, as indicators that the predisposition has begun to be partially expressed, or as signs of a substantial risk of becoming schizophrenic. Geneticists use the construct of incomplete penetrance to describe the fact that possession of a genetic predisposition conveys no certainty that the genotype will be expressed phenotypically. Even if it were clear that cognitive alterations mark a genetically transmitted vulnerability, there is currently no clear theory to predict whether cognitive markers of vulnerability should be expected among half of the first-degree relatives of schizophrenics (who may share a predisposition to schizophrenia), among 10% of relatives (who are at actual risk), or among some intermediate number of relatives. If cognitive alterations mark the true risk of developing schizophrenia, then they should deviate in only a very small proportion of healthy adult relatives of patients, since many adult relatives will have passed the age of risk.

In sum, studies involving first degree relatives of schizophrenics have two primary limitations. First, the power of the designs may be relatively weak, especially when mean comparison procedures are used. The conservatism of the designs is augmented if the vulnerability of a substantial proportion of the probands originated from nongenetic factors that did not affect their relatives. The second limitation is that it remains unclear whether cognitive vulnerability markers detect a predisposition or an actual risk for schizophrenia. Our ability to generate specific predictions is limited by this gap in knowledge and theory. Nonetheless, studies of biological relatives of patients enhance our understanding of possible familial sources for vulnerability to schizophrenia, especially when combined with data garnered from other research approaches.

Remitted Patients

Studies of "remitted" patients, whose positive symptoms have substantially subsided, represent another research approach. This strategy solves one of the problems that plagues research on healthy relatives of schizophrenics. Remitted schizophrenics are certain to possess a predisposition to the disorder. The conceptual difficulties involved in studying recovered schizophrenics are of a different type. They include controversy about whether markers should be construed as signs of a continuous disease process, whether cognitive alterations may be residual effects of the illness, and whether they may be affected by medication. Again, however, our confidence that a measure represents a vulnerability marker is enhanced when results dovetail with findings from other "high risk" research strategies.

Schizophrenia-Related Personality Features

Individuals who display schizophrenia-related personality features have also been studied in the effort to identify vulnerability markers. Interest has focused on personality characteristics that are theoretically related to schizophrenia-proneness and that might be signs that vulnerability has begun to be phenotypically expressed. The Chapmans' research on "hypothetically psychosis-prone" college students is the most programmatic example of this method (Allen, Chapman, Chapman, Vuchetich & Frost, 1987; Chapman, Edell & Chapman, 1980; Martin & Chapman, 1982). The Chapmans and others (Clementz, Grove, Katsanis & Iacono, 1991; Rosenbaum, Shore & Chapin, 1988; Lenzenweger & Loranger, 1989; Simons, MacMillan & Ireland, 1982) have examined young people who are characterized by anhedonia, perceptual aberrations, and impulsive nonconformity. It still remains to be determined whether these individuals will eventually go on to develop schizophrenia. If they do, the next issue that will need to be addressed is whether the deviant personality features that have been identified merely reflect vulnerability. An alternative possibility is that they may be prodromal signs which indicate that the person has already entered a schizophrenic episode.

Cognitive Vulnerability Markers

Because comprehensive reviews of research on cognitive markers of vulnerability to schizophrenia have recently been presented (Nuechterlein & Dawson, 1984; Nuechterlein & Zaucha, 1990; Spring, Lemon & Fergeson, 1990), the most promising markers will only be briefly discussed here. Some of the challenges involved in this research will be noted, including the issues of heterogeneity, specificity, and the likely nature of the processing deviation.

Reaction Time and Preparatory Set

Widely used to study attention in high risk populations, reaction time (RT) tasks have generated three primary measures. The first is simple *response speed.* The second is *cross-modal retardation,* which quantifies the lengthening of response speed for sequential stimuli that are in different modalities (cross-modal), as compared to sequential stimuli that are in the same modality (ipsimodal). A pronounced slowing of RT to cross-modal as compared to ipsimodal sequences is inferred to index problems in shifting attention across modalities (Sutton & Zubin, 1965).

The third measure is the magnitude of *crossover* in the set reaction time paradigm (Rodnick & Shakow, 1940) that compares response speeds to stimuli presented after sequences of identical preparatory intervals (regular condition) or different preparatory intervals (irregular condition). For schizophrenics, the curves for the regular and irregular conditions tend to converge (crossover) when the waiting interval reaches seven seconds. In other words, the facilitation of RT for regular trials is no longer found among schizophrenics when preparatory intervals last seven seconds or longer. For normal controls, the waiting interval that yields a crossover pattern is generally much longer (*e.g.*, 20 sec). These data have been interpreted to mean that schizophrenic individuals have difficulty maintaining a preparatory set and profiting from regularity of information. A slightly different interpretation is that redundancy produces adverse effects on the information processing of schizophrenics by evoking an active inhibitory process (Steffy & Galbraith, 1974).

Slowed reaction time is a nonspecific phenomenon that probably manifests disturbed motor functions in schizophrenia, even beyond its association with attention (Levin, Yurgelun-Todd & Craft, 1989). Response speed is slowed in many psychiatric and organic disorders, including tardive dyskinesia (King, 1991), and its responsivity to neuroleptic medications is unclear. Some evidence suggests that antipsychotic drugs speed reaction time (Spohn, Lacoursiere, Thompson & Coyne, 1977); some findings suggest an absence of effect (Held, Cromwell, Frank & Fann, 1970); and other results even suggest that high-dose neuroleptics may slow RT by inducing motor side effects (Spohn, Coyne, Lacoursiere, Mazur & Haynes, 1985).

Slowing of response speed was proposed as a vulnerability marker when Marcus (1973) found that children of schizophrenic mothers responded significantly more slowly than controls. Moreover, their RT remained significantly slowed under reward and informational conditions designed to facilitate performance. Adult siblings of patients also responded more slowly than controls in a choice RT study by Wood and Cook (1979). When Van Dyke, Rosenthal and Rasmussen (1975) examined all four combinations of being reared by a schizophrenic and being biologically related to a schizophrenic parent, they found RT slowing among adults who had been raised by a schizophrenic parent, regardless of whether they had a schizophrenic biological relative. RT slowing among groups at high risk for schizophrenia is not a robust phenomenon, however. Differences between high risk groups and controls have not reached significance in several studies, including those of Asarnow, Steffy, MacCrimmon and Cleghorn (1978); DeAmicis and Cromwell (1979); Phipps-Yonas (1984); Spring (1980); and Rosenbaum *et al.* (1988). Moreover, in several of these studies, the mean RTs of relatives and controls were virtually identical (DeAmicis & Cromwell, 1979; Phipps-Yonas, 1984; Spring, 1980).

Findings concerning the potential of cross-modal retardation as a vulnerability marker have not been promising (cf. Phipps-Yonas, 1984; Spring, 1980). Results for RT crossover in high risk individuals have been more encouraging, although not entirely consistent. Asarnow *et al.* (1978) and Marcus (1973) failed to find early crossover in children of schizophrenics, and Van Dyke *et al.* (1976) failed to find the pattern in adult relatives. On the other hand, DeAmicis and Cromwell (1979) found greater than normal crossover among relatives of a group of process schizophrenic probands who were preselected for high magnitudes of crossover. In addition, Rosenbaum *et al.* (1988) and Simons *et al.* (1982) found early crossover among college students who manifested schizotypal personality features, and Bohannon and Strauss (1983) found that outpatients continued to manifest the phenomenon. The reason for these discrepant findings is difficult to discern since similar procedures have been used in the studies that generated either positive or negative results.

In sum, response slowing and difficulties in maintaining a preparatory set to respond each holds some modest degree of promise as a vulnerability marker. Given the limitations of both indices, it might be wise to use them in combination, as was done in Rodnick and Shakow's (1940) original set index.

Vigilance

High risk research on vigilance has been more extensive than on any other attentional parameter, and findings are very promising. Most studies involve variants of the Continuous Performance Test (CPT), a visual sustained attention test first introduced by Rosvold and Mirsky (1956) to assess brain damage. In contrast to the original paradigm, which used a single clearly focused target, many current versions of the task increase its information processing demands by adding a memory load (to detect a sequence of targets), by blurring or masking (perceptually degrading) test stimuli, or by introducing distracting noises or lights. The number of hits (correctly detected targets) or d′ (an index of the ability to discriminate targets from nontargets, independent of biases to respond yes or no) are the measures usually reported. One often neglected index concerns the tendency for hits to decline over the course of testing (Nuechterlein, 1991). However, an examination of performance decrement is needed to discriminate whether lowered test scores arise chiefly from problems in maintaining attention over the course of an extended test session, or from difficulties in managing the moment-to-moment processing load of the task.

Only a subgroup of schizophrenic patients (40–50%) show CPT impairments (Orzack & Kornetsky, 1971; Walker, 1981), and these tend to have a family history of schizophrenia (Orzack & Kornetsky, 1971; Walker & Shaye, 1982). While hospitalized, the CPT impairment of this group of patients is more marked than that of either schizoaffective or affectively ill patients (Walker, 1981). Impaired CPT performance persists on demanding versions of the task even after psychotic symptoms have begun to subside (Wohlberg & Kornetsky, 1973). In fact, Asarnow and MacCrimmon (1978) found no significant differences in the number of targets missed by actively psychotic and clinically remitted schizophrenics.

Unless they are very young (Grunebaum, Weiss, Gallant & Cohler, 1974), children of schizophrenics are not impaired on the simplest versions of the CPT, even when stimuli are presented at a very rapid rate (Asarnow *et al.*, 1978; Erlenmeyer-Kimling & Cornblatt, 1987; Nuechterlein, 1983). More demanding versions of the task do detect impairments among older children at risk (Erlenmeyer-Kimling & Cornblatt, 1987; Nuechterlein & Zaucha, 1990). The data for specificity among high risk groups are not as clear-cut, as is the case for all vulnerability markers.

Generally, control groups at risk for other disorders fail to show vigilance impairments in comparison with normals who are not at risk. However, individuals at risk for other disorders can only inconsistently be differentiated from the target group at risk for schizophrenia (*e.g.*, Nuechterlein, 1983).

The nature of the problem that the CPT detects in high risk individuals is not yet well understood. The disturbance does not seem to involve difficulties sustaining attention over time, since the decrement in performance over time is no greater for children of schizophrenics than for controls (Nuechterlein, 1983). Deficits are elicited in a subset of vulnerable individuals whenever test conditions place high demands on moment-to-moment processing activity. An inherent reduction in information processing capacity would be one way to account for these findings (Nuechterlein & Dawson, 1984).

Sustained Selective Attention

The literature on sustained selective attenion in schizophrenia has recently been reviewed (Spring, Weinstein, Freeman & Thompson, 1991). Two main selective listening paradigms have been used to assess distractibility. Shadowing tasks require focused attention and continuous, immediate shadowing (repeating) of a primary message presented to one ear, with and without a simultaneous distracting message presented to the opposite ear. Instructions emphasize that distractors are to be ignored. Recall tasks require attention to a string of digits or words followed by recall of these stimuli. The main string of items is presented with interspersed distracting items, and subjects are told to ignore the distracting stimuli. Distractibility can be measured in two ways: by the degree to which distractors disrupt performance of the main task, and by the extent to which distractors intrude or are interjected into main channel responses.

Meta-analyses indicate that shadowing tasks elicit a large generalized deficit (Spring *et al.*, 1991). Even when shadowing without distraction, schizophrenic patients shadow less accurately than 89% of normal controls. Distraction widens this gap, however, leading the average schizophrenic patient to perform less well than 99.5% of normals. Although the data for intrusion errors are only inconsistently reported, effect size analyses indicate that the average schizophrenic exceeds 100% of normal controls on the frequency of these errors. In sum, these findings suggest both large generalized deficits in shadowing and differential deficits due to distractibility. When recall is used to assess selective attention, the findings are similar, although slightly less robust. Meta-analyses indicate that the schizophrenic's performance mean falls 1.57 standard deviations below the normal mean for recall without distraction, as compared to 1.92 standard deviations below the normal mean for recall with distraction. Stated differently, the average schizophrenic performs more poorly than 94% of normal controls at recalling words without distraction. With distraction, he or she performs worse than 97% of normals.

Evidence is mixed regarding the stability of distractibility across changes in clinical state. Two studies (Wohlberg & Kornetsky, 1973; Asarnow & MacCrimmon, 1978) found that a significant differential deficit due to distraction persisted in remitted schizophrenics. A third (Frame & Oltmanns, 1982) found that only a tendency toward distractibility (large in magnitude but nonsignificant with this small sample of 8) persisted.

Similarly, high risk children have only inconsistently been found significantly more distractible than controls. Heightened distractibility has emerged in some (Erlenmeyer-Kimling & Cornblatt, 1978; Cornblatt & Erlenmeyer-Kimling, 1984; Winters, Stone, Weintraub & Neale, 1981; Driscoll, 1984), but not all analyses (Cornblatt & Erlenmeyer-Kimling, 1985; Asarnow *et al.*, 1978).

Some evidence suggests that the stability of distractibility as well as its presence in individuals at risk for schizophrenia may depend upon which index is chosen to assess failures in selective attention. Spring, Lemon, Weinstein and Haskell (1989) administered nondistractor and distractor shadowing tasks that had been matched on psychometric discriminating power to a variety of samples at risk for schizophrenia. Results indicated that the tendency of distractors to disrupt the accuracy of shadowing appeared to be a state marker. Only schizophrenic patients who were hospitalized or just recently released from hospital showed a deterioration in performance when distraction was introduced. Stably remitted patients, first-degree relatives of patients and schizotypal college students shadowed as well with distraction as without it. On the other hand, although none of these samples made a very great number of intrusion errors (interjecting distracting phonemes into their shadowing responses), the number of intrusion errors was in each case greater than for normal controls. This finding might be interpreted to mean that a schizophrenic disturbance in pigeonholing and response selection is more trait-related than is the disturbance in stimulus selection.

Further, intrusion errors tended to occur immediately after the task required a shift in the to-be-shadowed ear, suggesting that distractibility may have been provoked by the requirement to shift attention and to re-establish a selective set. If so, it may be the case that failures in selective attention, like failures in vigilance, only become evident in high risk populations under high momentary processing demand, such as that required by the need to shift attention.

A similar interpretation may be offered for the finding that distraction selectively disrupts the primacy effect: the recall advantage that is usually found for items appearing early in a memory list (Winters *et al.*, 1981; Frame & Oltmanns, 1982). For vulnerable individuals, the capacity-demanding cognitive rehearsal that is needed to maintain early list items in working memory is disrupted by distraction. In contrast, the passive storage mechanism that retains later items demands little cognitive capacity, and is not disrupted by distraction. If some individuals vulnerable to schizophrenia allocate greater than normal capacity to processing distractors, the toll on performance is likely to be more evident for highly demanding cognitive activities. Cognitive operations that demand minimal mental effort can apparently proceed smoothly even when attention is allocated to processing distractors.

The Cognitive Substrate for Vulnerability to Schizophrenia

Reduced Capacity

Many cognitive anomalies observed among high risk individuals could be caused by a reduction in the total pool of available information processing capacity. This hypothesis has been advocated by Nuechterlein and Dawson (1984), who point out that the construct of reduced capacity makes theoretically meaningful the observation that schizophrenics and high risk individuals display generalized deficits in performance on a wide variety of tasks. If capacity becomes overloaded, many processing operations will be slowed and the most capacity demanding activities will be disrupted. Thus, a reduction in capacity could account for the findings that response speed is slowed, that the effortful processes of maintaining set are impaired, and that the active rehearsal operations required for recall memory are especially impaired.

A question may be raised about whether information processing capacity is truly or *inherently* reduced in vulnerable individuals, or whether it is only *apparently* reduced. An inherent reduction in processing capacity would be one likely outcome of a dementing or infectious process that resulted in diffuse insults to the brain. If a true reduction in capacity comprised the substrate for cognitive vulnerability markers, this would be entirely compatible with some biological explanations of the etiology of schizophrenia.

Alternatively, it remains possible that information processing capacity is not actually reduced, but that it only appears to be. A pseudo-reduction in capacity might come about in two ways. First, processing activities might be allocated to events other than the primary laboratory test procedure. Second, processing capacity might be taxed unduly because of a reliance on inefficient cognitive operations.

Altered Allocation Policies

To what events other than the laboratory procedures might capacity be allocated? There are at least three likely candidates: external distractors, physiological sensations, and thoughts or emotions.

At least some individuals predisposed to schizophrenia seem to allocate greater than normal capacity to processing external distractors. For example, remitted patients, biological relatives of schizophrenics and schizotypical individuals interject distractors during a dichotic listening test. Greater than normal allocation of attention to distractors might explain why at least some evidence suggests that the performance of children of schizophrenics is disrupted to a greater than normal extent by distraction. Also consistent is the observation that children who go on to develop adult schizophrenia are described clinically as highly distractible (Hartmann, Milofsky, Vaillant, Oldfield, Falks & Ducey, 1984).

It is actually a bit surprising that extraneous distractors presented during a laboratory task can command processing activity. Laboratory distractors are not usually of great personal salience to experimental subjects. One might guess that internal events such as physiological sensations, thoughts and emotions would represent more compelling sources of distraction, although this is admittedly a speculative suggestion. The adaptive, evolutionary function of autonomic activation is to alert the organism to events that require immediate attention and response. When arousal is persistent, and cannot be met with fight, flight or other behavioral responses, the sensations of autonomic activation begin to command processing activity even in healthy individuals (Mandler, 1979). Individuals prone to schizophrenia manifest several anomalies in autonomic nervous system functioning that might result in attention-commanding states of physiological arousal. These characteristics include proneness to chronic hyperarousal and strong reactivity to social stressors (Dawson & Nuechterlein, 1984; Szymanski, 1991). When processing resources are devoted to monitoring physiological arousal, the capacity available to be allocated to processing external stimuli should be reduced accordingly. In fact Gjerde (1983) proposed that all of the cognitive deficits observed in schizophrenic individuals could be attributed to more fundamental disturbances in physiological arousal. In addition to sensations of physiological events, it may be that thoughts and emotions command a disproportionate share of the processing capacity of vulnerable individuals. Indeed, some of the more distracting thoughts and emotions may arise in the process of interpreting unusual sensations. Maher (1974) and Zimbardo, Anderson and Kabot (1981) proposed that delusional thinking may emerge in an effort to explain sensory and emotional experiences that appear disproportionate to ongoing events. To the extent that preschizophrenic individuals experience persistent or intense sensations related to heightened arousal, they may be compelled to allocate processing capacity to interpreting these inner events.

In sum, although many findings on cognitive markers of vulnerability are consistent with the hypothesis of reduced capacity, it remains unclear whether capacity reductions are biologically or psychologically engendered. Capacity limitations could be rooted in brain dysfunction. Alternatively, they could arise because vulnerable individuals allocate proportionally less capacity to an experimental task and proportionally more capacity to processing distracting external and internal events.

Overreliance on Controlled Processing

Apparently reduced capacity in the presence of actually normal processing resources could also arise because of excessive reliance on inefficient cognitive strategies. For example, some cognitive operations that are ordinarily performed automatically, and out of awareness might among schizophrenics be driven consciously and effortfully. The distinction between automatic processing and controlled, sequential, effortful processing was proposed by Schneider and Shiffron (1977) and by Posner (1978). Controlled processes must be activated and monitored via conscious attention, are slow and demanding of information processing capacity, and interfere with other ongoing controlled processes. Automatic processes, by contrast, are rapid, unconscious, nondemanding of capacity and able to be executed simultaneously with other processing operations.

There is only anecdotal evidence to suggest that schizophrenic patients may overuse controlled processes when automatic processes would operate more efficiently. As one patient described this phenomenon,

> I'm not sure of my own movements any more. It's very hard to describe this but at times I'm not sure about even simple actions like sitting down . . . I found recently that I was thinking of myself doing things before I would do them. If I'm going to sit down for example, I've got to think of myself and almost see myself sitting down before I do it. It's the same with other things like washing, eating or even dressing—things that I have done at one time without even bothering or thinking at all . . . I take more time to do things because I am always conscious of what I am doing. If I could just stop noticing what I am doing, I would get things done a lot faster. I have to do everything step by step now, nothing is automatic.
>
> (McGhie, 1969)

Empirical research on automatic and controlled processing remains to be done with schizophrenic patients and vulnerable individuals. However, the hypothesis that schizophrenics may overtax cognitive capacity by overusing controlled processing operations is intriguing because it suggests a therapeutic intervention. Extended practice is one means by which cognitive operations become automatic and free up information processing capacity (Spelke, Hirst & Neisser, 1976). The most successful attention training programs have incorporated an emphasis on extensive cognitive rehearsal and practice, and some have reported promising results with schizophrenic individuals (*e.g.*, Wagner, 1968; Meichenbaum & Cameron, 1973; Adams, Malatesta, Brantley & Turkat, 1981).

DISCUSSION

A primary rationale for research on cognitive markers of vulnerability is the hope that markers will one day enable us to identify and preventively intervene with persons at risk for schizophrenia. This hope has been complicated by the possibility that there may be several types of schizophrenia that need to be predicted by different markers. Nonetheless, progress has been made and a group of potential vulnerability markers has withstood the test of replication.

One problematic issue concerns the specificity of the cognitive indicators as markers of risk for schizophrenia. Most of the laboratory tests reveal quantitative rather than qualitative differences among individuals at risk for different types of psychopathology, and there is overlap among the score distributions for various groups. It may be that the specificity we are seeking will come from combining the data from various markers. However, another possibility is that some markers may index a predisposition to certain symptoms that can cut across diagnostic lines. If, for example, some markers detect a proneness to develop delusions or hallucinations, then these indicators may prove to be symptom-specific, rather than diagnosis-specific.

It is presently unclear whether cognitive markers should be interpreted as indicators of the presence of a predisposition to schizophrenia, as signs that the predisposition has begun to be expressed, or as markers of actual risk for the disorder. In part, this uncertainty reflects the incompleteness of theories about the etiology of schizophrenia. We do not know, for example, whether information processing anomalies are part of the pathophysiological process that results in schizophrenia or whether they are linkage markers only indirectly associated with the etiology of schizophrenia. A major gap is the lack of models to explain how a problem in vigilance, selective attention or memory could give rise to the cognitive symptoms of schizophrenia. An additional gap concerns an understanding of other parameters, such as life stress, social supports, personality traits, or social competencies that may need to be entered into the prediction equation that predicts actual risk of illness (Spring & Coons, 1982).

An implicit belief among researchers who study cognitive vulnerability markers is that information processing anomalies are centrally involved in the causal pathway that culminates in schizophrenia. If so, then vulnerability research may provide information about fruitful pivot points for preventive or therapeutic intervention (Spring & Ravdin, 1992). Especially, if capacity limitations result from anomalous policies for allocating processing resources, or from overreliance on controlled processing strategies, then psychological interventions may prove to be highly fruitful.

REFERENCES

Adams, H. E., V. Malatesta, P. J. Brantley & I .D. Turkat. 1981. Modification of cognitive processes: a case study of schizophrenia. J. Consult. Clin. Psychol. **49:** 460–464.

Allen, J. J., L. J. Chapman, J. P. Chapman, J. P. Vuchetich & L. A. Frost. 1987. Prediction of psychoticlike symptoms in hypothetically psychosis-prone college students. J. Abnorm. Psychol. **96:** 83–88.

ANDREASON, N. C., M. FLAUM, V. W. SWAYZE, G. TYRRELL & S. ARNDT. 1990. Positive and negative symptoms in schizophrenia. Arch. Gen. Psychiatry **47:** 615–521.
ASARNOW, R. F. & D. J. MACCRIMMON. 1978. Residual performance deficit in clinically remitted schizophrenics: a marker of schizophrenia? J. Abnorm. Psychol. 1978, **87:** 597–608.
ASARNOW, R. F., R. A. STEFFY, D. J. MACCRIMMON & J. M. CLEGHORN. 1978. An attentional assessment of foster children at risk for schizophrenia. *In* The Nature of Schizophrenia: New Approaches to Research and Treatment. L. C. Wynne, R. L. Cromwell & S. Matthysse, Eds. 339–358. John Wiley & Sons. New York.
BATESON, G., D. D. JACKSON, J. HALEY & J. WEAKLAND. 1956. Toward a theory of schizophrenia. Science **1:** 251–264.
BERQUIER, A. & R. ASHTON. 1991. A selective review of possible neurological etiologies of schizophrenia. Clin. Psychol. Rev. **11:** 645–661.
BLEULER, M. 1978. The Schizophrenic Disorders. Trans. S. M. Clemens. Yale University Press. New Haven.
BOHANNON, W. E. & M. E. STRAUSS. 1983. Reaction time crossover in psychiatric outpatients. Psychiatry Res. **9:** 17–22.
BRENNER, H. D., S. KRAEMER, M. HERMANUTZ & B. HODEL. 1990. Cognitive treatment in schizophrenia. *In* Schizophrenia: Concepts, Vulnerability, and Intervention. E. R. Straube & K. Hahlweg, Eds. 161–191. Springer-Verlag. Heidelberg.
CANNON, T. D., S. A. MEDNICK & J. PARNAS. 1990. Antecedents of predominantly negative and predominantly positive-symptom schizophrenia in a high-risk population. Arch. Gen. Psychiatry **47:** 622–632.
CHAPMAN, L. J. & J. P. CHAPMAN. 1978. The measurement of differential deficit. J. Psychiatr. Res. **14:** 303–311.
CHAPMAN, L. J., W. S. EDELL & J. P. CHAPMAN. 1980. Physical anhedonia, perceptual aberrations, and psychosis proneness. Schizophr. Bull. **6:** 639–653.
CLEMENTZ, B. A., W. M. GROVE, J. KATSANIS & W. G. IACONO. 1991. Psychometric detection of schizotypy: perceptual aberration and physical anhedonia in relatives of schizophrenics. J. Abnorm. Psychol. **100:** 607–612.
COOK, W. L., D. A. KENNY & M. J. GOLDSTEIN. 1991. Parental affective style risk and the family system: a social relations model analysis. J. Abnorm. Psychol. **100:** 492–501.
CORNBLATT, B. & L. ERLENMEYER-KIMLING. 1985. Global attentional deviance as a marker of risk for schizophrenia: specificity and predictive validity. J. Abnorm. Psychol. **94:** 470–486.
CORNBLATT, B. & L. ERLENMEYER-KIMLING. 1984. Early attentional predictors of adolescent behavioral disturbances in children at risk for schizophrenia. *In* Children at Risk for Schizophrenia: a Longitudinal Perspective. N. F. Watt, E. J. Anthony, L. C. Wynne & J. E. Rolf, Eds. 198–211. Cambridge University Press. New York.
CROMWELL, R. L. & W. SPAULDING. 1978. How schizophrenics handle information. *In* Phenomenology and Treatment of Schizophrenia. W. E. Fann, I. Karacan, A. D. Pokorny, & R. L. Williams, Eds. 127–162. Spectrum Publications. New York.
CROW, T. J. 1985. Positive and negative schizophrenic symptoms and the role of dopamine. Br. J. Psychiatry **137:** 383–386.
CROW, T. J., C. D. FRITH, E. C. JOHNSTONE & D. G. C. OWENS. 1980. Schizophrenia and cerebral atrophy. Lancet **1:** 1129–1130.
DAVIS, K. L., R. S. KAHN, G. KO & M. DAVIDSON. 1991. Dopamine in schizophrenia: a review and reconceptualization. Am. J. Psychiatry **148:** 1474–1486.
DEAMICIS, L. A. & R. L. CROMWELL. 1979. Reaction time crossover in process schizophrenic patients, their relatives and control subjects. J. Nerv. Ment. Dis. **167:** 593–600.
DRISCOLL, R. M. 1984. Intentional and incidental learning in children vulnerable to psychopathology. *In* Children at Risk for Schizophrenia: a Longitudinal Perspective. N. F. Watt, E. J. Anthony, L. C. Wynne & J. E. Rolf, Eds. 320–326. Cambridge University Press. New York.
ERLENMEYER-KIMLING, L. 1987. Biological markers for liability to schizophrenia. *In* Biological Perspectives of Schizophrenia. H. Helmchen & F. A. Henn, Eds. 33–56. Wiley. Chichester, England.

ERLENMEYER-KIMLING, L. & B. A. CORNBLATT. 1987. High-risk research in schizophrenia: a summary of what has been learned. J. Psychiatry Res. **21:** 401–411.

ERLENMEYER-KIMLING, L., R. R. GOLDEN & B. A. CORNBLATT. 1989. A taxometric analysis of cognitive and neuromotor variables in children at risk for schizophrenia. J. Abnorm. Psychol. **98:** 203–208.

FARAONE, S. V. & M. T. TSUANG. 1985. Quantitative models of the genetic transmission of schizophrenia. Psychol. Bull. **98:** 41–66.

FISHER, J. D. & A. FARINA. 1979. Consequences of beliefs about the nature of mental disorders. J. Abnorm. Psychol. **88:** 320–327.

FRAME, C. L. & T. F. OLTMANNS. 1982. Serial recall by schizophrenic and affective patients during and after psychotic episodes. J. Abnorm. Psychol. **91:** 311–318.

GARVER, D. L. 1987. Methodological issues facing the interpretation of high-risk studies: biological heterogeneity. Schizophr. Bull. **13:** 525–529.

GERSHON, E. S., C. R. MERRILL, L. R. GOLDIN, L. E. DELISI, W. H. BERRETTINI & J. I. NURNBERGER. 1987. The role of molecular genetics in psychiatry. Biol. Psychiatry **22:** 1388–1405.

GJERDE, P. F. 1983. Attentional capacity dysfunction and arousal in schizophrenia. Psychol. Bull. **93:** 57–72.

GRUNEBAUM, H., J. L. WEISS, D. GALLANT & B. COHLER. 1974. Attention in young children of psychotic mothers. Am. J. Psychiatry **131:** 887–891.

HALFORD, W. K. & R. HAYES. 1991. Psychological rehabilitation of chronic schizophrenia patients: recent findings on social skills training and family psychoeducation. Clin. Psychol. Rev. **11:** 23–44.

HARTMANN, E., E. MILOFSKY, G. VAILLANT, M. OLDFIELD, R. FALKE & C. DUCEY. 1984. Vulnerability to schizophrenia. Arch. Gen. Psychiatry **41:** 1050–1056.

HARVEY, P., L. WINTERS, S. WEINTRAUB & J. M. NEALE. 1981. Distractibility in children vulnerable to psychopathology. J. Abnorm. Psychol. **90:** 298–304.

HELD, J. M., R. L. CROMWELL, E. T. FRANK & W. E. FANN. 1970. Effect of phenothiazines on reaction time in schizophrenics. J. Psychiatr. Res. **7:** 209–213.

HELMCHEN, H. 1988. Methodological and strategical considerations in schizophrenia research. Comp. Psychiatry **29:** 337–354.

HENN, F. A. & S. KETY. 1982. Introduction. *In* Schizophrenia as a Brain Disease. F. A. Henn & H. A. Nasrallah, Eds. Oxford Press. New York.

HOGARTY, G. E., C. M. ANDERSON, D. J. REISS, S. J. KORNBLITH, D. P. GREENWALD, R. F. ULRICH & M. CARTER. 1991. Family psychoeducation, social skills training and maintenance chemotherapy in the aftercare treatment of schizophrenia. II. Two year effects of a controlled study on relapse and adjustemnt. Arch. Gen. Psychiatry **48:** 340–347.

KAY, S. R. 1990. Significance of the positive-negative distinction in schizophrenia. Schizophr. Bull. **16:** 635–652.

KING, H. E. 1991. Psychomotor dysfunction in schizophrenia. *In* Neuropsychology, Psychophysiology and Information Processing. S. R. Steinhauer, J. H. Gruzelier & J. Zubin, Eds. 273–301. Elsevier. New York.

KRAEPELIN, E. 1919. *Dementia praecox*. Trans. R. Barclay. E. S. Livingston, Ltd. Edinburgh.

LENZENWEGER, M. F., R. H. DWORKIN & E. WETHINGTON. 1989. Models of positive and negative symptoms in schizophrenia: an empirical evaluation of latent structures. J. Abnorm. Psychol. **98:** 62–70.

LENZENWEGER, M. F. & A. W. LORANGER. 1989. Psychosis proneness and clinical psychopathology: examination of the correlates of schizotypy. J. Abnorm. Psychol. **98:** 3–8.

LEVIN, S., D. YURGELUN-TODD & S. CRAFT. 1989. Contributions of clinical neuropsychology to the study of schizophrenia. J. Abnorm. Psychol. **98:** 341–356.

LEWINE, R. R. J. 1990. A discriminant validity study of negative symptoms with a special focus on depression and antipsychotic medication. Am. J. Psychiatry **147:** 1463–1466.

LIBERMAN, R. P., C. J. WALLACE, C. E. VAUGHN, K. S. SNYDER & C. RUST. 1980. Social and family factors in the course of schizophrenia. *In* The Psychotherapy of Schizophrenia. J. S. Strauss, M. Bowers, T. W. Downey, S. Fleck, S. Jackson & I. Levine, Eds. 21–54. Plenum Medical Book Co. New York.

LIDZ, T., A. CORNELISON, S. FLECK & D. TERRY. 1957. The intrafamilial environment of schizophrenic patients: II. Marital schism and marital skew. Am. J. Psychiatry **114:** 241–248.

LIEM, J. H. 1974. Effects of verbal communications of parents and children: a comparison of normal and schizophrenic families. J. Consult. Clin. Psychol. **42:** 438–450.

MAHER, B. A. 1974. Delusional thinking and perceptual disorder. J. Individ. Psychol. **30:** 98–113.

MANDLER, G. 1979. Thought processes, consciousness and stress. *In* Human Stress and Cognition. V. Hamilton & D. M. Warburton, Eds. John Wiley & Sons. New York.

MARCUS, L. M. 1973. Studies of attention in children vulnerable to psychopathology. Diss. Abstr. Int. **33:** 5023–B.

MARTIN, E. M. & L. J. CHAPMAN. 1982. Communication effectiveness in psychosis-prone college students. J. Abnorm. Psychol. **91:** 420–425.

McGHIE, A. 1969. Pathology of Attention. Penguin Books. Baltimore.

MEICHENBAUM, D. & R. CAMERON. 1973. Training schizophrenics to talk to themselves: a means of developing attentional controls. Behav. Ther. **4:** 515–534.

MIKLOWITZ, D. J., A. M. STRACHAN, J. M. GOLDSTEIN, J. A. DOANE, K. S. SNYDER, G. E. HOGARTY & I. R. H. FALLOON. 1986. Expressed emotion and communication deviance in the families of schizophrenics. J. Abnorm. Psychol. **95:** 60–66.

MOISES, H. W., J. GELERNTER, L. A. GIUFFRA, V. ZARCONE, L. WETTERBERG, O. CIVELLI, K. K. KIDD & L. L. CAVALLI-SFORZA. 1991. No linkage between D2 dopamine receptor gene region and schizophrenia. Arch. Gen. Psychiatry **48:** 643–647.

NUECHTERLEIN, K. H. 1983. Signal detection in vigilance tasks and behavioral attributes among offspring of schizophrenic mothers and among hyperactive children. J. Abnorm. Psychol. **92:** 4–28.

NUECHTERLEIN, K. H. & M. E. DAWSON. 1984. Information processing and attentional functioning in the developmental course of schizophrenic disorders. Schizophr. Bull. **10:** 160–203.

NUECHTERLEIN, K. H. 1991. Vigilance in schizophrenia and related disorders. *In* Neuropsychology, Psychophysiology and Information Processing. S. R. Steinhauer, J. H. Gruzelier & J. Zubin, Eds. 397–433. Elsevier. New York.

NUECHTERLEIN, K. H. & K. M. ZAUCHA. 1990. Similarities between information-processing abnormalities of actively symptomatic schizophrenic patients and high-risk children. *In* Schizophrenia: Concepts, Vulnerability and Intervention. E. R. Straube & K. Hahlweg, Eds. 77–96. Springer-Verlag. Heidelberg.

ORZACK, M. H. & C. KORNETSKY. 1971. Environmental and familial predictors of attention behavior in chronic schizophrenics. J. Psychiatr. Res. **9:** 21–29.

PARDES, H., C. KAUFMANN, H. A. PINCUS & A. WEST. 1989. Genetics and psychiatry: past discoveries, current dilemmas and future directions. Am. J. Psychiatry **146:** 435–443.

PHIPPS-YONAS, S. 1984. Visual and auditory reaction time in children vulnerable to psychopathology. *In* Children at Risk for Schizophrenia: a Longitudinal Perspective. N. F. Watt, F. J. Anthony, L. C. Wynne & J. E. Rolf, Eds. 312–319. Cambridge University Press. New York.

POSNER, M. I. 1978. Chronometric Explorations of Mind. Lawrence Erlbaum Associates. Hillsdale, N.J.

REVELEY, A. M., M. A. REVELEY & R. M. MURRAY. 1984. Cerebral ventricular enlargement in non-genetic schizophrenia: a controlled twin study. Br. J. Psychiatry **144:** 89–93.

RODNICK, E. & D. SHAKOW. 1940. Set in the schizophrenic as measured by a composite reaction time index. Am. J. Psychiatry **97:** 214–225.

ROSENBAUM, G., D. L. SHORE & K. CHAPIN. 1988. Attentional deficit in schizophrenia and schizotypy: marker versus symptom variables. J. Abnorm. Psychol. **97:** 41–47.

ROSVOLD, H. E., A. MIRSKY, I. SARASON, E. D. BRANSOME & L. H. BECK. 1956. A continuous performance test of brain damage. J. Consult. Psychol. **20:** 343–350.

SACCUZZO, D. P. 1977. Bridges between schizophrenia and gerontology: generalized or specific deficits. Psychol. Bull. **84:** 595–600.

SCHNEIDER, W. & R. M. SHIFFRON. 1977. Controlled and automatic human information processing: 1. Detection, search and attention. Psychol. Rev. **84:** 1–66.

SIMONS, R. F., F. W. MACMILLAN & F. B. IRELAND. 1982. Reaction time crossover in preselected schizotypic subjects. J. Abnorm. Psychol. **91:** 414–419.

SPELKE, E., W. HIRST & V. NEISSER. 1976. Skills of divided attention. Cognition **4:** 215–230.

SPOHN, H. E., R. LACOURSIERE, K. THOMPSON & L. COYNE. 1977. Phenothiazine effects on psychological and psychophysiological dysfunction in chronic schizophrenics. Arch. Gen. Psychiatry **34:** 633–644.

SPOHN, H. E., L. COYNE, R. LACOURSIERE, D. MAZUR & K. HAYNES. 1985. Relation of neuroleptic dose and tardive dyskinesia to attention, information processing and psychophysiology in medicated schizophrenics. Arch. Gen. Psychiatry **42:** 849–859.

SPRING, B. J. 1980. Shift of attention in schizophrenics, siblings of schizophrenics, and depressed patients. J. Nerv. Ment. Dis. **168:** 133–140.

SPRING, B. & H. COONS. 1982. Stress as a precursor of schizophrenia. *In* Psychological Stress and Psychopathology. R. Neufeld, Ed. 13–53. McGraw Hill. New York.

SPRING, B., M. LEMON & P. FERGESON. 1990. Vulnerabilities to schizophrenia: information-processing markers. *In* Schizophrenia: Concepts, Vulnerability and Intervention. E. R. Straube & K. Hahlweg, Eds. 98–114. Springer-Verlag. Heidelberg.

SPRING, B., M. LEMON, L. WEINSTEIN & A. HASKELL. 1989. Distractibility in schizophrenia: state and trait aspects. Br. J. Psychiatry **155**(Suppl. 5), 63–68.

SPRING, B. & L. RAVDIN. 1992. Cognitive remediation in schizophrenia: Should we attempt it? Schizophr. Bull. **18:** 15–20.

SPRING, B., L. WEINSTEIN, R. FREEMAN & S. THOMPSON. 1991. Selective attention in schizophrenia. *In* Neuropsychology, Psychophysiology and Information Processing. S. R. Steinhauer, J. H. Gruzelier & J. Zubin, Eds. 371–396. Elsevier. New York.

SPRING, B. & J. ZUBIN. 1978. Attention and information processing as indicators of vulnerability to schizophrenic episodes. J. Psychiatr. Res. **14:** 289–302.

STEFFY, R. A. & K. A. GALBRAITH. 1974. A comparison of segmental set and inhibitory deficit explanations of the crossover pattern in process schizophrenic reaction time. J. Abnorm. Psychol. **83:** 227–233.

STRAUSS, M. E., W. E. BOHANNON, M. J. KAMINSKY & F. KHARABI. 1979. Simple reaction time and crossover in schizophrenic outpatients. Schizophr. Bull. **5:** 612–615.

SUTTON, S. & J. ZUBIN. 1965. Effect of sequence on reaction time in schizophrenia. *In* Behavior, Aging and the Nervous System. A. Welford & J. E. Birren, Eds. 1–36. Charles C. Thomas. Springfield, Illinois.

TORREY, E. F. 1983. Surviving Schizophrenia: a Family Manual. Harper & Row. New York.

VAN DYKE, J. L., D. ROSENTHAL & P. V. RASMUSSEN. 1975. Schizophrenia: effects of inheritance and rearing on reaction time. Can. J. Behav. Sci. **7:** 223–236.

VENTURA, J., K. H. NUECHTERLEIN, D. LUKOFF & J. P. HARDESTY. 1989. A prospective study of stressful life events and schizophrenic relapse. J. Abnorm. Psychology **98:** 407–411.

WAGNER, B. R. 1968. The training of attending and abstracting responses in chronic schizophrenics. J. Exper. Res. Pers. **3:** 77–88.

WALKER, E. 1981. Attentional and neuromotor functions of schizophrenics, schizoaffectives, and patients with other affective disorders. Arch. Gen. Psychiatry **38:** 1355–1358.

WALKER, E. & J. SHAYE. 1982. Familial schizophrenia: a predictor of neuromotor and attentional abnormalities in schizophrenia. Arch. Gen. Psychiatry **39:** 1153–1156.

WINTERS, K. C., A. A. STONE, S. WEINTRAUB & J. M. NEALE. 1981. Cognitive and attentional deficits in children vulnerable to psychopathology. J. Abnorm. Child. Psychol. **9:** 435–453.

WOHLBERG, G. W. & C. KORNETSKY. 1973. Sustained attention in remitted schizophrenics. Arch. Gen. Psychiatry **28:** 533–537.

WOOD, R. L. & M. COOK. 1979. Attentional deficit in the siblings of schizophrenics. Psychol. Med. **9:** 465–467.

WYNNE, L. C., I. RYCKOFF, J. DAY & S. HIRSCH. 1958. Pseudomutuality in the family relations of schizophrenics. Psychiatry **21:** 205–220.

ZIMBARDO, P. G., S. M. ANDERSON & L. G. KABOT. 1981. Induced hearing deficit generates experimental paranoia. Science **212:** 1529–1531.

P3 and Schizophrenia[a]

JUDITH M. FORD, ADOLF PFEFFERBAUM AND
WALTON ROTH

Department of Psychiatry 116A3
Veterans Affairs Medical Center
3801 Miranda Avenue
Palo Alto, California 94304
and
Department of Psychiatry and Behavioral Sciences
Stanford University School of Medicine
Stanford, California, 94305

Research using the P3 component of the event-related potential (ERP) to study schizophrenia has been both theoretically and empirically driven. Theoretical approaches have been based upon cognitive and brain structure abnormalities in schizophrenics, as well as their responses to neuroleptic medication. Empirical approaches have assessed the diagnostic utility of P3 and/or attempted to establish it as a genetic marker of the disease. The theoretical approaches have taken advantage of the unique properties of the ERP. It allows one to assess the allocation of attentional resources to ignored or distracting stimuli; observe the strategies used in selective attention; assess attention directed automatically and compare it to that directed effortfully; estimate the relative timing of stimulus evaluation processes; estimate the time taken to withhold a response; and, perhaps, observe regional anomalies in brain function.

THEORETICAL APPROACHES

Using P3 to Assess Cognitive Theories of Schizophrenia

Attention, Distraction, and Salience

Bleuler (1911) suggested that symptoms observed in schizophrenia may result from a cognitive deficit involving attention. More recent investigators have suggested that schizophrenics have difficulty in filtering out irrelevant information (Braff, Callaway & Naylor, 1977). Such a difficulty might result in an inability to focus attention and ignore irrelevant stimuli, both real and imagined, and be manifested as a thought disorder (McGhie, Chapman & Lawson, 1965). Research into the nature and extent of the inability of schizophrenics to ignore irrelevant information has examined three related aspects of this process: selective attention to relevant channels of stimuli, distraction by irrelevant stimuli, and salience of relevant and irrelevant stimuli.

[a] Supported by the Veterans Administration and National Institute of Mental Health, Mental Health Clinical Research Center (Bethesda, MD) Grant MH 30854 and National Institute of Mental Health Grant 40052.

Selective Attention. Although the focus of this review is the P3 component of the ERP, it is impossible to discuss selective attention without mentioning the earlier N1, Nd or processing negativity (PN). Baribeau, Picton, and Gosselin (1983) first used the dichotic target detection paradigm developed by Hillyard, Hink, Schwent, and Picton (1973) to study selective attention in schizophrenia. The rate of stimulation used in this paradigm makes it necessary to use a simple hierarchical strategy of actively excluding the irrelevant channel of input in order to perform the target detection task. Baribeau *et al.* (1983) used two rates of stimlation and found that only at the faster rate did schizophrenics (20 medicated)* have the same pattern of greater N1 amplitude to stimuli in the relevant channel as the controls. At the slower rate, the hierarchical strategy apparently broke down in the schizophrenics, as their N1s became equal in both channels.

Michie, Fox, Ward, Catts and McConaghy (1990) used a paradigm which added another level of selectivity to the attention task. Subjects were asked to detect a long duration tone of specific pitch presented to a specific location. The stimulus sequence was designed so that the task would be executed in a hierarchical manner—first select for location, then for pitch, and finally for duration. This hierarchical strategy used by the controls for detecting targets was revealed in the timing of the PN associated with each stimulus. The PN of schizophrenics (10 med) indicated that they selectively attended to location, but that the hierarchical strategy was not maintained to the next level of selectively attending for pitch. In addition, both Baribeau *et al.* (1983) and Michie *et al.* (1990) found that schizophrenics had smaller and delayed P3s to the target stimulus, indicating a deficit in maintaining attention and slowed target evaluation. The fact that Baribeau *et al.* used medicated patients and Michie *et al.* used unmedicated patients suggests that these deficits do not depend on medication status.

Distraction. The obverse of attending to something is being distracted away from it. A body of literature has focussed on demonstrating that schizophrenics' performance is more vulnerable to distraction than is controls' (Oltmanns & Neale, 1975). Grillon, Courchesne, Ameli, Geyer, and Braff (1990) predicted that schizophrenics would show increased distractibility due to their impaired ability to screen out irrelevant stimuli. Capitalizing upon the ERP's unique ability to assess the response to irrelevant stimuli, Grillon *et al.* (1990) demonstrated that schizophrenics (1 unmed/14 med) are more vulnerable to distraction than controls. Subjects were presented with a standard auditory oddball sequence and with a sequence to which novel distractors (a collection of unusual computer-generated sounds) were added. Abnormally small P3 responses to targets compared to distractors suggested that these resources were incorrectly apportioned between relevant and irrelevant stimuli. Smaller P3 responses to both the target and distractor suggested fewer attentional resources were available to the schizophrenics. Behavioral data were also consistent with schizophrenics being abnormally vulnerable to the effects of distractors.

Barrett, McCallum, and Pocock (1986) also observed smaller P3s to both targets and distractors in a group of schizophrenics (3 unmed/17 med), but did not find that the P3 to distractors was larger than P3 to targets. The task in this study was qualitatively different from that in the Grillon study (Grillon *et al.*, 1990) in that the distractors were less salient. Barrett *et al.* delivered

* Medication status of schizophrenic pateints will be detailed in this review, when available, either in the text or in parentheses.

standards and rares to each ear, but only the rares in one ear were designated as targets. In this way, the nontarget rare delivered to the unattended ear was a distractor, to which schizophrenics frequently made erroneous button presses. It is possible that if P3 had been computed only from those nontarget rare trials which distracted the patients, it would have been larger than P3 to the targets.

Salience. Germane to research on distraction by irrelevant stimuli is research on attention allocated to task-irrelevant stimuli. Again, the ERP is uniquely suited to inform us directly about the attention paid to such stimuli. Attention to irrelevant stimuli can be conceived of as attention automatically elicited by a very salient evoking stimulus. Our recent paper suggested that even the automatic, or passive, P3 is reduced in both medicated (13) and unmedicated (18) schizophrenics (Pfefferbaum, Ford, White & Roth, 1989), consistent with the original report of Roth and Cannon (1972) of smaller P3s to task-irrelevant stimuli in a group of mostly medicated schizophrenics (2 unmed/19 med).

When the timing and nature of stimuli are predictable, they are less salient, elicit less attention, and require less processing. Indeed, P3s elicited by predictable stimuli do not differ between schizophrenics and controls (Levit, Sutton & Zubin, 1973; Verleger & Cohen, 1978). These findings suggest that schizophrenics and controls allocate equivalent resources to the processing of predictable stimuli. The processing of unpredictable stimuli requires more attentional resources, which schizophrenics either have less of or allocate differently than controls. These alternatives can best be distinguished with ERP studies using attentional allocation paradigms developed in the Donchin laboratory (Isreal, Chesney, Wickens & Donchin, 1980).

Duncan, Perlstein, and Morihisa (1987b) tested schizophrenics (11 unmed/15 med) with auditory and visual sequences of stimuli at three levels of probability (p = 0.5, 0.3, 0.1). P3 was reduced in schizophrenics compared to controls, but only for the low probability auditory stimuli. Roth, Pfefferbaum, Kelly, Berger, and Kopell (1981) compared the P3 responses of schizophrenics (8 unmed/14 med) across four paradigms in which the target stimulus probability was p = 1.0, 0.33, 0.15 or 0.10. In the first paradigm, timing was uncertain but the identity of the next stimulus was 100% predictable. Only the last two paradigms with the low probability targets (0.15 and 0.10 probability) elicited P3s that differed between schizophrenics and controls. Unfortunately, they were also the only paradigms with an RT task, leaving open the question of whether P3 reduction in schizophrenia was due to the imposition of an RT task or to the use of unpredictable targets. In a recent study (Pfefferbaum *et al.*, 1989), we used the same type of stimulus that Roth *et al.* (1981) had used in their first two paradigms, and like them had no RT requirement. This time the presence of frequent background stimuli gave the infrequent stimulus a 0.15 probability. Lowering the probability of the loud noise stimulus by adding frequently occurring background tone pips resulted in a P3 reduction in schizophrenics compared to controls. In another study (Roth, Goodale & Pfefferbaum, 1991) we again found that without background stimuli, the P3 elicited by a loud noise is not reduced in magnitude in schizophrenics despite long interstimulus intervals (12–17 sec) and an RT task requirement. A comparison of these studies suggests background stimuli are more important than long ISIs or task requirements in eliciting reduced P3 from schizophrenics. This may be because background stimuli engender uncertainty about the identity of the next stimulus, and it is the response to uncertainty which distinguishes schizophrenics from normals.

Insensitivity to Context

The above discussion suggests that when there is a context against which to compare an event, electrophysiological responses to that event differ between schizophrenics and controls. Two types of context have been investigated with ERPs—local probability structure of tone sequences and semantic.

Probability Context. From the foregoing, we might expect that schizophrenics and controls would differ in sensitivity to probability of events. It might be predicted that P3s elicited by highly probable events will not differ between schizophrenics and controls, and that the P3 amplitude difference between schizophrenics and controls will increase as events become less probable. Squires, Wickens, Squires, and Donchin (1976) published data demonstrating that P3 was sensitive to global as well as local sequential probability of events. Not only was P3 larger to a tone that occurred on 30% of the trials compared to 50%, but P3 was also larger to a tone that had not occurred for 5 trials compared to a tone that had just occurred on the previous trial. Duncan-Johnson, Roth, and Kopell (1984) used this analysis technique to test the hypothesis that P3 amplitude reduction was due to a failure of the schizophrenics (16 unmed/27 med) to estimate the local probability of stimuli. They did not vary global probability in this study, but analyzed the data for the effects of local probability. They found that the P3 amplitude and RTs of schizophrenics showed the same dependence as controls on the sequence of preceding stimuli in spite of an overall reduction in P3 amplitude and delay in RT. They concluded that smaller P3s observed in schizophrenics were not due to their abnormal estimations of sequential probabilities. This suggests the dependence of P3 amplitude reduction in schizophrenia on the presence of background stimuli is not due to alterations in local probability, but perhaps to global probability.

Semantic Context. Linguistic research has indicated that schizophrenics have fewer constraints on word selection. Maher (1972) showed that it is harder for normals to guess the words from passages of schizophrenic speech than normal speech. When schizophrenics are asked to fill in deleted words from normal speech, their responses are dictated by the immediately preceding word rather than the sentence as a whole. Kutas and Hillyard (1984) observed a large, late negative component (N400) whose amplitude reflected the semantic expectancy of the final word in the sentence. When the final word was unrelated to the sentence, there was a large N400. Several laboratories have recorded N400s from schizophrenics. Andrews, Mitchell, Fox, Catts, Ward, and McConaghy (1990) found that when the subjects were engaged in a semantic task, the incongruous word elicited larger N400s in controls than schizophrenics. However, when engaged in a physical match task, both groups had equivalent N400 amplitudes. In a word pair lexical decision task, Hokama, Koyama, Miyatani, Miyazato, Ogura, Nageishi *et al.* (1990) showed that when the second word was unrelated to the first, it elicited a larger N400 in controls than in schizophrenics (7 unmed). A similar result was obtained by Holinger and Wagman (1990) in 5 medicated thought-disordered schizophrenic patents performing a semantic matching task. Adams, Faux, McCarley, Marcy, and Shenton (1989) found reduced N400 in only two of the four schizophrenics (all med) studied, although the N400 in the composite average was reduced in schizophrenics compared to the controls. Even though no task instructions were mentioned by Adams *et al.*, it is reasonable to assume that subjects were instructed to read the visually presented sentences. We might conclude from these preliminary studies that the impact of semantic context as reflected in N400 amplitude is somewhat reduced in schizophrenics, and may interact with task demands but not with medication status.

Automatic vs Controlled Processing

One of the interesting aspects of the study by Duncan-Johnson *et al.* (1984) mentioned above is that it might represent a comparison of automatic and controlled processing in the same task. The formation of trial-to-trial expectancies from a random tone sequence might be considered an automatic process (Hasher & Zacks, 1979), while the allocation of attentional resources to identify and respond to a target stimulus might be considered a controlled or effortful process. Thus, the findings of Duncan-Johnson *et al.* might be interpreted as meaning that schizophrenics are similar to normals in automatic, but differ in controlled aspects of processing. This is consistent with recent data from our laboratory (Pfefferbaum *et al.*, 1989). Eye blinks elicited automatically by intense noise bursts were normal in both medicated and unmedicated schizophrenics, while the P3 was considerably smaller but not later. We interpreted this to mean that the initial registration that "something important happened" (*i.e.*, automatic processing) is normal in the schizophrenics, but subsequent processing is deficient.

Motivation/Task Performance

Schizophrenics are often slow to respond verbally or with a button press. While such an observation invites speculation about slowed cognition, P3 latency provides a direct test of it. Kutas, McCarthy and Donchin (1977) laid the groundwork for a decade of P3 studies by showing that P3 latency is an indication of the relative speed of stimulus evaluation processes and is unaffected by response-related processes. With few exceptions (Pfefferbaum, Christensen, Ford & Kopell, 1986), P3 latency has been demonstrated to be a veridical index of the timing of stimulus-related cognitive events. RTs collected in ERP studies have been consistently longer in schizophrenics, while P3 latency may or may not be later. Some studies report significantly increased P3 latency in medicated patients (Baribeau *et al.*, 1983; Blackwood, Whalley, Christie, Blackburn, St Clair & McInnes, 1987; Pfefferbaum *et al.*, 1989; Romani, Merello, Gozzoli, Zerbi, Grassi & Cosi, 1987), in unmedicated patients (Kutcher, Blackwood, St Clair, Gaskell & Muir, 1987; Michie *et al.*, 1990), and in a mixture of medicated and unmedicated patients (Duncan *et al.*, 1987b; Pfefferbaum, Wenegrat, Ford, Roth & Kopell, 1984; Roth, Horvath, Pfefferbaum, Tinklenberg, Mezzich & Kopell, 1979). Others have not found significant P3 slowing in medicated (Levit *et al.*, 1973; Verleger & Cohen, 1978), unmedicated (Brecher & Begleiter, 1983; Brecher, Porjesz & Begleiter, 1987a; Brecher, Porjesz & Begleiter, 1987b; Pfefferbaum *et al.*, 1989), or in a mixture of medicated and unmedicated patients (Barrett *et al.*, 1986; Duncan-Johnson *et al.*, 1984; Grillon *et al.*, 1990; Roth & Cannon, 1972; Roth, Horvath, Pfefferbaum & Kopell, 1980; Roth, Pfefferbaum, Horvath, Berger & Kopell, 1980; Roth *et al.*, 1981). Considering only those studies with uniform medication status, auditory P3s were more likely than visual P3s to be delayed in schizophrenics. One exception is our study (Pfefferbaum *et al.*, 1989) in which both visual and auditory P3s were delayed in medicated schizophrenics. Romani *et al.* (1987) found that delayed P3s were consistently associated with poor neuropsychological performance, suggesting that some kind of general cognitive ability affects whether P3 is delayed in schizophrenia. Studies finding delayed RT but no delayed P3 suggest that these schizophrenics are not slow to evaluate the events they encounter but are slow to make a response.

Although several reports mentioned above show that P3 is reduced even when there is no call for motivated effort (Pfefferbaum *et al.*, 1989; Roth *et al.*, 1980), the possibility remains that schizophrenics have reduced P3s because they are not motivated to allocate resources to the task (Roth, Tecce, Pfefferbaum, Rosenbloom & Callaway, 1984). Certainly, most ERP studies involving a task find both P3 amplitude and performance decrements in schizophrenic patients. Some studies have directly addressed the question of whether P3 reduction in schizophrenics is due to a lack of motivation by equating performance on individual trials while recording P3. Roth *et al.* (1980) sorted trials according to whether they were associated with short (100–286 msec) or long (287–600 msec) RTs. While the short RT trials had larger P3s than the long RT trials, schizophrenics (8 unmed/7 med) had smaller P3s than controls, regardless of RT range. Although RT was more variable in schizophrenics than in controls, single-trial latency adjustment of P3 did not eliminate the group differences. It has also been shown that when performance is equated by averaging only correct trials, P3 is still reduced in schizophrenics compared to controls (Steinhauer & Zubin, 1982).

Brecher and Begleiter (1983) attempted to manipulate the incentive value of correct vs fast stimulus identification, by offering on different runs, $1 for either correct or fast responses. P3 was reduced in schizophrenics (14 unmed) regardless of incentive. P3 was larger to the rewarded stimulus than the unrewarded stimulus for controls but not for schizophrenics. Although both controls and schizophrenics made very few errors, schizophrenics had fewer correct responses. Errors were especially pronounced on trials with fast RTs for schizophrenics when fast performance was rewarded, suggesting that schizophrenics, more than controls, traded accuracy for speed.

The lack of responsiveness of visual P3 to task manipulations was found in this same group of 14 unmedicated schizophrenics performing a different task (Brecher *et al.*, 1987a), and it was suggested that schizophrenics have a relatively invariant P3 response. While this may be true for the visual P3 elicited during an effortful task, it is not true for the response of the auditory P3 during the automatic extraction of local probability information (Duncan-Johnson *et al.*, 1984).

Using P3 to Assess Structural and Biochemical Theories of Schizophrenia

Structural Abnormalities

While there may be many generators of P3, two strong candidates for important roles are the hippocampus deep in the temporal lobe (Halgren, Squires, Wilson, Rohrbaugh, Babb & Crandall, 1980; Okada, Kaufman & Williamson, 1983) and the frontal lobes (McCarthy, 1985). Both locations have also been suggested as sites of pathology in schizophrenia (Bogerts, Meertz & Schonfeldt-Bausch, 1985; Brown, Colter, Corsellis, Crow, Firth, Jagoe *et al.*, 1986; Buchsbaum, Ingvar, Kessler, Waters, Cappelletti, Van Kammen *et al.*, 1982; Crow, Colter & Brown, 1988; Ingvar & Franzen, 1974; Kovelman & Scheibel, 1984). Neuropsychological and behavioral data further have been interpreted to indicate disturbances in the left hemisphere (Gur, 1978; Hammond & Gruzelier, 1978). The McCarley laboratory has attempted to find electrophysiological evidence for left temporal lobe deficits by recording P3 from electrodes placed over the right (T4) and left (T3) temporal scalp recording sites (McCarley, Faux, Shenton, LeMay, Cane, Ballinger *et al.*, 1989). They have published many papers (reviewed in McCarley *et al.*, 1989) showing P3 reduction at left temporal scalp electrode sites in schizophrenics. Recently, they related structural brain changes as measured by CT to

ERP measures. In a group of 9 medicated schizophrenics and 9 controls, they found that left sylvian fissure enlargement was highly correlated with reduction of the auditory oddball P3 recorded over the left temporal scalp (T3). Both were correlated with positive symptoms in the schizophrenics. While other laboratories have not reported attempts to correlate regional CT values with asymmetrical P3s, we (Pfefferbaum *et al.*, 1989; Pfefferbaum, Ford, White, Roth & Mathalon, 1991) specifically tested the hypothesis that P3 recorded at T3 would be smaller than that recorded at T4 in schizophrenics but not in controls. Using a noncephalic reference, we found no support for this hypothesis in either medicated or unmedicated groups of schizophrenics. Michie *et al.* (1990) found no support for asymmetry of the auditory P3 in 10 unmedicated schizophrenics when recording from T3 and T4. Koga, Hashiguchi, Ogino, Chiba, Taguchi, and Hori (1987) have also recorded P3 bilaterally off the midline, while presenting auditory and visual, linguistic and nonlinguistic stimuli to 25 medicated schizophrenics. Although not specifically testing for group differences at T3 and T4, they found that P3s recorded from multiple electrodes over the right and left hemispheres, including T3 and T4, were reduced in schizophrenics compared to controls, and more significant differences between patients and controls were found in P3 recorded over the left hemisphere. Although they did not record P3 over left and right temporal areas, Kemali, Galderisi, Maj, Mucci, Cesarelli, and D'Ambra (1988) found no evidence for lateral asymmetry (P3 and P4) in either 14 unmedicated schizophrenics or controls. While lateralized topographic differences in P3 amplitude remain to be confirmed, many investigators have noted that the anterior-posterior midline scalp distribution of P3 is different in schizophrenics, being somewhat more frontally maximal than in controls (Miyazato, Ogura, Miyatani, Nashiro, Fukao & Hokama, 1989; Pfefferbaum *et al.*, 1989).

Neuroleptic and Clinical Correlates of P3

The dopamine theory of schizophrenia suggests that schizophrenia is related to a relative excess of dopamine (DA). Most neuroleptic medications block postsynaptic DA receptor sites and presumably in that way ameliorate symptoms. If P3 amplitude reduction in schizophrenia reflects the clinical state of a patient, P3 should be larger when the patient is symptomatically improved on neuroleptic medication. Longitudinal designs are rare, however, though one has been reported by Duncan, Morihisa, Fawcett, and Kirch (1987a). Furthermore, it is often clinically impractical to randomly assign patients to on- and off-medication groups since only those who are able or willing to be taken off their medication can be studied in either the medicated or unmedicated state. Many patients can only be studied in the medicated state.

Several cross-sectional studies have reported no difference in P3 between groups of medicated and unmedicated patients (Pass, Klorman, Salzman, Klein & Kaskey, 1980; Pfefferbaum *et al.*, 1989; Roth *et al.*, 1981). Others have found larger P3s in medicated patients (Josiassen, Shagass, Straumanis & Roemer, 1984; Roth *et al.*, 1979). Pfefferbaum *et al.* (1989) compared the auditory and visual P3s of medicated and unmedicated schizophrenics. Compared with controls, both groups had smaller auditory P3s at Pz. The medicated schizophrenics had normal size auditory P3s at the frontal sites. While the visual P3 was somewhat smaller in the schizophrenics, the difference did not reach the a priori level of significance ($p < 0.05$). In a longitudinal design, Duncan *et al.* (1987a) studied 7 schizophrenics on and off medication. Patients who exhibited clinical improvement on the BPRS (positive and negative symptoms) with medication showed an increase in visual, but not auditory, P3 amplitude.* Using only the auditory ERP, Blackwood *et al.* (1987) studied the same 14 patients before neuroleptic treatment began, 1 week after treatment started, and 4 weeks after treatment started. While N1 amplitude was reduced with treatment, P3 was not affected.

The early work of Levit *et al.* (1973) and the more recent work of Duncan *et al.* (1987a) suggest that any effects of medication on P3 amplitude may be mediated by a medication-related improvement in clinical symptoms. In a recent review of the literature, Pritchard suggested that P3 amplitude reduction in schizophrenia may represent a marker of the deficit state of schizophrenia as indexed by negative symptoms (Pritchard, 1986). Since then, several attempts have been made to test that hypothesis. Kemali *et al.* (1988) demonstrated that in 14 unmedicated schizophrenics, the visual P3 amplitude was negatively correlated with negative symptoms as measured by the Scale for the Assessment of Negative Symptoms (SANS) (Andreasen & Olsen, 1982). Ward, Catts, Fox, Michie, and McConaghy (1991) found that the auditory P3 was also negatively related to the alogia scale of the SANS in 10 unmedicated schizophrenics. Our recent study (Pfefferbaum *et al.*, 1989) showed that both the auditory and visual P3 amplitudes in unmedicated schizophrenics negatively correlated with negative symptoms as assessed by the Brief Psychiatric Rating Scale (BPRS) (motor retardation, blunted affect, mannerisms and posturing, and emotional withdrawal). This is consistent with the proposal by Roth and Tinklenberg (1982), who argued that P3 is the ERP component most analogous to the skin conductance response, which like P3 is negatively correlated with negative symptoms (Bernstein, 1987).

Although they did not specifically test the relationship between negative symptoms and P3 amplitude, Brecher and Begleiter (1983) found no relationship between total BPRS or the anergia score of the BPRS and the visual P3 amplitude in a group of unmedicated schizophrenics. Barrett *et al.* (1986) noticed no relationship between auditory P3 amplitude and positive symptomatology, but did not attempt to correlate amplitudes and symptoms. Blackwood *et al.* (1987) found that while BPRS scores improved with neuroleptic medication, auditory P3 amplitudes did not. However, auditory P3 amplitude was significantly correlated with "depressive mood" ($r = -0.47$, $p < 0.05$) and motor retardation ($r = -0.44$, $p < 0.05$). No correlation of auditory P3 amplitude with a composite score of negative symptomatology was made.

Summary of Theoretical Work

Reduction of P3 amplitude in schizophrenics is robust. Neuroleptic treatment does *not* affect it unless there is clinical improvement with medication; P3 reduction does not depend upon task relevance, nor does it always correlate with poor performance; and it may depend upon stimulus certainty, global probability, and context. In the auditory modality, to elicit a P3 that is reduced in schizophrenia, stimuli should be salient and their identity made uncertain, perhaps by the introduction of frequently occurring background stimuli. Whether P3s elicited by auditory and visual stimuli reflect different processes in schizophrenics is unsettled. Cer-

* This might be due to patients better fixating and focussing their eyes on the visual stimuli when their clinical state has improved; with auditory stimuli presented via headphones, no improvement in peripheral selection is possible.

tainly both are reduced, but perhaps for different reasons. Both the auditory and visual P3 amplitudes may be sensitive to negative symptoms, with only visual P3 being sensitive to overall clinical improvement.

The ERP research reviewed above has added to our understanding of the psychopathology of schizophrenia. Before the application of ERPs we did not know that schizophrenics use different hierarchical strategies in selective attention; that schizophrenics do not differentiate distracting and task-relevant stimuli as well as controls; and that slowed reactions in schizophrenics to a visual stimulus probably are not due to slowed stimulus evaluation. Further, questions regarding the total amount of processing resources available to schizophrenics can be addressed using ERP paradigms (Isreal *et al.*, 1980).

Yet to be ascertained is how diagnostic any of these cognitive ERP findings may be. In reporting results, very few investigators report how well controls and schizophrenics can be discriminated by single variables or combinations of variables. In fact, rarely are the cognitively-sensitive paradigms that are described above used in any attempts to establish the ERP as a diagnostic tool in psychiatry.

EMPIRICAL APPROACHES

Using P3 to Diagnose Schizophrenia

For a diagnostic test to be useful, it should be both sensitive and specific. The relative importance given each depends on the ultimate application of the test. For example, a diagnostic test for schizophrenia might be the auditory P3, which seems to be reliably reduced in amplitude in schizophrenics. The sensitivity of P3 would be estimated as the percentage of schizophrenics with P3s smaller than the age-corrected P3 amplitude cutoff. The specificity of P3 would be estimated as the percentage of nonschizophrenic psychiatric patients who have P3 amplitudes larger than the cutoff. For a screening test (*i.e.*, to rule out schizophrenia), sensitivity is important; for a definitive test (*i.e.*, to establish the presence of schizophrenia), specificity is emphasized; for a differential diagnosis, both sensitivity and specificity are required.

Sensitivity

Most studies of P3 in schizophrenia were designed to demonstrate *group* differences. Few have established a P3 amplitude cutoff and determined how many schizophrenics fall below it. One exception is Roth and Cannon (1972), who could classify 16/21 controls and 19/21 schizophrenics with a cutoff of 3.2 uV. Another is a study by Faux, Torello, McCarley, Shenton, and Duffy (1988) which established that a P3 recorded over the left temporal scalp electrode site (T3) that is smaller than 2 uV is diagnostic for schizophrenia. They correctly classified 9/11 medicated schizophrenics and 7/9 controls. Of course, the exact cutoff depends upon the paradigm, and in these examples they may be in the noise range. Barrett *et al.* (1986) used a discriminant analysis to correctly classify 85% of the schizophrenics (3 unmed/17 med) and 95% of the controls. They suggested that the source of the differences lay in a protracted positive potential shift which overlapped P3 and was present in waveforms of controls to frequent and infrequent stimuli.

Specificity

In 1986, Pfefferbaum (Shagass, Roth, Duncan & Pfefferbaum, 1986) listed studies done through 1984 that compared P3 in schizophrenics and depressed patients. Of the five (Levit *et al.*, 1973; Pfefferbaum *et al.*, 1984; Roth *et al.*, 1981; Shagass, Roemer, Straumanis & Amadeo, 1978; Steinhauer & Zubin, 1982), only Pfefferbaum *et al.* (1984), using both auditory and visual modalities, found no difference between schizophrenics and depressed patients. Since then, Blackwood *et al.* (1987) have compared auditory P3s in schizophrenics and depressed patients and have also found P3 amplitude to be reduced in both groups in the acute phase. Following treatment and clinical improvement, P3 normalized in the depressed, but remained reduced in the schizophrenics, suggesting that the auditory P3 reduction in schizophrenics is an enduring trait, while in depressed patients it is a function of clinical status. Kutcher *et al.* (1987) compared patients with borderline personality disorder (BPD), nonborderline personality disorder, depression, and schizophrenia using the auditory P3. Both the amplitude and latency of the auditory P3 differentiated the BPD and schizophrenic patients from the other patients and controls, but did not differentiate the BPD and schizophrenics from each other. In one of the earliest papers using P3 to study schizophrenia, Levit *et al.* (1973) compared the P3s of schizophrenics, psychotic depressives, and controls. Naive observers were given the task of indicating which waveform was the "most different" in a set of waveforms containing one from each group of subjects. Twenty-one of 24 observers picked out the schizophrenic waveform as being "most different." Brecher *et al.* (1987b) were unable to demonstrate specificity in comparing schizophrenics and alcoholics to controls. P3 was sensitive to both schizophrenia and alcoholism, failing to distinguish the patient groups from each other.

The E. Roy John laboratory (John, Prichep, Fridman & Easton, 1988) has advocated a "neurometric" approach to diagnosis and been able to distinguish schizophrenics from controls and from other psychiatric patients (affective disorders, alcoholism, dementia) using a number of components including the P3 (Prichep, John & Easton, 1989). Sandman, Gerner, O'Halloran, and Isenart (1987) compared the auditory P3s from depressed, alcoholic, and schizophrenic patients. Using the information from the amplitude of P1, N1, P2, N2, and P3, and the latency of N2, 57% of the schizophrenics were correctly classified where 33% would have been by chance. Very few (1 each) of the alcoholic and depressed patients were misclassified as schizophrenics.

Signal Detection Analysis of Sensitivity and Specificity

Our own data originally reported in 1984 (Pfefferbaum *et al.*, 1984) is pertinent to discussions of sensitivity and specificity. For this review, we have applied a signal detection analysis to these data to determine how sensitive and specific auditory P3 amplitude reduction is to the diagnosis of schizophrenia. The P3 amplitudes of 20 schizophrenic, 34 depressed, 37 demented, and 9 nondemented patients were first statistically corrected for the effects of age using the P3 amplitudes from 115 control subjects, and were expressed as z-scores. An iterative procedure (Kraemer, 1988) allowed us to identify the sensitivity and specificity of P3 amplitude using various cutoffs (*e.g.*, 1.6, 1.8, 2.0 standard deviations). This procedure also estimated the quality of the sensitivity and specificity of P3 amplitude reduction by taking into account how many correct diagnoses would have been made by chance. The analysis demonstrated that 100% of the patients in our sample with P3 amplitudes exceeding 1.6 standard deviations are *not* schizophrenic. FIGURE 1 compares the receiver operating characteristics (ROC) and the quality of the receiver operating characteristics (QROC) for the auditory P3 amplitude. The perpendicular appearance of the QROC curve indicates that a criterion P3 amplitude level can be established, above which one can rule out the diagnosis of schizophrenic (+ 1.6 SD in this case), but no criterion amplitude level can reasonably establish a positive diagnosis of schizophrenia. In contrast, this approach was useful in establishing a P3 latency cutoff which was specific (albeit with low sensitivity) for the diagnosis of dementia (Pfefferbaum, Ford & Kraemer, 1990).

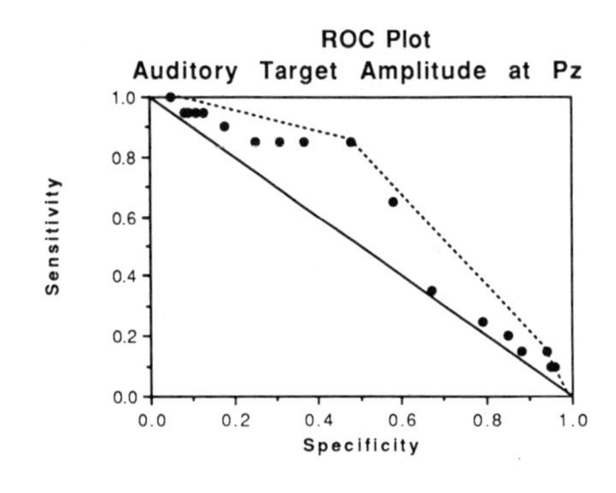

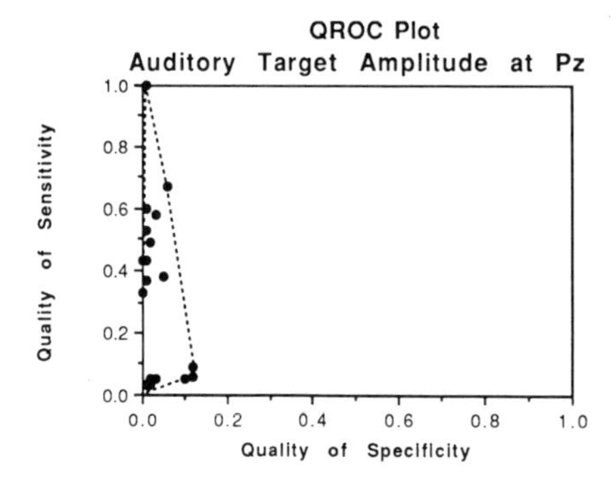

FIGURE 1. *Top:* receiver operating characteristics (ROC), a graph of sensitivity vs specificity using various cutoffs for the auditory P3 amplitude. *Bottom:* quality receiver operating characteristics (QROC), a graph of quality of sensitivity vs quality of specificity using various cutoffs for the auditory P3 amplitude. Note that sensitivity is high but specificity is not.

Using P3 as a Genetic Marker for Schizophrenia

The desire to predict who will and who will not develop schizophrenia has motivated research on young people who are at risk for the disease by virtue of family history or "preschizophrenic" personality. In both cases, the subjects studied are not yet psychotic and thus the investigators are better able to distinguish precursor traits from the consequences of schizophrenia. The strength of both models lies in the potential to follow these subjects longitudinally to determine the validity and power of the predictive variable. This, of course, necessitates the inclusion of large numbers of subjects due to the low probability of developing the disease even in the high-risk population, and attrition of the original sample over time.

There is considerable evidence for familial factors in the transmission of schizophrenia, though in most studies, genetic transmission has not been proved. Children of schizophrenic mothers have been observed to be underactive and hypotonic as infants (Fish & Alpert, 1962), signs which others have related to later diagnoses of schizophrenia and schizophrenia spectrum disorder (Parnas, Schulsinger, Schulsinger, Mednick & Teasdale, 1982). Later in childhood and adolescence, deficits in focussed attention, concentration, and alertness often precede a diagnosis of schizophrenia (Parnas *et al.*, 1982). Nuechterlein and Dawson (1984) noted that the continuous performance task (CPT) reveals deficits in vigilance performance both in subjects with a positive family history and in subjects with a schizotypal profile. Furthermore, the performance deficit on the CPT depended on the processing load. With its sensitivity to focussed attention, distraction, and attentional capacity, the ERP is a potentially powerful tool for studying populations at risk for schizophrenia.

The likelihood of schizophrenia is estimated at 12–14% among the offspring of schizophrenic parents and 8–10% among siblings (Gottesman, 1978). Thus 80–90% of a high-risk sample will not develop schizophrenia. However, the genetic susceptibility (and perhaps a biologic marker such as reduced P3 amplitude) might be inherited with a higher frequency than 12–14%, with the disease manifesting itself only with certain environmental interactions. Thus, genotypically positive individuals with a positive biological marker could be phenotypically negative for schizophrenia. Furthermore, schizophrenia may present as a "spectrum" of disorders running from schizophrenia to schizotypal personality to borderline personality to anhedonia, with the degree of symptoms depending on both the penetrance of the gene(s) responsible for the disorder and environmental interactions. If so, a continuous predictor variable like P3 amplitude, might be sensitive to the degree of manifestation of the spectrum disease symptoms. However, Kutcher, Blackwood, Gaskell, Muir, and St Clair (1989) found that auditory oddball P3 amplitude could not differentiate between borderline and schizotypal subjects, but could differentiate them both from controls. While the auditory oddball was not a sufficient stimulus to elicit P3s sensitive to anhedonia (Miller, 1986; Ward, Catts, Armstrong & McConaghy, 1984), tones warning of high- and low-pleasure slides are able to do so (Simons, 1982). Although Ward *et al.* found that the oddball P3 did not distinguish between anhedonics and controls, it did distinguish nonpsychotic college students characterized as "allusive thinkers" from those without this kind of thinking (Ward *et al.*, 1984).

An important consideration in this work is the establishment of specificity. Children of parents with other mental disorders (*i.e.*, alcoholism) also demonstrate reduced P3 amplitudes relative to children of normal parents (Begleiter & Porjesz, 1984). While several investigators have recorded P3 from children at-risk for developing schizophrenia (Herman, Mirsky, Ricks & Gallant, 1977; Itil, Hsu, Saletu & Mednick, 1974; Saletu, Saletu, Marasa, Mednick & Schulsinger, 1975), only the Friedman laboratory has included children who are at risk for other psychiatric disorders. Unfortunately, they found few ERP differences between the children of normal parents, affectively disordered parents, and schizophrenic parents using either a CPT (Friedman, Cornblatt, Vaughan & Erlenmeyer-Kimling,

1986) or an auditory oddball paradigm to elicit P3 (Friedman, Cornblatt, Vaughan & Erlenmeyer-Kimling, 1988). More disappointing was the failure to find a subgroup of outliers who might have been most vulnerable. It is possible that the ERP is *not* a genetic index of risk, or that the sample of 31 children at risk for schizophrenia was not large enough to detect any effect since only a fraction of the children will carry the gene(s). Follow-up of their original sample would be informative.

Using adult siblings of schizophrenics and a more difficult syllable discrimination task, Saitoh, Niwa, Hiramatsu, Kameyama, Rymar, and Itoh (1984) had better success in distinguishing high-risk subjects from controls with P3 amplitude. Their P3 amplitudes were similar to their unmedicated schizophrenic siblings, which the authors interpreted as reflecting "a genetic predisposition to schizophrenia." The siblings were free of any history of psychiatric disease even though some of them were beyond the peak age of risk (18–40 years, mean = 28.5). While phenotypically negative for schizophrenia, perhaps they were genotypically positive. Interpretation of these results would be facilitated by knowledge of the general heritability of P3 amplitude across siblings, independent of the heritability of schizophrenia.

Summary of Empirical Work

Attempts have been made to establish P3 as a diagnostic tool and a marker for schizophrenia. While P3 might be useful in ruling out a diagnosis of schizophrenia, P3 amplitude reduction is not specific to schizophrenia. Its lack of specificity is most likely due to the influence of a general deficit state, associated with psychological disorder, that cuts across many psychiatric diagnoses. The ultimate success of P3 as a marker for either biological or psychometric risk for developing schizophrenia awaits analysis of follow-up data in the coming years.

ACKNOWLEDGMENTS

The authors wish to thank Margaret J. Rosenbloom and Rudolf Cohen for a critical reading of the manuscript, and Patricia M. White for help with the signal detection analysis of the data.

REFERENCES

ADAMS, J., S. FAUX, R. W. MCCARLEY, B. MARCY & M. SHENTON. 1989. The N400 and language processing in schizophrenia. *In* International Conference on Event-Related Potentials of the Brain (EPIC IX)—Poster Session 3. 8–9. Noordwijk, The Netherlands. Not published.

ANDREASEN, N. & S. OLSEN. 1982. Negative v positive schizophrenia: definition and validation. Arch. Gen. Psychiatry **39:** 789–794.

ANDREWS, S., P. MITCHELL, A. M. FOX, S. V. CATTS, P. B. WARD & N. MCCONAGHY. 1990. ERP indices of semantic processing in schizophrenia. *In* Psychophysiological Brain Research, Vol. 2. C. H. M. Brunia, A. W. K. Gaillard & A. Kok, Eds. 187–192. Tilburg University Press. Tilburg, The Netherlands.

BARIBEAU, J., T. W. PICTON & J. Y. GOSSELIN. 1983. Schizophrenia: a neuro-physiological evaluation of abnormal information processing. Science **219:** 874–877.

BARRETT, K., W. C. MCCALLUM & P. V. POCOCK. 1986. Brain indicators of altered attention and information processing in schizophrenic patients. Br. J. Psychiatry **148:** 414–420.

BEGLEITER, H. & B. PORJESZ. 1984. Event-related brain potentials in boys at risk for alcoholism. Science **225:** 1493–1496.

BERNSTEIN, A. S. 1987. Orienting response research in schizophrenia: Where we have come and where we might go? Schizophr. Bull. **13:** 623–641.

BLACKWOOD, D., L. WHALLEY, J. CHRISTIE, I. BLACKBURN, D. ST CLAIR & A. MCINNES. 1987. Changes in auditory P3 event-related potential in schizophrenia and depression. Br. J. Psychiatry **150:** 154–160.

BLEULER, E. 1911. Dementia praecox or the group of schizophrenias. Translated, 1950 by J. Zinkin. International Universities Press, Inc. New York.

BOGERTS, B., E. MEERTZ & R. SCHONFELDT-BAUSCH. 1985. Basal ganglia and limbic system pathology in schizophrenia. Arch. Gen. Psychiatry **42:** 784–791.

BRAFF, D. L., E. CALLAWAY & H. NAYLOR. 1977. Very short-term memory function in schizophrenia: defective short time constant information processing in schizophrenia. Arch. Gen. Psychiatry **34:** 25–30.

BRECHER, M. & H. BEGLEITER. 1983. Event-related brain potentials to high incentive stimuli in unmedicated schizophrenic patients. Biol. Psychiatry **18:** 661–674.

BRECHER, M., B. PORJESZ & H. BEGLEITER. 1987a. Late positive component amplitude in schizophrenics and alcoholics in two different paradigms. Biol. Psychiatry **22:** 848–856.

BRECHER, M., B. PORJESZ & H. BEGLEITER. 1987b. The N2 component of the event-related potential in schizophrenic patients. Electroencephalogr. Clin. Neurophysiol. **66:** 369–375.

BROWN, R., N. COLTER, N. CORSELLIS, T. J. CROW, C. D. FRITH, R. JAGOE, E. C. JOHNSTONE & L. MARSH. 1986. Postmortem evidence of structural brain changes in schizophrenia: differences in brain weight, temporal horn area and parahippocampal gyrus compared with affective disorder. Arch. Gen. Psychiatry **43:** 36–42.

BUCHSBAUM, M. S., D. H. INGVAR, R. M. KESSLER, R. N. WATERS, J. CAPPELLETTI, D. P. VAN KAMMEN, A. C. KING & J. L. JOHNSON. 1982. Cerebral glucography with positron tomography in normals and patients with schizophrenia. Arch. Gen. Psychiatry **39:** 251–259.

CROW, T. J., N. COLTER & R. BROWN. 1988. Lateralized asymmetry of temporal horn enlargement in schizophrenia. Schizophr. Res. **1:** 155–156.

DUNCAN, C. C., J. MORIHISA, R. W. FAWCETT & D. G. KIRCH. 1987a. P300 in schizophrenia: state or trait marker? Psychopharmacol. Bull. **23:** 497–501.

DUNCAN, C. C., W. M. PERLSTEIN & J. M. MORIHISA. 1987b. The P300 metric in schizophrenia: effects of probabiligy and modality. *In* Current Trends in Event-Related Potential Research (EEG Suppl. 40). R. Johnson Jr., J. W. Rohrbaugh & R. Parasuraman, Eds. 670–674. Elsevier Science Publishers B.V. Amsterdam.

DUNCAN-JOHNSON, C. C., W. T. ROTH & B. S. KOPELL. 1984. Effects of stimulus sequence on P300 and reaction time in schizophrenics: a preliminary report. Ann. N. Y. Acad. Sci. **425:** 570–571.

FAUX, S., M. TORELLO, R. MCCARLEY, M. SHENTON & F. DUFFY. 1988. Schizophrenia: confirmation and statistical validation of temporal region deficit in P300 topography. Biol. Psychiatry **23:** 776–790.

FISH, B. & M. ALPERT. 1962. Abnormal states of consciousness and muscle tone in infants born to schizophrenic mothers. Am. J. Psychiatry **119:** 439–445.

FRIEDMAN, D., B. CORNBLATT, H. VAUGHAN & L. ERLENMEYER-KIMLING. 1988. Auditory event related potentials in children at risk for schizophrenia: the complete initial sample. Psychiatry Res. **26:** 203–221.

FRIEDMAN, D., B. CORNBLATT, H. J. VAUGHAN & L. ERLENMEYER-KIMLING. 1986. Event-related potentials in children at risk for schizophrenia during two versions of the continuous performance test. Psychiatry Res. **18:** 161–177.

GOTTESMAN, I. 1978. Schizophrenia and genetics: Where are we? Are you sure? *In* The Nature of Schizophrenia: New Approaches to Research and Treatment. L. C. Wynne, R. L. Cromwell & S. Matthysse, Eds. 59–69. John Wiley and Sons. New York.

GRILLON, C., E. COURCHESNE, R. AMELI, M. A. GEYER & D. L. BRAFF. 1990. Increased distractibility in schizophrenic patients—electrophysiologic and behavioral evidence. Arch. Gen. Psychiatry **47**(2), 171–179.

GUR, R. E. 1978. Left hemisphere dysfunction and left hemisphere overactivation in schizophrenia. J. Abnorm. Psychol. **87:** 226–238.

HALGREN, E., N. SQUIRES, C. WILSON, J. ROHRBAUGH, T. BABB & P. CRANDALL. 1980. Endogenous potentials generated in the human hippocampal formation and amygdala by infrequent events. Science **210:** 803–805.

HAMMOND, N. & J. GRUZELIER. 1978. Laterality, attention and rate effects in auditory temporal discrimination of chronic schizophrenics. Quart. J. Exp. Psychol. **30:** 91–103.

HASHER, L. & R. T. ZACKS. 1979. Automatic and effortful processes in memory. J. Exp. Psychol. **108:** 356–388.

HERMAN, J., A. MIRSKY, N. RICKS & D. GALLANT. 1977. Behavioral and electrographic measures of attention in children at risk for schizophrenia. J. Abnorm. Psychol. **86:** 27–33.

HILLYARD, S. A., R. HINK, V. L. SCHWENT & T. W. PICTON. 1973. Electrical signs of selective attention in the human brain. Science **182:** 177–180.

HOKAMA, H., S. KOYAMA, M. MIYATANI, Y. MIYAZATO, C. OGURA, Y. NAGEISHI & M. SHIMOKOCHI. 1990. Abnormality of contextual effect on ERPs in schizophrenia: N400 topography. *In* Psychophysiological Brain Research, Vol 2. C. H. M. Brunia, A. W. K. Gaillard & A. Kok, Eds. 213–217. Tilburg University Press. Tilburg, The Netherlands.

HOLINGER, D. P. & A. M. WAGMAN. 1990. Brain potentials and cognitive dysfunction in schizophrenia (abs.). *In* Annual Meeting of Society for Research in Psychopathology, Denver, Co.

INGVAR, D. H. & G. FRANZEN. 1974. Abnormalities of cerebral blood flow distribution in patients with chronic schizophrenia. Acta Psychiatr. Scand. **50:** 425–462.

ISREAL, J. B., G. L. CHESNEY, C. D. WICKENS & E. DONCHIN. 1980. P300 and tracking difficulty: evidence for multiple resources in dual-task performance. Psychophysiology **17**(3), 259–273.

ITIL, T., W. HSU, B. SALETU & S. A. MEDNICK. 1974. Auditory evoked potential investigations in children at high risk for schizophrenia. Am. J. Psychiatry **131:** 892–900.

JOHN, E. R., L. S. PRICHEP, J. FRIDMAN & P. EASTON. 1988. Neurometrics: computer-assisted differential diagnosis of brain dysfunctions. Science **239:** 162–169.

JOSIASSEN, R. C., C. SHAGASS, J. J. STRAUMANIS & R. A. ROEMER. 1984. Psychiatric drugs and the somatosensory P400 wave. Psychiatry Res. **11:** 151–162.

KEMALI, D., S. GALDERISI, M. MAJ, A. MUCCI, M. CESARELLI & L. D'AMBRA. 1988. Event-related potentials in schizophrenic patients: clinical and neurophysiological correlates. Res. Commun. Psychol. Psychiatry Behav. **13:** 3–16.

KOGA, Y., K. HASHIGUCHI, K. OGINO, C. CHIBA, H. TAGUCHI & K. HORI. 1987. Language information processing in schizophrenia. *In* Current Trends in Event-Related Potential Research (EEG Suppl 40). R. J. Johnson, J. W. Rohrbaugh & R. Parasuraman, Eds. 705–711. Elsevier Science Publishers B.V. Biomedical Division. Amsterdam.

KOVELMAN, J. A. & A. B. SCHEIBEL. 1984. A neurohistological correlate of schizophrenia. Biol. Psychiatry **19:** 1601–1621.

KRAEMER, H. C. 1988. Assessments of 2 × 2 associations: generalization of signal-detection methodology. Am. Statistician **42:** 37–49.

KUTAS, M. & S. A. HILLYARD. 1984. Brain potentials during reading reflect word expectancy and semantic association. Nature **307:** 161–163.

KUTAS, M., G. MCCARTHY & E. DONCHIN. 1977. Augmenting mental chronometry: the P300 as a measure of stimulus evaluation time. Science **197:** 792–795.

KUTCHER, S. D. BLACKWOOD, D. GASKELL, W. MUIR & D. ST CLAIR. 1989. Auditory P300 does not differentiate borderline personality disorder from schizotypal personality disorder. Biol. Psychiatry **26:** 766–774.

KUTCHER, S., D. BLACKWOOD, D. ST CLAIR, D. GASKELL & W. MUIR. 1987. Auditory P300 in borderline personality disorder and schizophrenia. Arch. Gen. Psychiatry **44:** 645–650.

LEVIT, R., S. SUTTON & J. ZUBIN. 1973. Evoked potential correlates of information processing in psychiatric patients. Psychol. Med. **3:** 487–494.

MAHER, B. 1972. The language of schizophrenia: a review and interpretation. Br. J. Psychiatry **120:** 3–17.

MCCARLEY, R. W., S. FAUX, M. SHENTON, M. LEMAY, M. CANE, R. BALLINGER & F. H. DUFFY. 1989. CT abnormalities in schizophrenia. Arch. Gen. Psychiatry **46:** 698–708.

MCCARTHY, G. 1985. Intracranial recordings of endogenous ERPs in humans. Electroencephalogr. Clin. Neurophysiol. **61:** S11.

McGhie, A., J. Chapman & J. Lawson. 1965. The effect of distraction on schizophrenic performance, I: perception and immediate memory. Br. J. Psychiatry **11:** 383–390.

Michie, P. T., A. M. Fox, P. B. Ward, S. V. Catts & N. McConaghy. 1990. Event-related potential indices of selective attention and cortical lateralization in schizophrenia. Psychophysiology **27**(2), 209–227.

Miller, G. S. 1986. Information processing deficits in anhedonia and perceptual aberration: a psychophysiological analysis. Biol. Psychiatry **21:** 100–115.

Miyazato, Y., C. Ogura, M. Miyatani, S. Nashiro, K. Fukao & H. Hokama. 1989. Dynamic event-related potential topography in schizophrenics. *In* International Conference on Event-Related Potentials of the Brain (EPIC IX)—Poster Session 3. 84–85. Noordwijk, The Netherlands. Not published.

Nuechterlein, K. H. & M. E. Dawson. 1984. Information processing and attentional functioning in the developmental course of schizophrenic disorders. Schizophr. Bull. **10:** 160–203.

Okada, Y. C., L. Kaufman & S. J. Williamson. 1983. The hippocampal formation as a source of the slow endogenous potentials. Electroencephalogr. Clin. Neurophysiol. **55:** 417–426.

Oltmanns, T. F. & J. M. Neale. 1975. Schizophrenic performance when distractors are present: attentional deficit or differential task difficulty. J. Abnorm. Psychol. **84:** 205–209.

Parnas, J., F. Schulsinger, H. Schulsinger, S. A. Mednick & T. W. Teasdale. 1982. Behavioral precursors of schizophrenia spectrum: a prospective study. Arch. Gen. Psychiatry **39:** 658–664.

Pass, H., R. Klorman, L. Salzman, R. Klein & G. Kaskey. 1980. The late positive component of the evoked response in acute schizophrenics during a test of sustained attention. Biol. Psychiatry **15:** 9–20.

Pfefferbaum, A., C. Christensen, J. M. Ford & B. S. Kopell. 1986. Apparent response incompatibility effects on P3 latency depend on the task. Electroencephalogr. Clin. Neurophysiol. **64:** 424–437.

Pfefferbaum, A., J. M. Ford & H. C. Kraemer. 1990. Clinical utility of long latency 'cognitive' event-related potentials (P3): the cons. Electroencephalogr. Clin. Neurophysiol. **76:** 6–12.

Pfefferbaum, A., J. M. Ford, P. White & W. T. Roth. 1989. P3 in schizophrenia is affected by stimulus modality, response requirements, medication status and negative symptoms. Arch. Gen. Psychiatry **46:** 1035–1046.

Pfefferbaum, A., J. M. Ford, P. M. White, W. T. Roth & D. H. Mathalon. 1991. Is there P300 asymmetry in schizophrenia?—Reply. Arch. Gen. Psychiatry **48**(4), 381–383.

Pfefferbaum, A., B. Wenegrat, J. Ford, W. T. Roth & B. S. Kopell. 1984. Clinical application of the P3 component of event-related potentials: II. Dementia, depression and schizophrenia. Electroencephalogr. Clin. Neurophysiol. **59:** 104–124.

Prichep, L. S., E. R. John & P. Easton. 1989. Neurometric evoked potential factors in normals and psychiatric patients. *In* International Conference on Event-Related Potentials of the Brain (EPIC IX)—Poster Session 3. 101. Noordwijik, The Netherlands. Not published.

Pritchard, W. S. 1986. Cognitive event-related potential correlates of schizophrenia. Psychol. Bull. **100:** 43–66.

Romani, A., S. Merello, L. Gozzoli, F. Zerbi, M. Grassi & V. Cosi. 1987. P300 and CT scan in patients with chronic schizophrenia. Br. J. Psychiatry **151:** 506–513.

Roth, W. T. & E. H. Cannon. 1972. Some features of the auditory evoked response in schizophrenics. Arch. Gen. Psychiatry **27:** 466–471.

Roth, W. T., J. Goodale & A. Pfefferbaum. 1991. Auditory event-related potentials and electrodermal activity in medicated and unmedicated schizophrenics. Biol. Psychiatry **29**(6), 585–599.

Roth, W. T., T. B. Horvath, A. Pfefferbaum & B. S. Kopell. 1980. Event related potentials in schizophrenics. Electroencephalogr. Clin. Neurophysiol. **48:** 127–139.

Roth, W. T., T. B. Horvath, A. Pfefferbaum, J. R. Tinklenberg, J. E. Mezzich & B. S. Kopell. 1979. Late event-related potentials and schizophrenia. *In* Evoked Brain Potentials and Behavior. H. Begleiter, Ed. Vol. 2: 499–515. Plenum Publishing. New York.

Roth, W. T., A. Pfefferbaum, T. B. Horvath, P. A. Berger & B. S. Kopell. 1980.

P3 reduction in auditory evoked potentials of schizophrenics. Electroencephalogr. Clin. Neurophysiol. **49:** 497–505.

ROTH, W. T., A. PFEFFERBAUM, A. F. KELLY, P. A. BERGER & B. S. KOPELL. 1981. Auditory event-related potentials in schizophrenia and depression. Psychiatry Res. **4:** 199–212.

ROTH, W. T., J. J. TECCE, A. PFEFFERBAUM, M. ROSENBLOOM & E. CALLAWAY. 1984. ERPs and psychopathology: behavior Process Issues. *In* Brain and Information: Event-Related Potentials. R. Karrer, J. Cohen & P. Tueting, Eds. 496–522. New York Academy of Sciences. New York.

ROTH, W. T. & J. TINKLENBERG. 1982. A convergence of findings in the psychophysiology of schizophrenia. Psychopharmacol. Bull. **18:** 78–83.

SAITOH, O., S. NIWA, K. HIRAMATSU, T. KAMEYAMA, K. RYMAR & K. ITOH. 1984. Abnormalities in late positive components of event-related potentials may reflect a genetic predisposition to schizophrenia. Biol. Psychiatry **19:** 293–303.

SALETU, B., M. SALETU, J. MARASA, S. MEDNICK & F. SCHULSINGER. 1975. Acoustic evoked potentials in offspring of schizophrenic mothers ("high risk" children for schizophrenia). Clin. Electroencephalogr. **6:** 92–102.

SANDMAN, C., R. GERNER, J. O'HALLORAN & R. ISENART. 1987. Event-related potentials and item recognition in depressed, schizophrenic and alcoholic patients. Int. J. Psychophysiol. **5:** 215–225.

SHAGASS, C., W. T. ROTH, D. D. DUNCAN & A. PFEFFERBAUM. 1986. P3 latency and amplitude abnormality in mental disorders. *In* Cerebral Psychophysiology Studies in Event-Related Potentials. C. McCallum, R. Zappoli, F. Denoth & M. Timsit-Berthier, Eds. 425–429. Elsevier Science Publishers B.V. Biomedical Division. Amsterdam.

SHAGASS, C., R. ROEMER, J. J. STRAUMANIS & M. AMADEO. 1978. Evoked potential correlates of psychosis. Biol. Psychiatry **13:** 163–184.

SIMONS, R. F. 1982. Physical anhedonia and future psychopathology: an electrocortical continuity? Psychophysiology **19:** 433–441.

SQUIRES, K. C., C. WICKENS, N. K. SQUIRES & E. DONCHIN. 1976. The effect of stimulus sequence on the waveform of the cortical event-related potential. Science **193:** 1142–1146.

STEINHAUER, S. & J. ZUBIN. 1982. Vulnerability to schizophrenia: information processing in the pupil and event related potential. *In* Biological Markers in Psychiatry and Neurology. I. Hanin & E. Usdin, Eds. Pergamon Press. Oxford.

VERLEGER, R. & J. COHEN. 1978. Effects of certainty, modality shift and guess outcome on evoked potentials and reaction times in chronic schizophrenics. Psychol. Med. **8:** 81–93.

WARD, P. B., S. V. CATTS, M. S. ARMSTRONG & N. MCCONAGHY. 1984. P300 and psychiatric vulnerability in university students. *In* Brain and Information Processing: Event-Related Potentials. R. Karrer, J. Cohen & P. Teuting. Eds. 645–652. New York Academy of Sciences. New York.

WARD, P. B., S. V. CATTS, A. M. FOX, P. T. MICHIE & N. MCCONAGHY. 1991. Auditory selective attention and event-related potentials in schizophrenia. Br. J. Psychiatry **158:** 534–539.

The Modality Shift Effect

Further Explorations at the Crossroads

RUDOLF COHEN[a] AND FRED RIST[b]

[a]*Sozialwissenschaftliche Fakultät*
University of Konstanz
Postfach 5560
D-7750 Konstanz, Germany
and
[b]*Zentralinstitut für Seelische Gesundheit*
Postfach 120 122
D-6800 Mannheim, Germany

INTRODUCTION

"Modality Shift at the Crossroads" was the title of a paper by Sutton, Spring and Tueting (1978). This paper is a beautiful example of the clear and candid style that characterizes Sutton's contributions to psychopathology. In a painstaking discussion of experimental findings the authors search for the most promising path to a better understanding of what had been called the Modality Shift Effect. This differential response pattern was first described by Sutton, Hakerem, Zubin and Portnoy (1961) and further explored in a comprehensive review by Sutton and Zubin (1965). Various additional data were available in 1978, when Sutton's scrutiny disclosed so many inconsistencies that no final conclusion appeared viable. He strongly invited the reader to pursue the search that he had begun. This contribution is an attempt to do so.

The term Modality Shift Effect (MSE) refers to a peculiar finding from simple Reaction Time (RT) tasks which require the subject to make the same motor response to each of the stimuli in a random series of lights and tones with variable intervals. In such tasks, which we will refer to as standard conditions, RTs were found to be longer when the imperative stimulus reflected a shift in modality than when a stimulus was repeated (FIG. 1a). This difference is more pronounced in subjects with a schizophrenic disorder than in normal subjects. We shall call this increment of the difference in the RTs of patients with schizophrenia a "differential" Modality Shift Effect (MSE_D). We will speak of a Modality Shift Effect (MSE) without any further specification only when referring to the more ubiquitous decrease in RT after a repetition of the preceding stimulus.

For several reasons the MSE_D appears to be particularly interesting for psychopathological research: 1) the effect is quite consistent; we located seven reports in the literature (cf. Rist & Cohen, 1991) and had two additional sets of data available that derived from standard conditions. In all nine sets of data from three different laboratories, the MSE—at least in the RTs to tones—was significantly larger in subjects with a schizophrenic disorder than in healthy subjects; 2) the effect is apparently not a by-product of neuroleptic and/or anticholinergic medication: all the early studies by Sutton and his colleagues (Sutton *et al.*, 1961; Sutton & Zubin, 1965; Waldbaum, Sutton & Kerr, 1975) were carried out with patients off medication for at least two weeks, before neuroleptic medication had become a standard diet of patients with schizophrenia. Moreover, none of the later studies found any correlation between medication status and the MSE_D; 3) there is nothing in the task to cue the subject that the experimenter is interested primarily in the difference between ipsi- and crossmodal RTs. Thus, it is unlikely that intentions to present oneself in a certain way could have produced the effect; 4) Sutton's finding is the highly exceptional case where the responses of patients with a schizophrenic disorder reveal not less but more sensitivity to the differences in experimental conditions, *i.e.*, between repetitions and alternations of stimuli. Therefore, the effect cannot be an artifact of poor reliability, deriving from a nonspecific lack of attention, motivation, skill or cooperation.

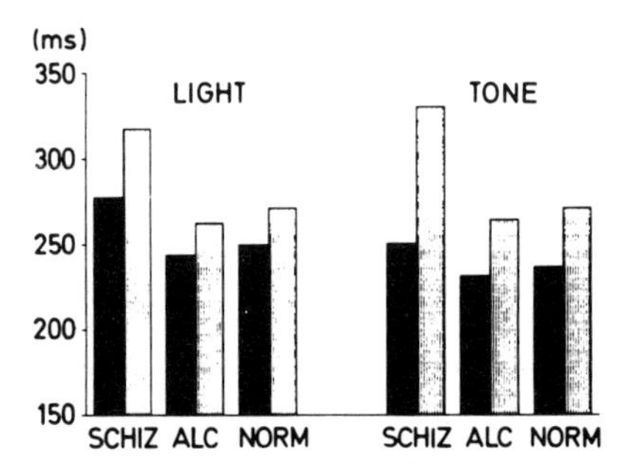

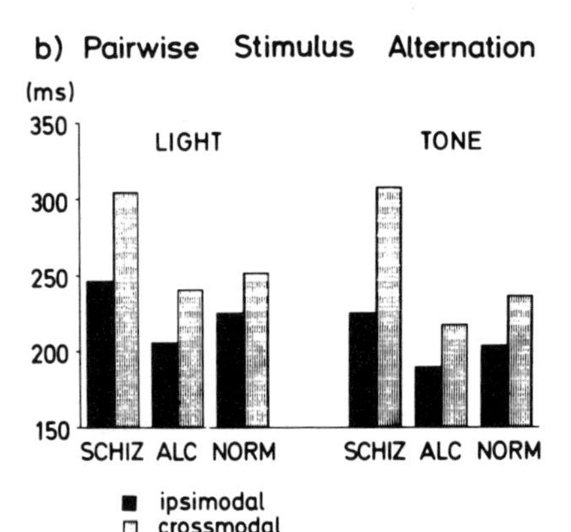

FIGURE 1. Simple reaction times (ms) to light and tone stimuli from STUDY II. **(a)** *Standard Condition* with keypress to each light and tone. **(b)** *Pairwise Stimulus Alternation* with stimuli presented in fixed order of alternating pairs of two lights and two tones.

In the first half of this paper we shall report some basic findings concerning the MSE_D, generally in support of earlier conclusions of Spring and Zubin (1977) and Mannuzza (1980). In the second half we shall turn to some more recent studies that might help to exclude some of the alternatives at the crossroads.

Throughout the paper we shall refer to two studies from our laboratory, in which the MSE_D was studied under various experimental conditions. In these studies, chronic or subchronic patients with a schizophrenic disorder were compared to alcoholic patients at least six weeks after withdrawal and to healthy volunteers. All patients were inpatients from the Psychiatrisches Landeskrankenhaus Reichenau, Konstanz, diagnosed according to DSM III criteria. Most schizophrenic patients were under neuroleptic maintenance medication. All subjects were male; the groups were matched for age (mean: 35 ys, range: 20–56 ys) and education. Thirty-two schizophrenic patients, 16 alcoholic and 16 normal subjects took part in STUDY I, and there were 18 subjects per group in STUDY II.

In both studies the duration of all stimuli was 80 ms. Stimuli were presented in a Bernoulli sequence with intertrial intervals varying randomly from 1.7 to 2.3 s. The visual stimuli were bright squares on a videoscreen (100 ftL at 1.5 m) presented in a dimly lit room. The auditory stimuli (STUDY I: 200 Hz, 65 db; STUDY II: 800 Hz, 73 db) were delivered through loudspeakers located on both sides of the screen. In STUDY I the location of the stimulus source varied randomly: tones were presented through the left or the right speaker, and the bright squares on the right or the left half of the video screen. In STUDY II the intensity of the tones was adjusted to the intensity of the lights according to the average judgment of 12 normals.

In STUDY I, RTs were collected under three experimental conditions: In the *Standard Condition* subjects were asked to quickly press the same button to each stimulus independent of location and modality. Under *Pairwise Response Alternation* subjects had to alternate between responding to two consecutive stimuli and withholding the response to the following two stimuli, again independent of location and modality. Under *Hemifield Selection* subjects had again to respond to both light and sound stimuli, but under this condition only to stimuli from one or the other of the two hemifields. Some results from these conditions were presented earlier by Rist and Cohen (1987).

STUDY II was carried out in collaboration with Thomas Finger. Similar to STUDY I, the *Standard Condition* required subjects to quickly press the same button to each stimulus. Under *Pairwise Stimulus Alternation* lights and tones were presented in a perfectly predictable order with two lights always following two tones etcetera (LLTTLLTT etc.). Under *Distinct Pairs* all tones were presented in distinct pairs separated by intervals of five seconds. In one block, pairs were always preceded by an identical tone (ipsimodal pairs), and in another block, pairs were always preceded by a light. Finally, a *Choice Reaction Time* condition was introduced mapping light and tone stimuli to key presses with the right or the left hand, respectively.

In both studies, event-related potentials (ERPs) were collected from frontal (Fz), vertex (Cz) and parietal (Pz) sites with Ag/AgCl electrodes referred to linked earlobes. Bipolar EOG was recorded from above and below the right eye. Amplifiers were set to a high-frequency cut-off of 35 Hz and a time constant of 10 s. The sampling rate was 100 Hz. Trials with eye movements were rejected when the EOG signal surpassed a threshold of 80 μV, and trials with blink artifacts were corrected according to the linear regression method of Berg (1986). Epoch duration was 980 ms in STUDY I (250 ms baseline) and 800 ms in STUDY II (100 ms baseline).

In all analyses of variance, the Greenhouse-Geisser correction was applied where appropriate. Mean RTs were computed using the median RTs of the individual subjects per experimental condition.

Basic Characteristics of the MSE$_D$

Stimulus Modality and the MSE$_D$

In two early papers, Sutton and his colleagues (Sutton *et al.*, 1961; Sutton & Zubin, 1965) reported a significant MSE$_D$ only for tones but not for lights. There are a number of later studies that found a significant MSE_D also for visual RTs (Mannuzza, Kietzman, Berenhaus, Ramsey, Zubin & Sutton, 1984; Spring, 1980; Rey & Oldigs, 1980; Waldbaum *et al.*, 1975) but in eight of the nine studies (cf. Rist & Cohen, 1991) comparing the MSE_D for auditory and visual RTs from the Standard Condition this difference between groups was larger for RTs to tones. However, the difference in discriminative power is usually small or moderate and has rarely been tested for statistical significance. We found a significant triple interaction of Group × Modality Sequence (ipsimodal/crossmodal) × Modality ($F(2.51) = 4.29$, $p = 0.02$) pointing to such a difference in discriminative power only in our STUDY II. It was in this study that we had taken special precautions to equate the subjective intensities of the stimuli. Rey, Beck, Morstadt and Oldigs (1987) found a MSE_D in the RTs to tones also in random sequences with tactile stimuli to the finger tips instead of light stimuli. Crossmodal retardation was not significant in any of the other combinations of tones, lights and tactile stimuli.

In their first report, Sutton *et al.* (1961) questioned whether the greater lengthening of the crossmodal reaction times for patients with a schizophrenic disorder is really due to the modality shift or simply to the fact that the previous stimulus was "different." In 1965, Sutton and Zubin reported a "MSE_D" within the auditory modality from an experiment in which two tones of different pitch and two lights of different color were presented in random order. Sutton and his colleagues suggested that ". . . the crossmodality effect . . . has nothing to do with sensory modality per se, but rather . . . (with) the magnitude of difference between the stimuli" (Sutton *et al.*, 1978; p. 263). They were certainly right as far as auditory stimuli are concerned; for other modalities no comparable results have been reported. In a study by Rist and Thurm (1984) only tones were presented, a high pitched tone to one ear and a low pitched tone to the other. Nevertheless, the interaction Group × Sequence (ipsimodal/crossmodal) for this differential shift effect proved to be significant. In the hemifield selection condition of STUDY I, the same tone was presented to one ear or the other, randomly interchanged with lights of different locations. Defining "modality shift" as a shift in location, we found the critical interaction even for this differential shift effect to be highly significant in the RTs to tones ($F(2,45) = 9.02, p < 0.01$) but not to lights ($F(2,45) = 0.68$). Apparently, reactions to tones are particularly sensitive to disparities between consecutive stimuli, a shift in modality being only one possible contrast.

The MSE_D as a Consequence of RT Retardation

Sutton and his colleagues were always well aware of the disquieting possibility that the MSE_D is an artifact of longer RTs (in patients with a schizophrenic disorder) leading to numerically larger differences between ipsi- and crossmodal RTs. As discussed by Chapman and Chapman (1989), this possibility is more than difficult to eliminate. In some studies they hoped to ward off this possible artifact through analyses of covariance, partialling out ipsi- from crossmodal RTs when testing for group differences. In their first report (Sutton *et al.*, 1961) they used another, possibly more convincing approach: for every subject they performed a Mann-Whitney U-test to compare the cross- and ipsimodal RTs of a given modality. The z-transformed U-scores for RTs to tones were significantly higher in schizophrenic than in normal subjects. This result was replicated ($p < 0.01$) in our STUDY I, but not in our STUDY II. The group difference for visual RTs never reached an acceptable level of significance.

Adjusting the MSE index with respect to overall reaction speed also provided evidence that the larger MSE in the auditory RTs of schizophrenic patients is not (only) an artifact of their longer RTs. Across seven sets of data from different experimental conditions, the critical F-values for the MSE_D based on the common MSE index (difference between ipsi- and crossmodal median RTs) varied between 2.41 and 8.35. When these median RTs were divided by the median RT from both conditions, the critical values varied from 2.41 to 8.92. The differences were negligible and the correlation (rho) of the F-values across the seven sets of data for the auditory MSE_D was 0.97. Furthermore, there was no indication that the critical F-ratios for the interaction Group × Sequence from the different conditions of STUDY I and STUDY II varied systematically with increases or decreases in the average RTs (from 231 ms to 453 ms) that had been introduced through different task demands. Moreover, within subjects Spring (1980) found no correlation between overall reaction time and MSE. The intraindividual correlations in her STUDY across ten blocks of trials varied from −0.50 to 0.39, with a mean value of zero.

In still another approach, Manuzza *et al.* (1984) compared their schizophrenic patients with a subgroup of normal subjects whose average geometric mean of the ipsimodal RTs was similar to those of the patients with schizophrenia. Even with this extremely strict control for possible artifacts through differences in average RT, the MSE_D proved to be highly significant for both lights and tones.

Taken together, these results provide strong evidence that the MSE_D does not simply result from the fact that patients with a schizophrenic disorder have longer RTs than normals and that larger differences between RTs are more likely with longer RTs.

The Influence of Different Task Demands on the MSE_D

As cited by Sutton and Zubin (1965), Wundt had already noted "a peculiar unrest" when the same response was required in random series of lights and tones. In later years, O. H. Mowrer and co-workers (cited in Sutton & Zubin, 1965) provided striking examples for the impact of central rather than peripheral "preparatory sets" on the retardation of simple reaction times when changes in modality are expected. Longer RTs after cross- than ipsimodal stimuli, in particular to auditory stimuli, are not specific to patients with a schizophrenic disorder but are also found in normal subjects. In all of our studies and in most reports from the literature the main effect of Modality Sequence (repetition vs alternation) accounts for up to about eight times more of the variance than the interaction Group × Modality Sequence. The increment in this universal effect is frequently larger in not only schizophrenic than in normal subjects, but is similarly increased in depressive patients (Spring, 1980). Alcoholic inpatients after the withdrawal phase are usually in the normal range (STUDY I; STUDY II; Rey *et al.*, 1987; Rist & Thurm, 1984).

Not only has a MSE been found in all groups studied, also the rules that determine the size of the MSE seem to be quite similar in schizophrenics, other patients and healthy subjects.

In STUDY I, we compared the Standard Condition to a variant that required the subjects to respond only to stimuli from one of the two hemifields (Hemifield Selection) and to another task where they had to alternate between responding and nonresponding to two consecutive stimuli in a row (Pairwise Response Alternation). For RTs to tones the MSE was largest in the Standard Condition ($F(1,61) = 135.27$; $p < 0.001$) and smallest under Pairwise Response Alternation ($F(1,61) =$

64.25; $p < 0.001$). In STUDY II, the Standard Condition was compared both to a choice RT variant requiring different responses to lights and tones and to a variant of the Standard Condition, with the stimuli presented in a perfectly predictable order as alternating pairs of lights and tones (Pairwise Stimulus Alternation). In this study, the MSE in the RTs to tones was again largest in the Standard Condition ($F(1,51) = 77.95$; $p < 0.001$) and smallest under Pairwise Stimulus Alternation ($F(1,51) = 41.85$; $p < 0.001$). In none of these studies did a change in task demands differentially effect the MSE of schizophrenic patients or controls.

More recently, we asked subjects to count the stimuli in the RT task either of one modality only, or across both lights and tones (unpublished). Again, there was no significant interaction of Group × Sequence × Condition. Counting stimuli in one modality reduced the MSE in that modality due to a shortening of the (cross-modal) RTs whenever a shift back allowed counting to continue. Further modifications of the tasks will be discussed in later sections.

The reliability of a subject's difference between cross- and ipsimodal RTs is modest. The retest reliability in STUDY I ($N = 64$) was 0.65 for the Standard Condition, the correlation of the differences from visual and auditory RTs was $r = 0.47$, and the intercorrelations of the differences in auditory RTs between the three experimental conditions varied from 0.49 to 0.65.

Considering the robustness of the MSE_D across experimental modifications of task demands and the similarity in the responses of normal and schizophrenic subjects vis-a-vis these experimental manipulations, it appears promising to first describe in more detail the rules that determine the overall effect before exploring possible factors responsible for the increased MSE in subjects with a schizophrenic disorder.

The MSE and First Order Sequential Effects in Choice Reaction Times

The basic pattern of a MSE corresponds to the first order effect in choice reaction time tasks: longer RTs after alternations than after repetitions. This is the most common finding, if the response-stimulus interval (RSI) is at least 500 ms (*e.g.*, Laming, 1968; Remington, 1969; Falmagne, Cohen & Dwivedi, 1975; for discrepant findings see Kirby, 1980; Luce, 1986), which has always been the case in modality shift experiments.

In a noteworthy series of studies this increase of RTs to alternations has been parallelled in event-related potentials (ERPs) by an increase in the amplitude of the P300 and the parietal positive Slow Wave (SW_{pos}) (Squires, Wickens, Squires & Donchin, 1976; Johnson & Donchin, 1980, 1982; Ford, Duncan-Johnson, Pfefferbaum & Kopell, 1982; Duncan-Johnson, Roth & Kopell, 1984). In contrast to standard MSE experiments, these studies confronted subjects with a discriminative task requiring them either to silently count one of two tones presented in random order ("oddball") or to perform a choice RT task. In other studies subjects were either informed on every trial about the stimulus to appear next or had to guess its modality (Levit, Sutton & Zubin, 1973; Verleger & Cohen, 1978). In these two studies P300 amplitudes were larger after cross- than after ipsimodal stimuli and larger in normals than in patients with a schizophrenic disorder. But despite larger MSE in RTs, the difference in P300 amplitude was smaller in patients with schizophrenia than in normals.

In our studies there was a dissociation of first order effects in RT and late positivities when ERPs were collected during a simple RT task with identical task requirements for all stimuli. Under such conditions the difference between ipsi-

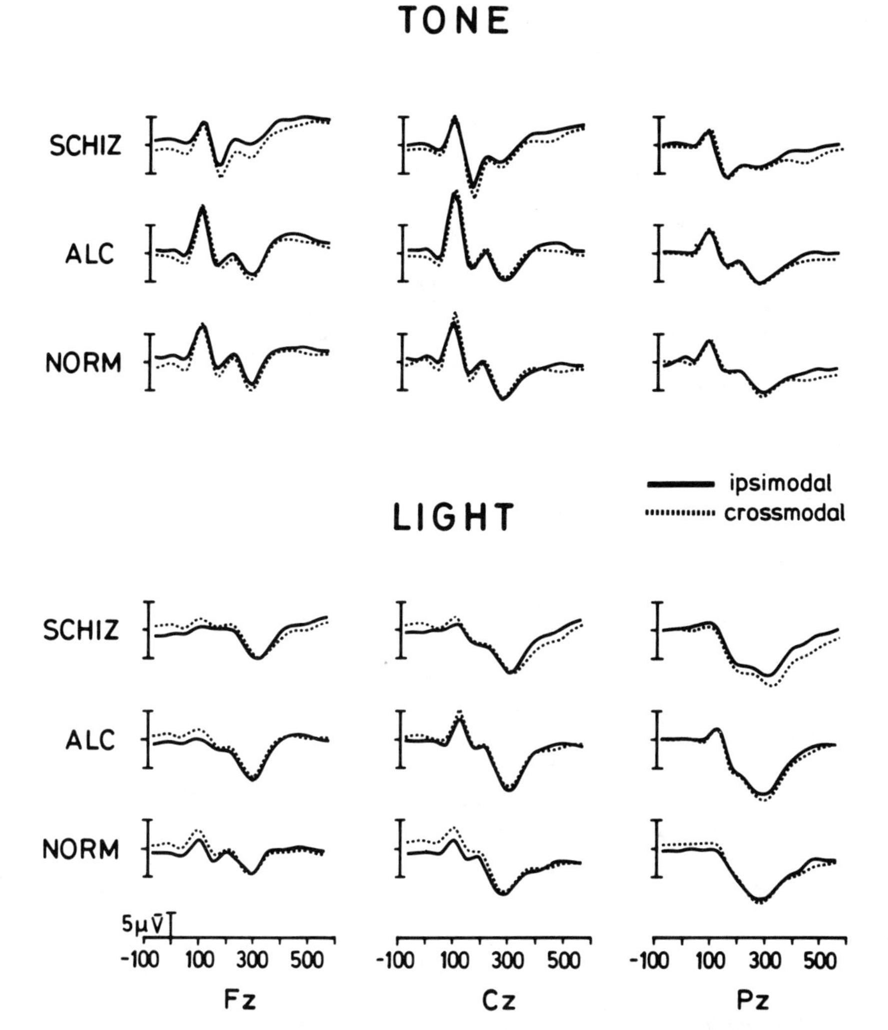

FIGURE 2. Event-related potentials to light and tone stimuli from STUDY II, recorded from Fz, Cz and Pz to linked ears. Negative deflections are pictured *upward*; the stimulus occurred at time zero.

and crossmodal P300 amplitudes is negligible. FIGURE 2 shows the grand averages for ipsimodal (repetitions) and crossmodal (alternation) stimuli from the Standard Condition in STUDY II. Similar to STUDY I and for both modalities, only the parietal Slow Wave was larger after alternations than after repetitions ($F \leq 8.76$, $p < 0.001$).

Actually, the MSE in RT is seen without any indication of a P300 in the ERPs when stimuli are presented in a fully predictable order. In one condition of STUDY

II we presented lights and tones in "pairwise stimulus alternation (LLTTLLTT etc.)." In a different condition we had one block of trials with distinct pairs of the same tone (TT.TT.TT), while in another block the tone was always preceded by a light (LT.LT.LT etc.). As one would expect for a variable so sensitive to the effects of subjective probability, the introduction of perfect predictability eliminated nearly all indications of a parietal P300 component. It also eliminated the differences between ipsi- and crossmodal sequences in the parietal Slow Wave, but no attenuation of the RT effects MSE and MSE_D occurred. We conclude that those sequential expectancies and stimulus evaluation processes that are reflected in the late positive components (cf. Johnson, 1986; Ruchkin & Sutton, 1983; Ruchkin, Sutton & Mahaffey, 1987) are not necessary and immediate prerequisites for the sequential effect in RTs we set out to explore.

In both STUDY I and STUDY II, differences between ipsi- and crossmodal ERPs were highly significant for the auditory N1 component. The auditory N1 is smaller after ipsi- than after crossmodal tones, perhaps due to the possibility of a superimposed processing negativity (Näätänen, 1987). Unfortunately, our data do not allow us to determine the possible contribution of processing negativity to this difference. We shall later comment on systematic differences in baselines, which in both studies turned out to be significantly more negative whenever the preceding stimulus was a tone. All of these differences between ipsi- and crossmodal ERPs were found to be significant only as main effects. Thus they are poor candidates to explain the MSE_D.

With one exception, the differences between the ERP scores of schizophrenic and control subjects were also found to be significant only as main effects and all of them pertained exclusively to auditory ERPs. In line with the literature (cf. Cohen, 1990), patients with schizophrenia had both smaller N1 and smaller P3 amplitudes ($F \geq 5.09$, $p < 0.001$). In STUDY II (FIG. 2), they also had larger P2 amplitudes ($F(2,51) = 11.85$, $p < 0.001$) which may well be the result of larger overlapping processing negativities in the controls. Also in STUDY II their ERPs showed larger frontal ($F(2,51) = 4.11$, $p = 0.02$) and central $F(2,51) = 7.56$, $p < 0.001$) negative Slow Waves. It is this late fronto-central negativity for which the only interaction of Group × Sequence was found. In STUDY I this late fronto-central negativity was larger following crossmodal tones. This effect was more evident in patients with a schizophrenic disorder than in controls ($F(1,62) = 8.37$, $p < 0.01$). In STUDY II the corresponding trend did not reach an acceptable level of significance.

The MSE and Higher Order Sequential Effects in Choice Reaction Times

"Higher order effects" usually refer to the influence of up to about four preceding stimulus-response cycles on the most recent response. A common way to present these effects is the repetition-alternation curve (see FIG. 3). In this figure the left portion depicts the RTs after a repetition and the right portion the RTs after an alternation. The data are presented separately for subsets with, from left to right, a decreasing number and proximity of prior repetitions. In several studies (Laming, 1968; Remington, 1969; Falmagne *et al.*, 1975; Soetens, Boer & Hueting, 1985) choice RTs have been found to be longest whenever a run of either repetitions or alternations is discontinued; the longer the run the shorter the choice RTs, as if some of the processes in stimulus-response coding or mapping proceed faster or bypass certain critical stages (cf. Rabbitt & Vyas, 1979). In the repetition-alternation curve this is seen as a positive slope in the repetition (left side of FIG. 3) and a negative slope in the alternation curve (right side of FIG. 3). Only for very short intervals between stimuli, a positive slope appears also in the alternation curve (cf. Kirby, 1980; Soetens *et al.*, 1985).

The higher order effects in simple RTs from the Standard Condition in STUDY II (FIG. 3) correspond to this picture known from choice RTs. In this analysis RT data from lights and tones have been combined, since the sequential trends did not interact with modality. The linear trends were significant for both the repetition and the alternation curve, with an additional cubic trend in the repetition curve. These gradients are steeper in the schizophrenic than in either of the two control groups leading to a significant interaction Group × Linear Trend for the alternation ($F(2,51) = 4.92$, $p = 0.01$) and a corresponding tendency for the repetition curve ($F(2,51) = 2.75$, $p = 0.07$). In line with a suggestion of Sutton and Zubin (1965; p. 595), "that . . . influences over the whole series of trials act to 'over-influence' . . . the schizophrenic subjects . . . (and not only) events which are contiguous in time" the largest difference between groups is seen after a long run of stimuli from the other modality.

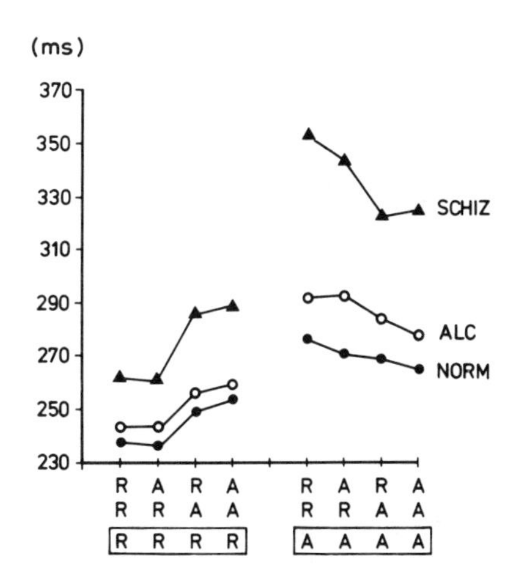

FIGURE 3. Simple reaction times to higher order sequences of alternations (A) and repetitions (R) of light and tone stimuli from the *Standard Condition* of STUDY II. RTs to light and tone stimuli were combined. The *left hand* shows RTs to stimulus repetitions (light-light and tone-tone) preceded by two repetitions and/or alternations. The *right hand* shows RTs to stimulus alternations (light-tone and tone-light).

Trends quite similar to the higher order effects in choice RT have been reported for the P300/SW_{pos} amplitudes from ERPs collected during oddball tasks (*e.g.*, Ford *et al.*, 1982; Johnson & Donchin, 1982; Squires *et al.*, 1976). These effects are usually depicted as tree graphs. FIGURE 4 gives an example for auditory ERPs from our Standard Condition in STUDY II. Analyses of the factor scores from these ERPs closely correspond to the trends in RTs shown in FIGURE 3. In the repetition curve to lights, linear increases were found for the P300 across all leads and the parietal Slow Wave. For tones they were significant only for the frontal P3a. In the alternation curve to lights, linear trends indicated a significant decrease for the parietal P300 and the positive Slow Wave. Similar trends were seen for the P3b and the parietal Slow Wave to tones.

Interestingly, a similar pattern was found for the N1 component (Cz) with a linear increase in the repetition curve to tones and a linear decrease in the alternation curve for both tones and lights. With the exception of a linear increase of the

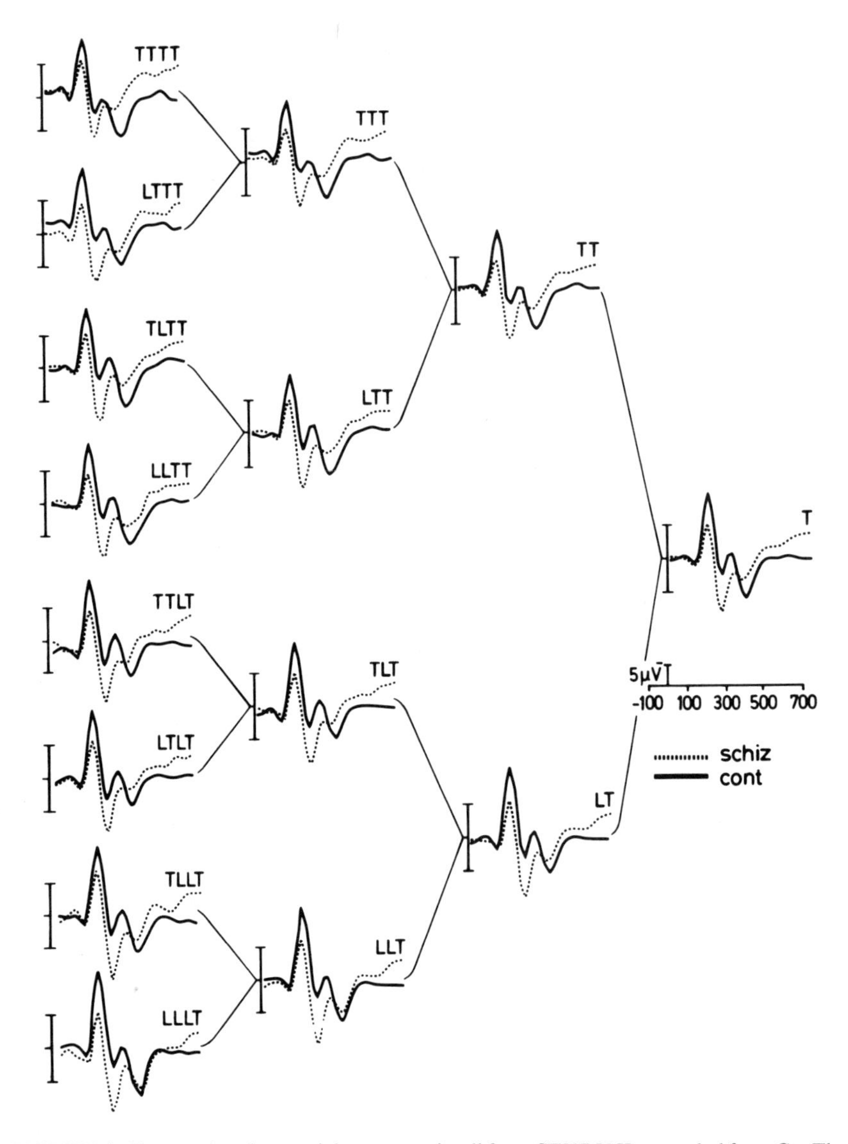

FIGURE 4. Event-related potentials to tone stimuli from STUDY II, recorded from Cz. The ERPs are displayed for all possible sequences of light and tone stimuli up to sequences of four stimuli. Negative deflections are pictured *upward*; the stimulus occurred at time zero.

N1(Cz) to lights in the repetition curve, which we find ourselves unable to interpret, all trends of ERP component scores are in parallel to the higher order curves for RTs. Also all the trends found for late positivities correspond to the sequential trends repeatedly found in oddball studies (cf. Donchin, Karis, Bashore, Coles & Gratton, 1986; Pritchard, 1981).

In contrast to the RT findings, none of the ERP components showed any significant interaction of Group with these higher order sequential effects. Surprising as this might appear at first glance, it is entirely in line with an earlier report on higher order sequential effects of the P300 amplitude by Duncan-Johnson *et al.* (1984). This correspondence encourages us to conclude that the cognitive processes leading to the formation of expectancies determining sequential effects in RT and oddball tasks follow the same rules in patients with schizophrenia as in normals. It is only in RTs that we find patients with a schizophrenic disorder showing a stronger influence of the preceding three or four stimulus-response cycles than nonpsychotic controls.

In Search of an Explanation

"Expectancy" as Explanatory Concept

The most common concepts used to explain the sequential RT effects in modality shift experiments are (1) expectation, preparation or strategies on the one hand, and (2) trace or automatic facilitation on the other. Sutton and his colleagues (Sutton *et al.*, 1978) found considerable problems with both of them.

In 1977 Spring and Zubin posited that it appears impossible to maintain a simple expectancy or strategy formulation to explain the MSE. In the RT literature (*e.g.*, Kirby, 1980; Luce, 1986) such a formulation usually implies some sort of differential preparation with a benefit if the next stimulus happens to concur with the expectation and with costs if it does not. More specifically, Näätänen and Merisalo (1977)) defined "expectation" as the subjective probability of the immediate delivery of an imperative stimulus. In this sense, expectation is a crucial determinant of "preparation," which in turn consists of carrying out everything of a response that can be carried out in advance (Näätänen & Merisalo, 1977).

Donchin *et al.* (1986) as well as other ERP workers typically use the term "expectancy" in a somewhat different way, to refer to largely automatic "subroutines" or "algorithms" involved in the updating of working memory without any obligatory connections to the motor system. The term had been used by Sutton and Zubin (1965) in a similar way that was closely related to their later use of the term "trace:" "Human beings . . . develop certain 'expectancies,' either consciously or unconsciously. These expectancies may be in the nature of facilitatory or inhibitory pathways established by previous experience" (Sutton & Zubin, 1965; p. 565). In later years they used "expectancy" more in line with the RT literature, and as we shall use it here, to imply some sort of differential preparation for a specific stimulus-response cycle.

The following four sets of results are at variance with an "expectancy" interpretation of the MSE:

1) When asking their subjects to guess what the next stimulus would be, Sutton *et al.* (1978) did not find any indication of a differential bias in these predictions. If anything, normals seemed more biased to expect a repetition than patients with schizophrenia. Even more detrimental to an expectancy formulation, they found

the MSE_D to be significant not for the incorrectly but only for the correctly predicted trials. However, correct guesses should have no costs but benefits only.

2) To Sutton *et al.* (1978; p. 264) "the most devastating blow to an expectancy hypothesis" came from a study by Waldbaum *et al.* (1975), when the experimenter told the subject whether the next stimulus would be a light or a tone and had the subject confirm this information. Under such conditions of certainty the MSE_D was even more pronounced than when the subject had no prior knowledge of the upcoming stimulus. Spring (1980) also found a strong MSE_D both when the experimenter or when an actual light or tone stimulus informed the subject about the next stimulus.

3) In STUDY II we followed a suggestion by Muriel Hammer to control for the effects of expectancy: lights and tones were presented as alternating pairs (LLTTLL etc.) in a perfectly predictable sequence. A comparison with the Standard Condition (see FIG. 1b) clearly showed in both studies that this introduction of certainty had no effect on the MSE_D. In a replication of this study predictability even enhanced the overall MSE for both modalities. In another experimental condition of STUDY II we presented distinct pairs of stimuli. In one block of trials each pair consisted of two identical tones (repetition), in another block of trials the tone was always preceded by a light (alternation). RTs to the second stimuli in these pairs, always tone stimuli, were compared with the auditory RTs from the Standard Condition. Again, the experimental condition did not interact with sequential effects: predictability did not modify the overall MSE or the MSE_D.

4) Finally, as reported in the last section, the P300 components of the ERPs to ipsi- and crossmodal stimuli in random series of lights and tones requiring the same motor response gave no indication of stronger "surprise reactions" (Donchin, 1981) after crossmodal stimuli. Only when discriminatory responses are required have differential P300s been found, but these were not more pronounced in schizophrenic than in healthy subjects (Duncan-Johnson *et al.*, 1984; Verleger & Cohen, 1978). A more detailed comparison of our data with the results from Levit *et al.* (1973) or Verleger and Cohen (1978) is not possible, since their subjects had to guess which stimulus was to be presented next.

The first three of these sets of results consistently indicate that the longer RTs to crossmodal stimuli cannot be the result of incorrect preparations for a specific stimulus-response cycle. The fourth set of results indicates that the differential RT effect is not accompanied by differential P300 effects. This could have been expected on the basis of a number of earlier studies that reported P300 to reflect differences in "unexpectedness" (Tueting, Sutton & Zubin, 1971), "subjective probabilities" or "expectancies", independent of response selection or motor preparation (cf. Donchin *et al.*, 1986; Johnson, 1986; Ruchkin, Munson, Mahaffey & Sutton, 1982). Whether the term "expectancy" has been used with reference to processes of memory updating and stimulus evaluation as in ERP research or to processes of guessing or response preparation as in the studies with experimental variations of stimulus predictability, the concept "expectancy" appeared inadequate or misleading when used to explain the RT effect.

"Trace" as Explanatory Concept

Realizing the problems with an "expectancy" formulation, Sutton and Zubin repeatedly turned to a "trace" formulation to explain the MSE. In these formulations, quite similar to physiological suppositions of Sokolov and the most recent work of Näätänen (in press) on mismatch and processing negativities in the auditory ERP, the term "trace" referred to some sort of neural activity in sensory pathways and/or short-term memory. Sutton *et al.* (1961, p. 232) speculated that "the occurrence of relevant stimuli in a given sensory modality predisposes the organism to be maximally ready for further stimuli in the same modality." Elaborating this idea, Zubin's (1975) "neural trace theory" suggests: "as the number of . . . trials increases, trace residues of each trial . . . summate and simulate the effect of increased intensity. . . . Of course, if the interval between stimuli is long enough, this temporary trace would vanish and leave no pool of excited neurons as a residue" (p. 152). In addition, he proposes "that the immediate effect of a stimulus persists longer in the case of schizophrenics. For this reason, RT in a crossmodal shift takes longer, since the effect of the prior stimulus persists, interfering with the response to the shift" (p. 164).

To Sutton and his colleagues the main problem with such a concept of residual activity in sensory pathways to explain the MSE_D arose from the fact that "as we went from experiment to experiment, we were able to shift the intertrial interval with impunity, making it quite long in some cases. A trace should decrease in proportion to the intertrial interval" (Sutton *et al.*, 1978, p. 264). One might add, that there is no physiological evidence for traces in afferent pathways to persist for periods longer than a second, which is the shortest interstimulus interval in modality shift experiments. Even the evidence from RT studies pertains only to intervals of up to about 500 ms (cf. Kirby, 1980; Luce, 1986; Soetens *et al.*, 1985). Hence, the only report of a reduction in the MSE with an increase in the intertrial interval from 2.5 to 7.5 s (Rey & Oldigs, 1980) probably reflected some other phenomenon such as the ease of motor preparation with only two intertrial intervals.

Another argument against a trace formulation stems from various findings showing that stimulation immediately preceding an imperative stimulus does elicit a MSE only if it requires a motor response:

1) Rist and Thurm (1984) postponed every second stimulus in a random series of lights and tones with an average intertrial interval of two seconds and defined them as "warning signals" for the imperative stimulus 500 to 1000 ms later. A motor response was only required to the imperative stimuli. Neither for lights nor for tones did these warning stimuli elicit a MSE, while the preceding imperative stimuli some 2000 to 3500 ms earlier clearly did. Zubin's trace model would have predicted a stronger effect of the immediately preceding stimuli. This finding corresponds well with results from Spring (1980) and Waldbaum *et al.* (1975) who reported a strong MSE_D even though every imperative stimulus was preceded by an intermittent verbal or ipsimodal warning signal informing the subject about the modality of the upcoming stimulus.

2) When subjects had to respond to two consecutive stimuli but not respond to the next two (Pairwise Response Alternation), most of the variance was accounted for by the interaction between stimulus (repetition/alternation) and response sequence ($F(1,61) = 91.95$ for lights, $F(1,61) = 124.17$ for tones, $p < 0.001$). The MSE is significantly reduced if the preceding stimulus does not require a motor response. The crossmodal RTs to lights are even slightly shorter than the ipsimodal RTs if no response was given in the preceding trial. The MSE to tones is more robust and, although drastically reduced, still significant for patients with schizophrenia even when the preceding stimulus required no motor response.

3) In another unpublished study from our laboratory, we had subjects respond only to the offset of stimuli after two or four seconds duration. This prolonged stimulation, we assumed, would mask the postulated traces from preceding stimuli, separated by another two or four seconds from the onset of the immediate stimulus.

Independent of stimulus and interval duration, a similarly strong overall MSE was found in both groups. If the MSE was an aftermath of residual activity in sensory pathways, it should have disappeared at least under some of these conditions.

In line with Spring's (1980) conclusion from her experiment with verbal and ipsimodal warning signals that "cross-modal retardation is influenced primarily by the sequence of responded-to stimuli" (p. 139), these studies inadvertently demonstrate that the concept of neural traces in sensory pathways cannot be used to explain the MSE. Only if "traces" are defined as referring not to afferent pathways well described by neuroanatomical or neurophysiological research, but to entire stimulus-response cycles in substantially more enigmatic systems of the brain, can we readily bring together the existing findings under the heading of a trace concept. With such a definition it even appears that those results that had caused problems for a sensory trace formulation can be seen as buttressing the concept of traces for whole stimulus-response cycles or even for sequences of such cycles in the psychological pathways of short-term memory. The effect is only elicited by stimuli that required the same response while prolonged exposure or immediately preceding stimuli which do not require the same response have no influence. Posner (1978, p. 90) defined psychological pathways "as a set of internal codes and their connections that are activated automatically by the presentation of a stimulus." Quite similar to Zubin's neural trace theory he assumes: "once activated . . . pathways provide enhanced processing of stimuli that share the same pathway. When subjects . . . pay close attention to a stimulus, however, the facilitation is accompanied by an inhibition of stimuli that do not share the same pathway" (Posner, 1982, p. 173).

The duration of these traces can be assumed to be much longer than the neural activation in afferent pathways. Numerous RT studies by Bertelson, Rabbitt and others (cf. Kirby, 1980) of "identity" and "equivalence" responses have demonstrated particularly long duration of these effects under conditions of "incompatibility," *i.e.*, when two or more stimuli are arbitrarily assigned to different responses or mapped onto the same response. The latter type of incompatibility is a characteristic of all standard modality shift experiments: patently different stimuli are mapped onto the same motor response. With the typical stimuli of modality shift experiments it is quite difficult to imagine a prototypical exemplar embracing both kinds of stimuli. Subjects may flinch at the promiscuity of making the same response to a stimulus they had just made to a different one. Similar hesitations are well known from everyday life when one is to (correctly) address somebody with the same name one had used immediately before in (correctly) addressing somebody quite different.

Speculations about Determinants of the Differential Shift Effect

In all of our experiments, we had hoped to find a clue as to why the RTs of patients with a schizophrenic disorder are disproportionately prolonged relative to those of nonpsychotic controls when the stimulus sequence involves a shift of modality; or only a shift between patently different stimuli within the same modality. Why is the overall MSE and this differential retardation more pronounced in RTs to tones than in RTs to lights? As reported in previous sections, we found these effects to be (1) strikingly robust vis-à-vis various experimental manipulations, provided the preceding stimulus had required a motor response, and (2) not accompanied by corresponding differences between groups in the components of event-related potentials. This leaves us with the speculation that the differential effects might be a peculiarity of imperative stimuli and/or their correlation with voluntary motor responses. Pursuing this idea, two suggestions might be worth further examination. The first is that normals may be more likely than schizophrenic patients to develop an amodal code for lights and sounds that require the same response. The second is that the larger shift effect in RT to tones may reflect phasic activation through auditory stimulation. Schizophrenic patients might be more sensitive to this effect than controls.

The MSE_D and the Availability of Amodal Codes

In Posner's model of psychological pathways, an amodal code might allow some facilitation even across modalities. It is conceivable that patients with a schizophrenic disorder perform more on what Kurt Goldstein has called a concrete level, with different pathways for light and tone mapped onto the same motor response. In Posner's model, such concrete codes should lead to inhibition whenever a shift in modality occurs. Goldstein describes the patient with a concrete attitude as "stimulus-bound", " . . . confined to the immediate apprehension of the given thing or situation in its particular uniqueness . . . the individual is being shunted passively from one stimulus to the next" (Goldstein & Scherer, 1941, p. 3ff). In contrast, when in abstract mode, "we abstract common from particular properties, we are oriented in our action by a rather conceptual viewpoint, be it a category, a class, or a general meaning under which the particular (stimulus) before us falls" (Goldstein & Scheerer, 1941, p. 4f).

Such a concrete mode of processing might be enhanced and systematically elicited in normals when the experimenter, in order to reduce uncertainty about the forthcoming event, stresses the difference between modalities by informing the subject what the modality of the next stimulus will be. As mentioned above, an increase of the overall and/or differential MSE was found when either the experimenter told the subject prior to each trial whether the stimulus would be a tone or a light (Waldbaum *et al.*, 1975), or when we stressed the difference by ordering the stimuli according to modality in our Pairwise Stimulus Alternation condition.

The more concrete the stimulus-response codes, the more effort has to be invested whenever a shift to the other pathway is required. Since positive Slow Wave has been found to be larger in tasks that require greater effort (Ruchkin & Sutton, 1983), some evidence for such an extra effort might be seen in the finding that the positive Slow Wave component of the ERPs has been found to be significantly larger after alternations than after repetitions in both our studies. Moreover, in STUDY I positive Slow Wave showed an even larger difference between alternations and repetitions in patients with schizophrenia than in controls for both lights and tones ($F(1,62) \geq 9.08, p < 0.01$). We tend to interpret these findings as reflecting an extra effort to overcome the "glow" from previously activated pathways (Callaway & Naghdi, 1982).

The MSE and Phasic Activation through Auditory Stimulation

Part of the MSE_D phenomenon is that it is more easily provoked by a shift from light to tone than vice versa. This may reflect a particular sensitivity of schizophrenic patients to auditory stimulation. This speculation is mainly based on the finding of two quite reliable differences between the event-related potentials to lights and tones and between patients with schizophrenia and nonpsychotic controls: (1) In both our studies, for both groups and for all our experimental conditions we found the baseline to be significantly more negative whenever the preceding stimulus was a tone than when it was a light, *i.e.*, more negative baselines of the ERPs to ipsimodal tones and crossmodal lights. The baseline negativity of visual ERPs showed a linear increase with the number and proximity of preceding tones. (2) In both our studies the auditory ERPs of patients with schizophrenia had a significantly larger late negativity at frontal and central sites than nonpsychotic controls. Such negativities or differences in negativities are not found in visual ERPs (see FIG. 2).

Following Posner (1978) again, we interpret these negativities following an auditory stimulus as indicating an automatic increase of phasic alertness elicited by the previous auditory event, which may facilitate the response to the following stimulus. If the forthcoming stimulus happens to be a tone again, the phasic alertness would shorten the ipsimodal RT and thereby increase the MSE over and above the facilitation through activation of the same psychological pathway. If, on the other hand, it happens to be a light, the phasic alertness would shorten the crossmodal RT and thereby attenuate the difference between ipsi- and crossmodal RTs. Obviously, this facilitation cannot be the whole story considering the significant differential shift effect when the same tone was presented to one or the other ear (Rist & Thurm, 1984). What we suggest is an additive effect of pathway activation and phasic alertness through auditory stimulation to which patients with a schizophrenic disorder may be more sensitive than normals.

Unfortunately, our data do not allow a comparison between groups with respect to their ERP baseline. But if the more negative fronto-central Slow Waves in the auditory ERPs determine the baselines of the ERPs to the following stimuli, they might well provide a clue to why the MSE_D is more pronounced in the RTs to tones than in the RTs to lights.

But even if these late fronto-central negativities in the auditory ERPs of patients with schizophrenia were not related to the more negative baselines in the ERPs to subsequent stimuli, they clearly indicate a heightened cortical activation following auditory stimuli. This activation extends in time far beyond the motor response and might best be seen as a modality-dependent variant of the post-imperative negative variation (cf. Cohen, 1990). Such post-imperative negative variations with a fronto-central maximum often develop long after an imperative stimulus and are the only components that have been repeatedly found to be larger in the ERPs of patients with a schizophrenic disorder than for normals or most psychiatric controls. In a number of studies we have found these late fronto-central negativities to be considerably more pronounced after responses to auditory than to visual stimuli (Cohen, Rist, Finger & Mussgay, 1990). Most likely they reflect a basic uncertainty about response-outcome contingencies or the appropriateness of one's preceding response (cf. Rockstroh, Elbert, Canavan, Lutzenberger & Birbaumer, 1990). The late fronto-central Slow Wave in the auditory ERPs of schizophrenic patients may be conjectured to indicate a prolonged increase in cortical activation pointing to a basic uncertainty of these patients in the generation and monitoring of self-initiated actions. Such a deficit has been stressed by Frith and Done (Frith, 1987; Frith & Done, 1988) as a fundamental characteristic of schizophrenic disorders, possibly due to failures in the transmission of information from the prefrontal cortex to the hippocampal monitoring system. It could be potentiated through difficulties in discriminating between this basic irritation due to failures of internal feedback-loops and the phasic activation elicited through auditory stimulation. In turn, the difficulty to discriminate between these sources of activation might easily produce a state of heightened sensitivity vis-à-vis the alerting power of auditory stimulation.

We have, hopefully, moved forward a few steps from the crossroads detailed in the papers by Sutton and his colleagues. Both our propositions to interpret the MSE_D are certainly not mutually exclusive. We see them as woods in reachable distance that might be worth closer inspection to find a vantage point from which we may better understand the MSE_D. To summarize: (1) Sutton and his colleagues' revelation that expectancy may not be a viable formulation to explain the MSE_D is strongly supported. (2) The trace formulation also leads the wrong way if referring to sensory pathways, but is a promising route to pursue if referring to internal codes for complete stimulus-response cycles or even sequences of such cycles. (3) Subjects with a schizophrenic disorder might tend to process the stimuli in a more concrete code than normals with different pathways for lights and sounds. Furthermore, they may be particularly sensitive to the alerting power of auditory stimulation.

REFERENCES

Berg, P. 1986. The residual after correcting event-related potentials for blink artifacts. Psychophysiology **23:** 354–364.

Callaway, E. & S. Naghdi. 1982. An information processing model for schizophrenia. Arch. Gen. Psychiatry **39:** 339–347.

Chapman, L. J. & J. P. Chapman. 1989. Strategies for resolving the heterogeneity of schizophrenics and their relatives using cognitive measures. J. Abnorm. Psychol. **98:** 357–366.

Cohen, R. 1991. Event-related potentials and cognitive dysfunction in schizophrenia. *In* Search for the Causes of Schizophrenia. H. Häfner & W. F. Gattaz, Eds. Vol. 2: 342–360. Springer Verlag. Heidelberg.

Cohen, R., F. Rist, Th. Finger & L. Mussgay. 1990. Modality-dependent post-imperative negative variations to auditory stimuli in schizophrenic patients. *In* Psychophysiological Brain Research. C. H. M. Brunia, A. W. K. Gaillard, A. Kok, G. Mulder & M. N. Verbaten, Eds. 203–208. Tilburg University Press. Tilburg.

Donchin, E. 1981. Surprise! . . . Surprise? Psychophysiology **18:** 493–513.

Donchin, E., D. Karis, T. R. Bashore, M. G. H. Coles & G. Gratton. 1986. Cognitive psychophysiology and human information processing. *In* Psychophysiology: Systems, Processes and Applications. M. G. H. Coles, E. Donchin & S. W. Porges, Eds. 244–267. Guilford Press. New York.

Duncan-Johnson, C. C., W. T. Roth & B. S. Kopell. 1984. Effects of stimulus sequence on P300 and reaction time in schizophrenics. *In* Brain and Information: Event-Related Potentials. R. Karrer, J. Cohen & P. Tueting, Eds. Annals of the New York Academy of Science. Vol. 425: 570–577. New York Academy of Sciences. New York.

Falmagne, J. C., S. P. Cohen & A. Dwivedi. 1975. Two choice reactions as an ordered memory scanning process. *In* Attention and Performance. P. M. A. Rabbitt & S. Dornic, Eds. Vol. 4: 296–344. Academic Press. New York.

Ford, J. M., C. C. Duncan-Johnson, A. Pfefferbaum & B. S. Kopell. 1982. Expectancy for events in old age: stimulus sequence effects on P300 and reaction time. J. Gerontol. **37:** 696–704.

Frith, C. D. & D. J. Done. 1988. Towards a neuropsychology of schizophrenia. Br. J. Psychiatry **153:** 437–443.

Frith, C. D. 1987. The positive and negative symptoms of schizophrenia reflect impairments in the perception and initiation of action. Psychol. Med. **17:** 631–648.

Goldstein, K. & M. Scheerer. 1941. Abstract and concrete behavior. An experimental study with special tests. Psychol. Monogr. **53.**

JOHNSON, R. & E. DONCHIN. 1980. P300 and stimulus categorization: two plus one is not so different from one plus one. Psychophysiology **17:** 167–178.
JOHNSON, R. & E. DONCHIN. 1982. Sequential expectancies and decision making in a changing environment: an electrophysiological approach. Psychophysiology **19:** 183–200.
JOHNSON, R. 1986. A triarchic model of P300 amplitude. Psychophysiology **23:** 367–384.
KIRBY, N. 1980. Sequential effects in choice reaction time. *In* Reaction Times. A. T. Welford, Ed. 129–172. Academic Press. London.
LAMING, D. R. J. 1968. Information Theory of Choice Reaction Times. Academic Press. London.
LEVIT, R. A., S. SUTTON & J. ZUBIN. 1973. Evoked potential correlates of information processing in psychiatric patients. Psychol. Med. **3:** 487–494.
LUCE, I. 1986. Response Times. Their Role in Inferring Elementary Mental Organization. Oxford University Press. New York.
MANNUZZA, S., M. KIETZMAN, I. BERENHAUS, PH. RAMSEY, J. ZUBIN & S. SUTTON. 1984. The modality shift effect in schizophrenia: fact or artifact? Biol. Psychiatry **19:** 1317–1331.
NÄÄTÄNEN, R. 1990. The role of attention in auditory information processing as revealed by event-related potentials and other brain measures of cognitive function. Behav. Brain Sci. **13:** 201–288.
NÄÄTÄNEN, R. & A. MERISALO. 1977. Expectancy and preparation in simple reaction time. *In* Attention and Performance. S. Dornic, Ed. Vol. 6: 115–137. Academic Press. New York.
NÄÄTÄNEN, R. & T. W. PICTON. 1987. The N1 wave of the human electric and magnetic response to sound: a review and an analysis of the component structure. Psychophysiology **24:** 1–74.
POSNER, M. I. 1978. Chronometric Explorations of Mind. Lawrence Erlbaum. Hillsdale, NJ.
POSNER, M. I. 1982. Cumulative development and attentional theory. Am. Psychol. **37:** 168–179.
PRITCHARD, W. S. 1981. Psychophysiology of P300. Psychol. Bull. **89:** 506–540.
RABBITT, P. & S. VYAS. 1979. Memory and data-driven control of selective attention in continuous task. Can. J. Psychol. **33:** 71–87.
REMINGTON, R. J. 1969. Analysis of sequential effects in choice reaction times. J. Exp. Psychol. **82:** 250–257.
REY E. R., BECK, U., MORSTADT, E. & J. OLDIGS. 1987. Experimentelle Untersuchungen zu Aufmerksamkeitsstörungen Schizophrener. *In* Verhaltensmedizin: Ergebnisse und Perspektiven interdisziplinärer Forschung. W. D. Gerber, W. Miltner & K. Mayer, Eds. 537–562. VCH Verlagsgesellschaft. Weinheim.
REY, E. R. & J. OLDIGS. 1980. Experimental research of an attention deficit in schizophrenia. Eur. J. Behav. Anal. Mod. **4:** 127–140.
RIST, F. & R. COHEN. 1987. Effects of modality shift on event-related potentials and reaction time of chronic schizophrenics. *In* Current Trends in Event-Related Potential Research. R. Johnson, Jr., J. W. Rohrbaugh & R. Parasuraman, Eds. Electroencephalogr. Clin. Neurophysiol. Suppl. **40:** 738–745.
RIST, F. & I. THURM. 1984. Effects of intramodal and crossmodal stimulus diversity on the reaction time of chronic schizophrenics. J. Abnorm. Psychol. **93:** 331–338.
RIST, F. & R. COHEN. 1991. Sequential effects in the reaction times of schizophrenics: crossover and modality shift effects. *In* Experimental Psychopathology, Neuropsychology and Psychophysiology. J. Zubin, S. Steinhauer & J. Gruzelier, Eds. 241–271. Elsevier. Amsterdam.
ROCKSTROH, B., T. ELBERT, A. CANAVAN, W. LUTZENBERGER & N. BIRBAUMER. 1989. Slow Brain Potentials and Behavior. Urban & Schwartzenberg. Baltimore.
RUCHKIN, D. S. & S. SUTTON. 1983. Positive slow wave and P300: association and disassociation. *In* Tutorials in ERP Research: Endogenous Components. A. W. K. Gaillard & W. Ritter, Eds. 233–250. North Holland Publishing Company. Amsterdam.
RUCHKIN, D. S., R. MUNSON, D. MAHAFFEY & S. SUTTON. 1982. P300 and slow wave in a message consisting of two events. Psychophysiology **19:** 629–642.
RUCHKIN, D. S., S. SUTTON & D. MAHAFFEY. 1987. Functional differences between members of the P300 complex: P3e and P3b. Psychophysiology **24:** 87–103.

SOETENS, E., L. C. BOER & J. E. HUETING. 1985. Expectancy or automatic facilitation? Separating sequential effects in two-choice reaction time. J. Exp. Psychol. Hum. Percept. Perform. **11:** 598–616.

SPRING, B. & J. ZUBIN. 1977. Reaction time and attention in schizophrenia: a comment on K. H. Nuechterlein's critical evaluation of the data and theories. Schizophr. Bull. **3:** 437–444.

SPRING, B. 1980. Shift of attention in schizophrenics, siblings of schizophrenics, and depressed patients. J. Nerv. Ment. Dis. **168:** 133–140.

SQUIRES, K. C., C. WICKENS, N. K. SQUIRES & E. DONCHIN. 1976. The effect of stimulus sequence on the waveform of the cortical event-related potential. Science **193:** 1142–1146.

SUTTON, S. & J. ZUBIN. 1965. Effect of sequence on reaction time in schizophrenia. *In* Behavior, Aging, and the Nervous System. A. T. Welford & J. E. Birren, Eds. 562–579. Charles C. Thomas. Springfield, Ill.

SUTTON, S., B. SPRING & P. TUETING. 1978. Modality shift at the crossroads. *In* The Nature of Schizophrenia: New Approaches to Research and Treatment. L. Wynne, R. Cromwell & S. Matthysse, Eds. 262–269. Wiley. New York.

SUTTON, S., G. HAKEREM, J. ZUBIN & M. PORTNOY. 1961. The effect of shift of sensory modality on serial reaction time: a comparison of schizophrenics and normals. Am. J. Psychol. **74:** 224–232.

TUETING, P., S. SUTTON & J. ZUBIN. 1971. Quantitative evoked potential correlates of the probability of events. Psychophysiology **7:** 385–394.

VERLEGER, R. & R. COHEN. 1978. Effects of certainty, modality shift and guess outcome on evoked potentials and reaction times in chronic schizophrenics. Psychol. Med. **8:** 81–93.

WALDBAUM, J., S. SUTTON & J. KERR. 1975. Shift of sensory modality and reaction time in schizophrenia. *In* Experimental Approaches to Psychopathology. M. Kietzman, S. Sutton & J. Zubin, Eds. 167–177. Academic Press. New York.

ZUBIN, J. 1975. Problem of attention in schizophrenia. *In* Experimental Approaches to Psychopathology. M. Kietzman, S. Sutton & J. Zubin, Eds. 139–166. Academic Press. New York.

The Pupillary Response in Cognitive Psychophysiology and Schizophrenia[a]

STUART R. STEINHAUER[b] AND GAD HAKEREM[c]

[b]*Biometrics Research, 151R*
Department of Veterans Affairs Medical Center
Highland Drive
Pittsburgh, Pennsylvania 15206
and
[c]*Department of Psychology*
Queens College
City University of New York
Flushing, New York 11367

INTRODUCTION

The human pupillary response, whose final neural pathways are mediated through the autonomic nervous system, has been shown to reflect more centrally occurring processes when examined from the perspective of information processing activities. This paper will examine components of information processing as reflected in the pupillary response, and its relation to other neurophysiological signs of cognitive activity, most notably, event-related brain potentials. Both the constriction of the pupil to light (miosis), as well as the dilation (dilatation; mydriasis) resulting from information delivery, have provided useful adjuncts in the study of psychopathology, especially with reference to schizophrenia. The time course of pupillary dilation, reflected in the morphology of the pupillographic record, indicates a high degree of similarity among related family members. Findings from psychiatric patients and their relatives will be reviewed.

While the contributions of separate physiological components to the pupillary dilation response that relate to cognitive activity have never been adequately delineated, a review of findings indicates the possibility for experimental dissection of components. A model delineating likely psychophysiological contributions to the pupillary response, and approaches for testing the model, will be presented.

General Background and Historical Perspective

For centuries, the pupillary aperture has been thought of as a figurative window to the mind; with the advancement of medical sciences, the pupil began to serve as a literal window on brain function. In her 1958 monograph dealing primarily with pupillary dilatation, Loewenfeld (1958) cited nearly 1600 references, including 114 dated prior to 1830. Incidental observations of pupillary dilation associated with increased interest or arousal were well known, such as the use of belladonna to enlarge the pupil artifically as a cosmetic effect, and wearing of eyeshades to

[a] Supported by the Medical Research Service of the Department of Veterans Affairs, and National Institute of Mental Health No. 43615.

obscure any sudden dilatation for the poker player who might otherwise give away his hand.

During the early years of this century, aberrations in pupillary responsivity were carefully noted in psychotic patients (cf. Hakerem & Lidsky, 1975; Hess, 1972), especially by German psychiatrists such as Bumke (1904) and Bach (1908). These observations were followed up with studies by Lowenstein and Westphal (1933), Levine and Schilder (1942), and May (1948) in the third and fourth decades.

The earliest work in this century was performed by comparing the pupil with hand-held templates, or through laborious, often Herculean, photographic analyses. The first major leap in technology was provided by Otto Lowenstein's development of an electronic scanning pupillograph. Lowenstein, with Irene Loewenfeld, was responsible for many of the detailed investigations into the physiology of pupillary activity (*e.g.*, Loewenfeld, 1958, 1963; Lowenstein & Loewenfeld, 1952, 1962). By digitizing the analog output, it became possible to utilize computer averaging both in the recording, data storage, and analysis of pupillary activity (Hakerem, 1967). In the past two decades, infra-red TV-based systems, providing both analog and digital output, have permitted direct interface and control with laboratory minicomputers and even personal computers. Pupil diameter can now be measured with an accuracy of better than 0.025 mm at rates up to 60 times/sec. When data are averaged across multiple trials, background noise can be reduced, and changes of less than 0.01 mm can be detected. As an example of the technological effects, we once calculated that a research project investigating the reaction to light, which required seven months of data collection and analysis in 1962, could be performed today within a single afternoon. There is a caveat to such methodological speed. The generation of experiments may become more routinized rather than well conceived, a notion that would be strongly criticized by Samuel Sutton, who played an integral role in some of the research to be discussed below.

The re-emergence of pupillary studies among psychologists is related to a series of reports from several different laboratories in the early 1960s in the areas of experimental psychology and experimental psychopathology. The most polemic approach was generated by the initial papers of Eckhard Hess claiming pupillary dilation to positive affect stimuli and constriction to negative affect (Hess & Polt, 1960), which led to continuing controversies. Hess and Polt also began to report on pupillary dilation during mental activities (Hess & Polt, 1964), and this direction was taken up by Kahneman and Beatty (1966), representing a much stronger commitment to the developing concepts of cognitive psychology. A separate source of interest in pupillographics was being pursued in the domain of psychopathology. Leonard Rubin, at Eastern Psychiatric Research Institute in Philadelphia, was employing pupillary measurement to develop hypotheses of autonomic imbalance in psychiatric patients (Rubin, 1960, 1961, 1974; Rubin & Barry, 1968, 1972a, 1972b, 1976).

During the same period, both approaches, the psychopathological and the cognitive, were involved in the research program emerging from the Biometrics Research Unit at New York State Psychiatric Institute, involving Gad Hakerem, Samuel Sutton, and Joseph Zubin. Technological changes again played a substantial role in these developments. By the early 1960s, the Lowenstein pupillograph had replaced the camera in Hakerem's laboratory, and Manfred Clynes had introduced the first practical analog signal averager: the Computer of Average Transients (CAT 400). Clynes had been interested in using the device for physiological analysis of pupillary reactions as well as for electrophysiological analysis, and one of the earliest machines was made available to the Biometrics Research laboratories. Data were standardly recorded onto multichannel FM tape, along with trigger pulses preceding stimulus presentations. Off-line, the trigger pulse was used to initiate sampling with the CAT, and the resulting averages were plotted on paper. However, the same technology was appropriate for deriving evoked potentials, and Samuel Sutton's event-related potential laboratory was located directly across the hall. Tape recording allowed collection of data in either laboratory, but initially there was only the single averager. Thus began a history of the CAT being shuttled back and forth between laboratories, providing averaged pupillographic waveforms one day, averaged evoked potentials the next. Eventually, this traveling show resulted in the detection of the P300 component of the event-related potential (ERP), as it is known today, by Samuel Sutton and colleagues (Sutton, Braren, Zubin & John, 1965), and a keystone of cognitive psychophysiology was established. One of the initial pupillary studies involving cognitive processing activity demonstrated that visual stimuli at a 50% detection level (which were too weak to produce pupillary constriction) resulted in dilation when they were reported as seen, but not when they were missed, or when blanks were presented, or when no detection was required (Hakerem & Sutton, 1966). A further addendum to this story is that the continuing collaborations between the laboratories emphasized a number of parallels between pupillary dilation and P300 activity, so that eventually Hakerem and Sutton decided to add recording of the ERP to the pupillography laboratory. Some of the comparisons of pupillary and P300 findings are reviewed below.

Further study of constriction to light, and dilations evoked during cognitive tasks in normal subjects and schizophrenic patients has continued to be one of the primary aims of a second Biometrics Research Program that was established by Joseph Zubin in 1977, with Stuart Steinhauer, at the Highland Drive VA Medical Center in Pittsburgh. The research conducted in the Pittsburgh program was strongly influenced by the original findings in New York, and continued the collaborative efforts between the institutions (indeed, in order to be able to begin averaging procedures, Sutton loaned the original CAT 400 to the Pittsburgh laboratories for several years).

Cognitive Psychophysiology and Pupillography

Psychophysiological Measurement of Processing Effort, Capacity, and Information

Among those measures for which a correlate of both attentional effort and processing activities have been studied, perhaps the most widely emphasized is the pupillary dilation response (Beatty, 1982, 1986; Goldwater, 1972; Janisse, 1977). Pupil diameter enlarges with increasing effort during performance. This can be observed for purely mechanical effort, as when varying weights are picked up (Nunnally, Knott, Duchnowski & Parker, 1967) or even when a simple finger press occurs, in which both response preparation and execution contribute to the dilation (Richer, Silverman & Beatty, 1983). Mental effort has been manipulated by a number of means, including arithmetic problems of varying difficulty (often a typical "mental stress" paradigm), language-based tasks (including reading of material forward and backwards; Metalis, Rhoades, Hess & Petrovich, 1980), and especially the effect of increasing memory load during the digit span task, in which pupil diameter increases as the number of digits stored is increased (Kahneman & Beatty, 1966). Of special interest is that as maximum effective storage (judged by performance) is reached, pupillary dilation reaches a maximum (Peavler, 1974).

When memory is overloaded, the pupil may even decrease in diameter, suggesting that it is sensitive to both the extent of processing capacity as well as the breakdown of capacity (Poock, 1973). Kahneman (1973) relied heavily on results from pupillary experiments in the development of his treatise dealing with basic components of attention and effort.

Pupillary dilation can also be evoked by tasks in which there is little effort employed in recognizing a stimulus, but for which the "informational value" of the stimulus is high. Thus, simple click patterns show a quick habituation when the subject knows what each subsequent stimulus will be, but a clear dilation occurs to the clicks when the subject is asked to guess what stimulus pattern will occur (Hakerem, 1974). A variant of the missing stimulus paradigm first used by Sutton *et al.* (1967) for the study of P300 was also applied to pupillography. Findings indicated that when the subject was not certain whether a click would actually occur at a specific point in time, but the absence of a click indicated a particular outcome (*e.g.*, correct or incorrect, different amounts of monetary payoff), the "absence" of the stimulus itself elicited a pupillary dilation (Levine, 1969) which was related to the information conveyed by the stimulus absence.

Greater payoff will produce a dilation response of greater amplitude either when it is evoked (when the feedback cue is present) or emitted (when absence is the cue) (Steinhauer, 1982). FIGURE 1 presents average pupillary waveforms in which different amounts of money (0, 25, or 50 cents) were associated with each trial. The stimulus at S1 was a single 1 msec click. At S2, the informational stimulus was either the absence of a click (emitted) or an additional click identical to S1 (single click evoked). These events indicated whether the subject had won or lost money on those trials. The set of data on the bottom was recorded when a pair of clicks separated by 10 msec was presented instead at S2, indicating that no money was either won or lost.

Several aspects of the pupillary response are depicted in these data. First, regardless of whether the presence or absence of a click was used to provide critical feedback, the pupil showed a dilation reaching its widest diameter approximately 1200 msec after the time of the feedback. Second, the amplitude of dilation was related to the amount of money associated with the trial. Third, the extent of dilation was much greater when the amount of money involved was selected by the subject (Subject-Bet condition), rather than when the value was told to the subject as selected by a computer randomization prior to each trial (Computer-Bet condition). Finally, there was little dilation to conditions in which no money was involved, and the subject was merely told which stimulus would occur on the next trial (Certain condition). (Similar results were seen for the amplitude of the P300 component recorded simultaneously with the pupillary data.)

The phenomena of evoked and emitted pupillary responses directly parallel the initial findings for evoked (Sutton *et al.*, 1965) and emitted (Sutton, Tueting, Zubin & John, 1967) P300 components of event-related potentials, which were the templates for the design of several pupillary studies (Levine, 1969). A number of specific comparisons between ERP and pupillary data were subsequently conducted (Friedman, Hakerem, Sutton & Fleiss, 1973; Bock, 1976; Richer & Beatty, 1981; Steinhauer, 1982; Steinhauer & Zubin, 1982).

Following the demonstration by Patricia Tueting that P300 was inversely related to stimulus probability (Tueting, Sutton & Zubin, 1970), David Friedman simultaneously recorded pupil diameter and the vertex ERP during a guessing task, with the relative probabilities of two events changing across blocks (Friedman *et al.*, 1973). Friedman and colleagues compared pupillary and P300 data to the subjective probabilities (an interaction of the subject's guessing behavior and the stimulus

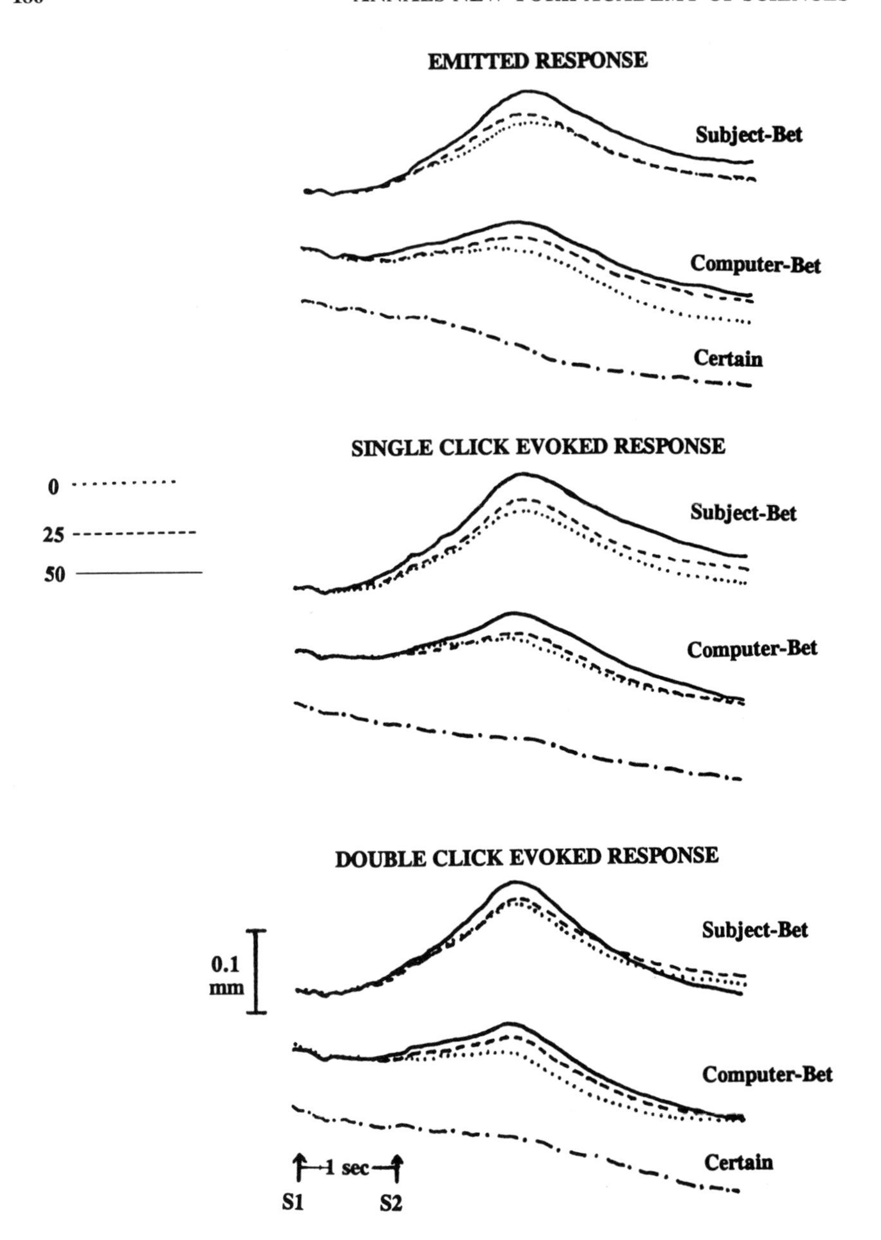

FIGURE 1. Pupillary dilation in 8 normal subjects during a task associated with monetary value and outcome. A single click at S1 was followed by one of three informational cues at S2: either no additional stimulus (emitted response), another identical click (single click evoked), or a pair of clicks (double click evoked). Little dilation was seen when the subject was told what each S2 would be (Certain). Pupillary dilation increased with the amount of money wagered on each trial (0, 25 or 50 cents), and was greater when the subject, rather than a computer, selected the values (adapted from Steinhauer, 1982).

probabilities), and found that the amplitudes of both dilation and P300 were inversely related to the subjective probabilities. Thus, larger amplitudes were seen for the least likely events. The same paradigm was employed by Bock (1976), who recorded pupillary and ERP data from monozygotic and dizygotic twin pairs, and from nontwin siblings.

A somewhat different approach was employed by Steinhauer and Zubin (1982), who used an auditory oddball procedure to record pupil diameter and ERPs from multiple scalp locations. The "oddball" in the typical experiment refers to the fact that two stimuli are presented with unequal probabilities. Unlike the tasks of Friedman *et al.* (1973) and Bock (1976), in which guessing was involved, the oddball task requires the subject to identify stimuli by either silently counting one of the stimuli, or by providing discriminative reactions (*e.g.*, a finger press) to one or both stimuli. It explicitly differs from the guessing paradigm both behaviorally and physiologically. Maximum P300 amplitudes are seen at vertex (Cz) during guessing paradigms, but are found at the midline parietal location (Pz) for post-hoc identification tasks such as the oddball paradigm. The guessing paradigm tends to produce larger pupillary dilations and P300 amplitudes than the oddball task, but at the cost of long intertrial intervals in order to provide prestimulus guesses and poststimulus reports by the subject. Thus, a limited number of individual trials (2–4/minute) can be obtained during test sessions, and this was problematic in our experiences testing schizophrenic patients. The guessing procedure could take several hours, and was especially fatiguing for patients. The oddball procedure, using continuous stimulus presentation at short intervals (*e.g.*, 1–3 sec intertrial intervals), greatly increases the number of trials recorded during a single session, and therefore results in less fatigue when employed with patients.

However, we did not wish to relinquish comparisons employed in the guessing task between "certain" and "uncertain" stimuli. In the ERP (Tueting *et al.*, 1970) and pupillary tasks (Friedman *et al.* 1973; Bock, 1976), subjects were recorded both during active guessing, and separately during blocks in which there was no guessing, but instead each stimulus was told to the subject before the trial began, and thus the subjects were "certain." In the certain condition, the stimuli elicit primarily sensory components (*e.g.*, N100, P200) and the pupil shows rapid habituation, with virtually no dilation (see FIG. 1, "Certain" conditions). P300 and the dilation response immediately emerge whenever guessing is reinstated. The difficulty in adopting the oddball paradigm was to incorporate the notion of "certainty" within the presentation of rare and frequent stimulus events. This was ultimately accomplished by the simple procedure of allowing only one rare stimulus (*e.g.*, a high tone) to occur at a time, without repetition. Furthermore, subjects were explicitly informed of this proviso—that two rare tones could not occur in a row. Consequently, even chronically hospitalized patients found it easy to answer the question "If you hear a high (rare) tone, what will the next sound be?" with the answer "a low." Thus, after every rare tone, the frequent tone could be predicted to occur with great cerainty (that is, a conditional probability of 1.0), while after any frequent tone, the subject was unsure whether the next sound would be one of the task-relevant rares or irrelevant frequents.

Average pupillary waveforms to these conditions indicated an inverse monotonic relationship with conditional event probability: the largest dilation occurred to the rare event, the least dilation to the predictable frequent tone, and an intermediate dilation to the unpredictable frequent tone (Steinhauer & Zubin, 1982; also Clark, Niemi & Balogh, 1985). Data from a recent sample of 19 normal control subjects, recorded in darkness, are shown in FIGURE 2. A similar relationship was observed between conditional probability and P300 amplitude (Steinhauer &

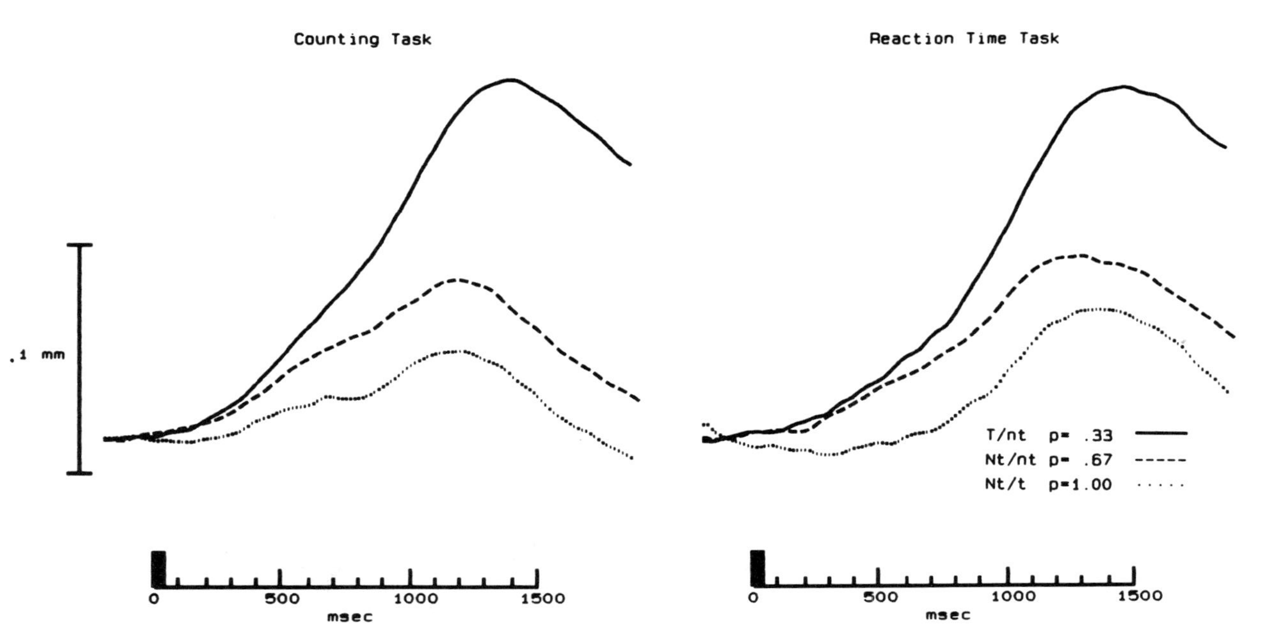

FIGURE 2. Average pupillary responses in 19 normal subjects. An infrequent high-pitched tone occurred one-third of the time after low-pitched tones [T/nt, $p = 0.33$]. On the other two-thirds of trials following a low tone, another low tone occurred [Nt/nt, $p = 0.67$]. After every high tone, only the low-pitched tone occurred [NT/t, $p = 1.00$]. Pupillary dilation was inversely related to these conditional probabilities both when the subjects counted only the high tones (Counting) and when the subjects made a different motor response to each high or low tone (Choice Reaction).

Zubin, 1982) and poststimulus cardiac acceleration (Steinhauer, Jennings, van Kammen & Zubin, in press), with parallel findings when reaction times, rather than counting, were required (Steinhauer, 1985). In FIGURE 2, the high probability events ($p = 0.67$ and $p = 1.00$) show greater dilation in the Choice Reaction task than for the Counting task. This is related to the fact that during Choice RT, every stimulus is relevant. Note, however, that for rare tones ($p = 0.33$), the dilations are similar across tasks (results for psychiatric patients are discussed below). Tonic diameters tend to decrease across experimental periods, reflecting habituation to the general test situation. Since the Counting task was always conducted first, it was not surprising to find that initial diameters were larger during the Counting than Choice Reaction task, which occurred later in the session. However, the lack of differences in dilation reinforce the notion that the amplitude of dilation is not strongly dependent on initial diameter, at least in the dark-adapted pupil.

While the majority of the studies incorporating both pupillary and ERP recording suggest strong parallels between the dilation and P300 responses, it would be erroneous to assume that the same process is always being manifest in the pupil and ERP. One class of exceptions will be noted. The presence of P300 has been related to information delivery (Sutton *et al.*, 1965), and loss of information, or equivocation, has been hypothesized as reducing P300 amplitude (Ruchkin & Sutton, 1978). Adams and Benson (1973) reported that decreasing differences in auditory stimulus intensity for feedback stimuli led to P300 reduction. Ruchkin and Sutton (1978) later interpreted the Adams and Benson study as follows: as the similarity between stimuli increased, and judgements of differences were made with less confidence, P300 decrement reflected loss of information or, in information theory terms, equivocation. In contrast, Kahneman & Beatty (1967) demonstrated that pupillary diameter increased, rather than decreased, as frequency differences between discriminative stimuli were reduced. The explanation for the dissociation between pupillary and ERP measures in these discrimination tasks is that in addition to processing of information, the pupil is also sensitive to effort, and the effort of making a difficult discrimination produces its own dilatation. This is presumably the source of the increased response in the Kahneman and Beatty (1967) experiment. Thus, it should be possible to demonstrate simultaneous dissociation of pupillary dilation and P300 amplitude in an auditory discrimination paradigm in which either frequency or intensity of stimuli are varied.

The activation of pupillary dilation as a prototypical component of the orienting reaction (Sokolov, 1963) has also led to some confusion in the interpretation of P300 mechanisms. Part of this confusion emanates from the fact that, as noted above, both pupillary diameter and P300 amplitude often covary in the same way in many studies. Combining these notions, Friedman (1978) suggested that since pupillary dilation was considered an aspect of orientation, and P300 and the dilation response were linked in several experiments, then a major aspect of P300 elicitation could be considered to be orienting activity. Subsequently, various discussions related orienting and P300 activity (*e.g.*, Donchin, Heffley, Hillyard, Loveless, Maltzman, Öhman, Rösler, Ruchkin & Siddle, 1984), and interpretations of orienting as a primary basis of P300 proliferated. Without detracting from those discussions, it seems worth noting that in them there was an unquestioned assumption that all signs of pupillary dilation in the tasks cited reflect orienting activity. Yet even the first association of pupil and P300 by Friedman and colleagues (1973) argued against this notion, since it was possible to eliminate dilation (and P300) merely by telling the subject which stimulus would be presented, yet dilation was elicited immediately when guessing was undertaken once again. In addition, the persistence, rather than habituation, of the dilation during task activity can be demonstrated. In an unpublished study, we (with David Friedman) once attempted to fatigue the pupil by continuing the guessing task for several hours: even after hundreds of trials, the subject (SRS) still showed dilation.

In dealing with the complexities of stimulus qualities which affect pupil diameter, it is worthwhile to take a brief look at one of the major controversies in pupillary research—the statements of Hess and colleagues (Hess & Polt, 1960; Hess, 1964) that positive affect is associated with dilation, while negative affect results in constriction. There have been many critical reviews of this research (*e.g.*, Janisse, 1977), as well as attempts by Hess and his students to justify the work (Hess, Beaver & Shrout, 1975). Two of the problems involved in using complex visual stimuli, which have usually been overlooked, will be mentioned.

The first consideration involves so-called control slides, which are typically presented before each stimulus slide. The notion in several studies was that the control and stimulus slides should be matched for brightness, so that no differential constriction to the slides could occur, and differences could only be attributable to the content of the target stimulus slide. This approach, however, takes a naive view of the physiology of the optic system, including the afferent pathway, even at the level of the retina. When stimuli of either different wavelengths or different intensities strike similar regions of the retina, they differentially stimulate receptors, which evoke pupillary constrictions. This was exquisitely demonstrated over two decades ago by Kohn and Clynes (1969): even matching for overall brightness did not eliminate sensory-related constrictions to the onset of different hues.

A second type of confounding is related to the pupillary constriction produced by the initial presentation of stimuli. This portion of the response was usually ignored by researchers employing pictures, who looked at average diameters over periods as long as ten seconds, rather than the specific dynamic responses to the pictures used. One exception to this was a study by Libby, Lacey, and Lacey (1973), whose data clearly showed the initial constriction resulting from stimulus presentations. In their study, pupillary dilation was most often seen to interesting pictures, and although several individuals were reported to show constriction, the unpleasant stimuli overall yielded larger dilations than pleasant stimuli—a finding totally at odds with the Hess formulation.

Our own research using complex pictorial stimuli was related to investigations of emotionality in neuropsychiatric patients. Patients with right hemisphere lesions have been reported to show reduced skin conductance responses to emotionally-laden images compared to patients with left hemisphere lesions (Morrow, Vrtunski, Kim & Boller, 1981). Pupillary reactions to the same stimuli were examined in 12 volunteer subjects. Each of 27 slides was preceded by a standard control slide, and 10 seconds of data were recorded at 50 msec intervals (Steinhauer, Boller, Zubin & Pearlman, 1983). Subsequently, it was possible to measure initial diameter, the point of maximum initial constriction which invariably followed slide onset, and subsequent maximum dilation. After each trial, the subject was asked to rate the slide on a five-point interval scale from very aversive to very pleasant. This in itself was unusual, since most previous studies had merely required subjects to look passively at stimuli and provide no judgements. In addition, a consensus judgement by laboratory staff was made for each stimulus. Responses were examined through a variety of techniques, including absolute maximum diameter, as well as maximum diameter after covarying for initial diameter or extent of constriction. Responses were analyzed first according to subjects' evaluations, and then according to the consensus evaluations. In all cases, the results were similar: the largest dilations were evoked by stimuli reported as most aversive or most pleasant, with smaller dilations to mildly unpleasant or pleasant stimuli, and the smallest dilation to neutral pictures. Thus, our own findings indicate that the level of emotional stimulation or interest, regardless of valence, is related to the pupillary dilation response, but the confounding effect of initial physiological reactions to visual stimuli must be carefully eliminated.

One of the more intriguing aspects of psychophysiological data is that there is clear evidence that familial similarity can be observed in tonic activity as well as in stimulus-evoked measures of cognitive activity (Boomsma & Gabrielli, 1985). Thus, Bock (1976), Surwillo (1980), Polich and Burns (1987), and Rogers and Deary (1991) have reported on event-related potential similarity in twins, and we have also noted high correlations for P300 amplitude even in nontwin siblings (Steinhauer, Hill & Zubin, 1986).

Patterns of pupillary dilation have also been examined among twin pairs in Hakerem's laboratory. Bock (1976) recorded pupillary dilation, comparing identical twins, fraternal twins, and nontwin siblings during a guessing task. Both objective numerical analyses of similarity, as well as judges' blind matching of pairs, indicated greater similarity of the pupil and ERP data for identical twins than for fraternal twins or nontwin siblings. In a recent dissertation, Gaudreau (1991) used a forced-choice procedure for matching pupillary waveforms, demonstrating significantly higher rates of matching identical compared to nontwin pairs across two different tasks. Studies of psychiatric patients and their relatives have also been carried out, and are described in the next section.

Psychopathology and Pupillary Motility

As noted at the beginning of this paper, pupillary abnormalities in psychiatric patients were well documented as early as the beginning of this century. With the advent of averaging techniques, it was possible to examine reactions of psychiatric patients with greater precision. In addition to the work of Rubin, already mentioned, Hakerem and colleagues in New York conducted a number of initial studies which indicated decreased light reactions and abnormal response latencies in schizophrenics (Hakerem & Lidsky, 1969; Hakerem, Sutton & Zubin, 1964; Lidsky, Hakerem & Sutton, 1971). In addition, patients exhibited difficulty in integrating irregular sequences of light pulses (Hakerem & Lidsky, 1975).

At the Pittsburgh laboratories, pupillary constriction to light has been recorded in schizophrenics and normals, examining intrasession variability and retest stability. In addition to replicating the widely reported finding of decreased constriction in schizophrenics, it has been possible to examine aspects of the response related to clinical state. In collaboration with Daniel P. van Kammen and Jeffrey L. Peters, twenty patients from the Schizophrenia Research Unit at the Highland Drive VA Medical Center in Pittsburgh were followed during initial neuroleptic treatment and subsequent (double-blind) drug-free withdrawal. Weekly recording of the pupillary light reaction was carried out for durations of up to six months. Stabilization on haloperidol resulted in a significantly larger amount of constriction (though relatively small) than during a subsequent drug-free period in patients. Thus, neuroleptic treatment appeared to normalize the response slightly, but generally the response remained below the mean for normals. During the drug-free period, the extent of constriction, already smaller in patients than normals, was found to be reduced even further as patients' psychosis ratings rose.

In addition, response characteristics were examined during haloperidol stabilization, prior to the subsequent drug-free phase. Compared to controls, schizophrenics tended to cease the active constriction process earlier after visual stimulation. Patients were grouped according to those who remained stable in the subsequent drug-free period vs those who would eventually relapse during the drug-free stage. The patients who would remain stable were observed to have reached the end of the constriction process significantly later—more like the normal response. This suggests that while amplitude of constriction may remain reduced, the latency to the end of constriction, while under haloperidol treatment, may predict the likelihood of relapse after withdrawal of medication.

Indicators of Vulnerability to Schizophrenia

A consortium of researchers led by Joseph Zubin, with the collaboration of Bonnie Spring and Samuel Sutton among others, worked on a conceptual framework and series of experimental approaches for dealing with the etiology of schizophrenia, as exemplified by the work on vulnerability (Zubin, Magaziner & Steinhauer, 1983; Zubin & Spring, 1977; Zubin & Steinhauer, 1981; Zubin, Steinhauer, Day & van Kammen, 1985). The vulnerability hypothesis postulates that the etiology of schizophrenia emerges from the interaction between genetic and other biological sources and environmental factors which influence the development of the individual. It is further postulated that schizophrenia, instead of being a chronically manifest disorder, is episodic in nature, but that the vulnerability to the disorder persists. A structure for interpreting the relationship of specific "markers" of vulnerability to schizophrenia has been developed (Zubin & Steinhauer, 1981), which is also applicable to other psychiatric disorders. Areas of research interest have included both psychosocial and neurophysiological aspects of schizophrenia, with the latter most relevant to the present discussion.

In contrast to a substantial literature dealing with pupillary light reactions in schizophrenics (and some discussion of the orienting response), there have been few studies of patients involving task-related dilation. Straube (1982) reported that schizophrenics exhibited larger dilations than controls during performance of the digit span task, which could be interpreted as an indication that patients employed greater effort than did controls. This represents an important finding which needs to be replicated. In our own laboratories, however, with different taks, we have observed decreased dilation amplitudes among schizophrenics compared to controls (Steinhauer, Hakerem & Spring, 1979; Steinhauer & Zubin, 1982).

In the New York Vulnerability Project (Steinhauer *et al.*, 1979), pupillary diameter and event-related potentials were recorded in 12 schizophrenic inpatients and 12 of their brothers, and in 15 control subjects during a task in which the subject guessed whether a sound or light would be presented on each trial, similar to the ERP paradigm originally employed by Levit, Sutton, and Zubin (1973). The schizophrenic patients showed decreased constriction to light as compared to controls, replicating previous studies. In addition, schizophrenics also showed greatly reduced dilations in response to unpredictable sounds, with no differences between correct and incorrect guesses. In contrast, controls and siblings showed vigorous dilations, which also differed following correct and incorrect guesses.

As noted earlier, pupillary and ERP (P300) amplitudes were found to be inversely related to the conditional probability of stimulus events during a counting task for normals. For inpatient schizophrenics, however, amplitudes were significantly smaller when compared to controls under all conditions, while responses for depressives tended to be intermediate (Steinhauer & Zubin, 1982; Steinhauer,

1985). FIGURE 3 illustrates grand means for a recent sample of 17 outpatient schizophrenics who were clinically stable at the time of testing. Note that while the amplitudes are greatly reduced compared to controls (FIG. 2), there is still some pupillary motility, suggesting that while neuroleptic treatment in these patients may affect amplitude, it is unlikely to account for the failure to differentiate between conditions.

In addition, the pattern of responses across conditions for patients suggested that schizophrenics responded more to a change in physical stimulus from trial to trial than to conditional probability of stimulus events, as was the case with normals. Specifically, more than half the patients showed a larger response to the frequent tone when it predictably followed an infrequent target, as compared to the same tone when it was not predictable, occurring with a decreased sequential probability (Steinhauer & Zubin, 1982). This is a pattern which was seen more recently among those brothers of patients where the brothers met criteria for Axis II schizotypal personality disorder (see FIG. 4), as described in greater detail below.

Data have been analyzed on a group of 33 brothers of schizophrenics, including brothers of the outpatient schizophrenic group reported above. The pattern of responding to stimulus change rather than event probability, which first was described for inpatient schizophrenics (Steinhauer & Zubin, 1982), was observed for many of the schizophrenic outpatient probands, but was not readily apparent in the averages for the entire sibling group. Moreover, siblings had significantly larger dilations than the patients for the Counting and Choice Reaction tasks. However, when the siblings were separated into groups according to Axis I and II diagnoses, the group of 5 brother meeting criteria for schizotypal personality disorder showed the abnormal pattern of responding—that is, they were hyperresponsive to stimulus change rather than conditional probability.

Data for five brothers are presented in FIGURE 4. All of these siblings were diagnosed as having schizotypal personality disorder. There are clear dilations present to all conditions. However, a careful inspection indicates that for the two types of frequent low tones, the subjects did not make use of the predictability of events. Instead, the subjects show greater dilation to the predictable low tones which followed high tones (dotted line) than to the unpredictable low tones (dashed lines). It is this pattern which we have interpreted as reflecting an overreaction to stimulus change, rather than to sequential probability. Thus, the pattern of responding appears to be a promising measure not only in schizophrenics, but also in their siblings.

Cardiac changes have also been observed to vary with probability changes in normals. Schizophrenics, compared to control subjects, showed reduced anticipatory deceleration as well as decreased post-stimulus acceleration during the counting task; patients also showed faster resting heart rate (Steinhauer, Jennings, van Kammen, & Zubin, in press).

These studies suggest that pupillary deviations are observed in schizophrenics for tasks in the information processing domain, but are not merely a reflection of a lack of effort or lack of potential pupillary change due to physiological limitations such as reduced tonic pupillary diameters. We believe that a pupillary study in which variations in effort, processing capacity, and information utilization are examined in the same patients will be of major importance in clarifying the contributions of each of these factors in the performance of schizophrenics.

Several questions are raised in reviewing the pupillary and other psychophysiological data from schizophrenic patients. (For a full review of pupillary reactions in schizophrenia, see Zahn, Frith & Steinhauer, 1991.) Schizophrenics show dimin-

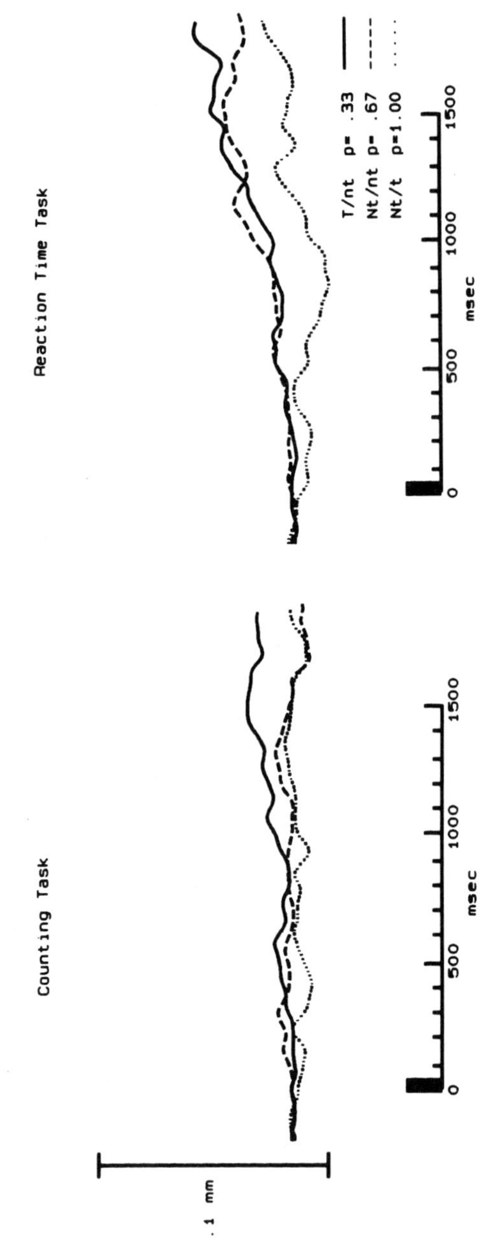

FIGURE 3. Pupillary responses of 17 outpatient schizophrenics during the Counting and Choice Reaction tasks.

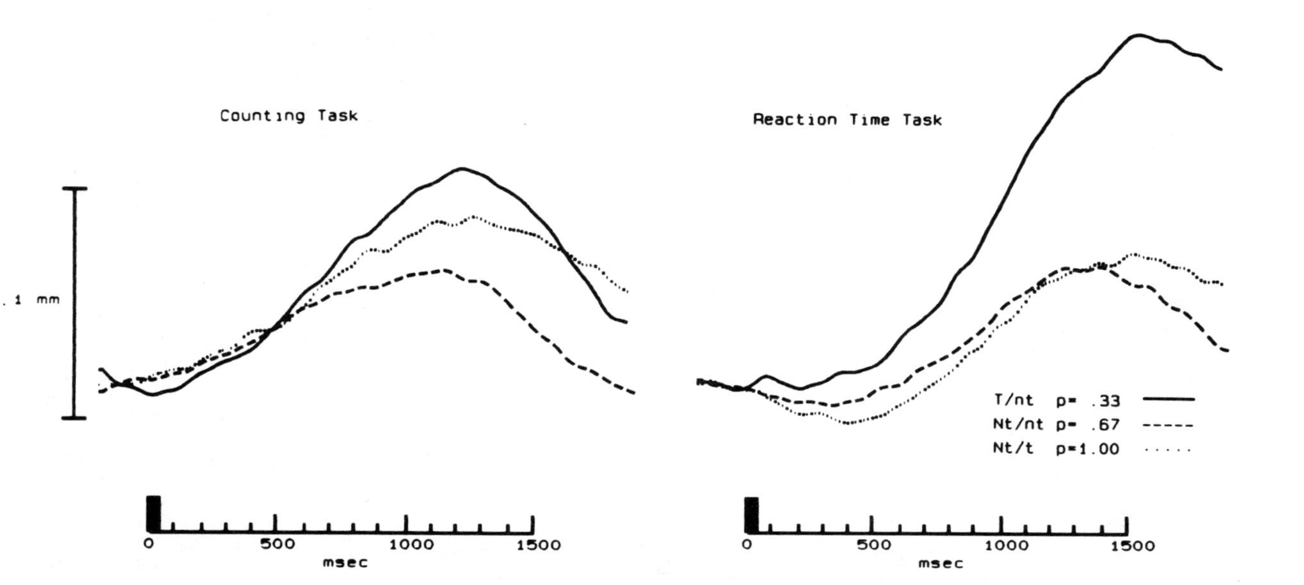

FIGURE 4. Pupillary response of five brothers of schizophrenic probands in the Counting and Choice Reaction tasks. All of these subjects met criteria for a diagnosis of schizotypal personality disorder. Note the greater dilation to frequent low tones that are predictable (*dotted line*) than unpredictable (*dashed line*) in both tasks.

ished response amplitudes for a wide variety of psychophysiological responses (for reviews, see Spohn & Patterson, 1979; Zahn, 1986). For example, the finding of a decreased amplitude for the P300 component of the ERP in schizophrenics is one of the most robust findings in the literature (cf. Roth, Tecce, Pfefferbaum, Rosenbloom & Callaway, 1984; Zahn, 1986; Zubin, Sutton & Steinhauer, 1986). It is not clear whether these indicators in schizophrenics point to a lack of capacity specific to information processing demands, or whether they signify more basic, general nonresponsiveness of physiological systems. What has often been overlooked is the question of whether or not the mechanics of the response system appear to be intact; it is possible to test the integrity of responses, independent of reactions elicited during information processing tasks. For example, the decreased pupillary constriction of schizophrenics in response to light has been reported even in patients who showed a normal constriction of the pupil during visual accommodation (Okada, Kase & Shintomi, 1978). Changes in bodily position (*e.g.*, going from a sitting to standing position) will normally increase heart rate, even for schizophrenics receiving neuroleptics (Castellani, Ziegler, van Kammen, Alexander, Siris & Lake, 1982). However, this has not yet been examined in cardiac studies of schizophrenics who at other times were also performing tasks. Even the P300 response, which usually is evoked during cognitive tasks (Donchin, 1979; Sutton, 1979; Sutton *et al.*, 1965) can be elicited without a task demand: Roth, Dorato, and Kopell (1984) have shown that very loud auditory stimuli evoke what appears to be a true, though involuntary, P300 response at the scalp. Thus, the integrity of physiological underpinnings of the pupillary dilation or "P300" systems could be tested even in patients who fail to show a large response during typical information processing tasks. Experiments are needed to determine whether some of the findings observed in schizophrenics may be attributed to either a general lack of ability to mobilize sufficient effort in approaching tasks, or to deficits primarily in processing capacity.

Further puzzles are suggested by findings of deviant response patterns in siblings of the patients. Are there only single measures which are likely to be markers, or are there patterns of indicators which characterize identifiable subgroups? Do those indicators which the siblings share with their proband brothers show an association of response deviance with history of psychopathology, or only to history of schizophrenia spectrum disorder, or is there familial association without regard to personal history?

It has also been suggested that schizophrenics may respond more than controls to tasks requiring physical or general mental effort. Do patients consistently show more of a pupillary dilation (and/or concomitant cardiac changes) during physical tasks, or during tasks requiring effort, such as the digit span (as indicated by the data of Straube, 1980)? If this were to be clearly established, it would negate the notion that patients are reticent to engage in tasks, and in fact would imply that some patients attempt to mobilize greater effort than controls in performance of tasks. If pupillary dilation in patients occurs during presentation of short digit strings, but shows a decrease in diameter when moderate string lengths are used, it would imply that the available capacity for storing information is decreased in the patients.

In contrast, decreased pupillary dilation is consistently observed in schizophrenic patients during tasks which require relatively little effort for adequate discrimination or performance, involving easily perceived and encoded stimuli. In such tasks, however, the dilation observed in normal subjects gives evidence that the information provided by the stimulus is analyzed in more complex forms than is demanded by the experimenter. For example, in our counting and choice reaction tasks, normal subjects show automatic encoding of sequential probabilities. The patients' responses imply that there is a difficulty in utilizing this information on event probability at a stage in information processing occurring after the registration of the stimulus.

What is the nature of the mechanisms which are responsible for these effects? The bases for many of the response deviations observed in schizophrenia tend to be unknown. Even for the normal subject's pupillary dilation response, for example, the relative contributions of the sympathetic and parasympathetic components of the autonomic nervous system are unclear, since dilation may result from sympathetic activation as well as reciprocal parasympathetic inhibition. Similarly, the reduced light reaction in patients could involve portions of decreased parasympathetic activity as well as increased sympathetic inhibition. A concern with determining the contributions of these systems, and review of the literature based on normal subjects, has led to a series of hypotheses regarding autonomic contributions to pupillary dynamics during information processing tasks. These are discussed in the next section.

Delineating Separate Autonomic Contributions to Pupillary Dynamics during Cognitive Processing

There has been only limited clarification of the mechanisms that are reflected in changes in pupillary diameter during performance of cognitive tasks, either from the conceptual or neurophysiological point of view. A model has been developed which proposes to evaluate the relative contributions of the two separate autonomic nervous system components in the expression of information-related changes in the pupillary system (Steinhauer, in preparation). This discussion summarizes the model, and methods for testing its validity.

A survey of the literature has indicated a number of trends which suggest that it may be possible to dissect out the components of the autonomic nervous system which are active in pupillary dilation and constriction in normals. Relevant facts include the following:

1) During a variety of cognitive processing tasks, when conducted in darkness with or without a required motor response, maximum pupillary dilation occurs with a peak latency of approximately 1200 msec. This response is also related to similar changes in P300 (Friedman *et al.*, 1973; Hakerem, 1974; Steinhauer & Zubin, 1982). Both types of changes occur even for responses emitted following stimulus absence, when that absence conveys information to the subject (Levine, 1969; Steinhauer, 1982; Sutton *et al.*, 1967).

2) In animal preparations, there is relatively little parasympathetic tone in darkness (Loewenfeld, 1958), so that most reflex dilatation in darkness has been attributed to sympathetic activity.

3) During tasks involving a clear motor-response component, carried out in the light, dilations tend to occur with an earlier peak (600–900 msec (Richer *et al.*, 1983)). Using auditory stimulation in light, Beatty (1989) has also demonstrated that an irrelevant stimulus in an attended channel produces an extremely small (approximately 0.01-mm) dilation with a peak latency of approximately 600 msec.

4) An auditory stimulus produces a dilation with a long latency (*i.e.*, 1200 msec) in darkness. As ambient light increases, an additional, earlier dilation can be seen. Shiga & Ohkubo (1979) observed only the long-latency peak during dark recording, but when they increased room illumination so that the initial diameter of the pupil was less than 4 mm in diameter, an earlier peak (700 msec) also appeared.

5) The pupillary light reflex can be *diminished* by psychosensory stimulation (Loewenfeld, 1958). We observed that even during a guessing task, resulting in a cognitive load, the light reaction (which reached minimum diameter at 800 msec) was reduced compared to conditions in which no task was imposed (Steinhauer *et al.*, 1979).

The above data can be interpreted as follows: the early dilation observed in tasks performed under light, and reduced light reactions for subjects under stress or while performing a task, reflect active inhibition of the parasympathetic efferent pathway (occurring maximally during 600–900 msec poststimulus). This is probably due to descending cortical and ascending reticular inhibitory inputs to the Edinger-Westphal complex of the oculomotor nucleus (N. III).

The major component of dilation elicited during cognitive processing activities occurs later, with a peak latency greater than 1100 msec. This is probably a primary sympathetic component, which is mediated through hypothalamic pathways via the cervical sympathetic chain.

We suggest that the typical pupillary dilation response during cognitive activation is the result of these two separate components, depicted in FIGURE 5. The first, early dilation component is associated with inhibition of the central parasympathetic pathway, leading to relaxation of the pupillary sphincter muscles. The sphincter pupillae form a band of muscles arranged in circular orientation around the pupillary margin; contraction of the muscles results in pupillary constriction. Given the well-known specificity of the parasympathetic system, the model assumes a relatively symmetrical onset and offset for this component. The second, later component is produced by direct sympathetic stimulation of the pupillary dilator muscles. The dilator pupillae are oriented in a radial fashion, so that contraction of the muscles enlarges the pupillary aperture. Once peak activity is reached, it is known that sympathetic activation takes longer to return to baseline (Loewenfeld, 1958; Lowenstein & Loewenfeld, 1962).

In FIGURE 5A, the dilation resulting from performance in a cognitive task has been modelled for recordings obtained in darkness. Dilation resulting from inhibition of parasympathetic centers is presumed to contribute a relatively small dilatory component, reaching its maximum at 600 msec. The sympathetic component increments somewhat later. The resulting waveform appears to be primarily sympathetic in nature, although a break at or slightly beyond 600 msec reflects the interaction of both components. Such waveforms are typically seen in guessing or oddball tasks recorded in darkness. For example, the data for control subjects performing the counting task seem to exhibit a similar pattern in the 0.67 and 1.00 probability conditions (FIG. 2).

During recording in bright ambient light conditions, the tonic pupil diameter would be decreased by parasympathetic stimulation. Thus, there should be a greater dilation produced by inhibition of this system. Consequently, in FIGURE 5B, the gain of the parasympathetic component has been increased, while the sympathetic component has been left constant. The resulting pupillographic record depicts both the later (sympathetic) dilation as well as an earlier peak, which reaches a maximum diameter between 600 and 1000 msec. This model fits well with the findings of earlier peaks for pupillary data recorded under normal room illumination (Richer *et al.*, 1983; Shiga & Ohkubo, 1979).

Moreover, we hypothesize that during performance of motor responses, the primary component of dilation associated with the motor response is parasympathetic inhibition having its effect on the early (600 msec) dilation. Richer *et al.* (1983) depict an early peak for averages time-locked to a motor response, which is entirely consistent with the present hypothesis. Consequently,

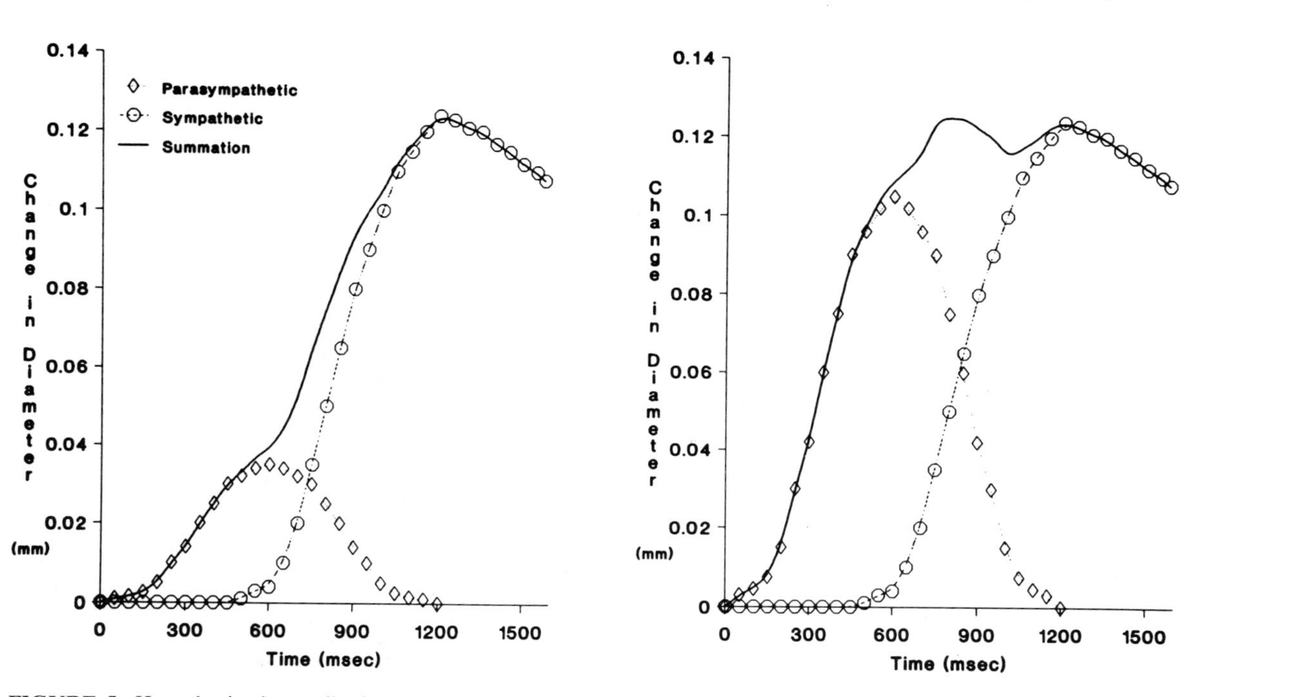

FIGURE 5. Hypothesized contributions of parasympathetic and sympathetic components of activation in the generation of pupillary dilation during cognitive tasks. In darkness **(A)**, there is minimal early dilation resulting from inhibition of parasympthetic pathways, and consequent relaxation of the sphincter pupillae. The primary dilation occurs later through sympathetic activation of the dilator pupillae. In light **(B)**, the greater tonic constriction of the sphincter muscles results in more extensive early dilation due to parasympathetic inhibition, which contributes to an additional early peak in the response.

the motor component should show only a minimum contribution to pupillary dilation responses recorded in darkness.

If we can demonstrate that these relations hold true across a variety of different tasks, then we will have established alternative procedures for testing the integrity of the separate sympathetic and parasympathetic components of the autonomic nervous system. This would be especially useful in examining the role of separate pathways in patient populations.

Studies by Beatty, Richer, and colleagues (Beatty, 1982; Richer *et al.*, 1983) have examined the contribution of motor responses in eliciting pupillary dilation, and we were inspired to investigate the effects of motor activity on the dilation produced during our own paradigms. We conducted a pilot experiment in which five subjects performed the counting task (Steinhauer & Zubin, 1982), with the addition that subjects were requested to clench their teeth when target tones occurred. We suspected that there might be different contributions in light and darkness, so the task was performed under both conditions of room lighting. For all subjects, we observed an extensive early dilation when the room was lit, but in darkness, we primarily observed only a later, lower amplitude dilation. We interpreted these findings as indicating differential activity contributing to the dilation response: a late sympathetic component in darkness, as well as an early dilation related to parasympathetic inhibition during the lighted room condition. The parasympathetic component, in this case, is presumably related to the motor activity required from the subject.

Several additional predictions emanate from the proposed model. If early dilations (600–900 msec peak) are related to parasympathetic inhibition, then this early component should be more prominent under all tasks performed in light, but less so (or not at all) in darkness. However, the inhibitory effect of ongoing cognitive activity should reduce the extent of pupillary constriction to light which is elicited simultaneously with the cognitive task. For example, the pupillary reaction to light pulses should be decreased during cognitive tasks such as the digit span. If motor activity results primarily in inhibition of the final parasympathetic pathway, then tasks involving motor responses should result in little early dilation activity when performed in darkness. Moreover, if some tasks primarily involving mental effort but no motor response (*i.e.*, digit span) show little dilation when performed in darkness, then it would be possible to identify such tasks with activity involving the parasympathetic rather than the sympathetic components of the autonomic nervous system; when both occur, it may be possible to estimate the change attributable to each component. Loewenfeld (personal communication) has pointed out that similar verification can be achieved by pharmacologically blocking postsynaptic activity of either the dilator or sphincter muscles of one iris; by comparing the response of the treated eye to the untreated eye (using binocular pupillographic recording) one can obtain a "pure" measure of the remaining autonomic component.

The investigation of the pupil as a window into the mechanisms of cognitive processes, and as a reflection of vulnerability to psychopathology, is proceeding. In a conversation with Sutton about the meaning and basis of such measures—why they should be so responsive, and what specific physiological changes they represented—he once remarked that aside from the new technologies, it may still take another century to understand the complexities of what we are measuring. The path along which he directed us involves the search for psychological and physiological understanding, keeping in sight the wonder of what these measures represent for the complexities of physiological and mental function.

ACKNOWLEDGMENTS

The authors acknowledge the major contributions of their late colleagues Samuel Sutton and Joseph Zubin to the establishment of the Biometrics Research Programs in New York and Pittsburgh, and to the work conducted in the pupillography laboratories of those programs, much of which forms the basis for this paper.

REFERENCES

Adams, J. C. & D. A. Benson. 1973. Task-contingent enhancement of the auditory evoked response. Electroencephalogr. Clin. Neurophysiol. **35:** 249–257.

Bach, L. 1908. Pupillenlehre. Anatomie, Physiologie und Pathologie. Methodik der Untersuching. Karger. Berlin.

Beatty, J. 1982. Phasic not tonic pupillary responses vary with auditory vigilance performance. Psychophysiology **19:** 167–172.

Beatty, J. 1982. Task-evoked pupillary responses, processing load, and the structure of processing resources. Psychol. Bull. **91:** 276–292.

Beatty, J. 1986. The pupillary system. *In* Psychophysiology: Systems, Process, and Applications. G. H. Coles, E. Donchin, & S. W. Porges, Eds. 43–50. Guilford Press. New York.

Beatty, J. 1989. Pupillometric signs of selective attention in man. *In* Neurophysiology and Psychology: Basic Mechanisms and Clinical Applications. E. Donchin, G. Galbraith, & M. L. Kietzman, Eds. 138–143. Academic Press. New York.

Bock, F. A. 1976. Pupillary Dilation and Vertex Evoked Potential Similarity in Monozygotic and Dizygotic Twins and Siblings. Doctoral dissertation, City University of New York. Diss. Abstr. Int. **36:** 6432B.

Boomsma, D. I. & W. F. Gabrielli, Jr. 1985. Behavioral genetic approaches to psychophysiological data. Psychophysiology **22:** 249–260.

Bumke, O. 1904. Die Pupillenstörunge, bei Geistes- und Nervenkrankheiten (Physiologie und Pathologie der Irisbewegungen). Fischer. Jena.

Castellani, S., M. G. Ziegler, D. P. van Kammen, P. E. Alexander, S. G. Siris & C. R. Lake. 1982. Plasma norepinephrine and dopamine-beta-hydroxylase activity in schizophrenia. Arch. Gen. Psychiatry **39:** 1145–1149.

Clark, W. R., C. A. Niemi & D. W. Balogh. 1985. Task-evoked pupillary response to a conditional probability counting task. Psychophysiology **22:** 586–587.

Donchin, E. 1979. Event-related potentials: a tool in the study of human information processing. *In* Evoked Brain Potentials and Behavior. H. Begleiter, Ed. 13–88. Plenum Press. New York.

Donchin, E., E. Heffley, S. A. Hillyard, N. Loveless, I. Maltzman, A. Öhman, F. Rösler, D. Ruchkin & D. Siddle. 1984. Cognition and event-related potentials: II. The orienting reflex and P300. *In* Brain and Information: Event-Related Potentials. R. Karrer, H. Cohen & P. Tueting, Eds. Annals of the New York Academy of Sciences. Vol. 424: 39–57). New York Academy of Sciences. New York.

Friedman, D. 1978. The late positive component and orienting behavior. *In* Multidisciplinary Perspectives in Event-Related Brain Potential Research. D. A. Otto, Ed. 178–180. Environmental Protection Agency. Washington, DC.

Friedman, D., G. Hakerem, S. Sutton & J. L. Fleiss. 1973. Effect of stimulus uncertainty on the pupillary dilation response and the vertex evoked potential. Electroencephalogr. Clin. Neurophysiol. **34:** 475–484.

Gaudreau, L. 1991. Event-Related Brain Potentials and Pupillary Responses Using a Cognitive Task in Monozygotic Twins. Unpublished doctoral dissertation, City University of New York.

Goldwater, B. 1972. Psychological significance of pupillary movements. Psychol. Bull. **77:** 340–355.

Hakerem, G. 1967. Pupillography. *In* A Manual of Psychophysiological Methods. P. H. Venables & I. Martin, Eds. 335–349. North-Holland Publishing Co. Amsterdam.

HAKEREM, G. 1974. Conceptual stimuli, pupillary dilation and evoked cortical potentials: a review of recent advances. *In* Pupillary Dynamics and Behavior. M.-P. Janisse, Ed. 135–158. Plenum Press. New York.
HAKEREM G. & A. LIDSKY. 1969. Pupillary reactions to sequences of light and variable dark pulses. Ann. N. Y. Acad. Sci. **156:** 951–958.
HAKEREM G. & A. LIDSKY. 1975. Characteristics of pupillary reactivity in psychiatric patients and normal controls. *In* Experimental Approaches to Psychopathology. M. L. Kietzman, S. Sutton & J. Zubin, Eds. 61–72. Academic Press. New York.
HAKEREM, G. & S. SUTTON. 1966. Pupillary response at visual threshold. Nature **212:** 485–486.
HAKEREM, G., S. SUTTON & J. ZUBIN. 1964. Pupillary reactions to light in schizophrenic patients and normals. Ann. N. Y. Acad. Sci. **105:** 820–831.
HESS, E. H. 1964. Attitude and pupil size. Sci. Am. **212:** 46–54.
HESS, E. H. 1972. Pupillometrics: a method of studying mental, emotional, and sensory processes. *In* Handbook of Psychophysiology. N. S. Greenfield & R. A. Sternbach, Eds. 491–531. Holt, Rinehart & Winston. New York.
HESS, E. H., P. W. BEAVER & P. E. SHROUT. 1975. Brightness contrast effects in a pupillometric experiment. Percept. Psychophys. **18:** 125–127.
HESS, E. H. & J. M. POLT. 1960. Pupil size as related to interest value of visual stimuli. Science **132:** 349–350.
HESS, E. H.& J. M. POLT. 1964. Pupil size in relation to mental activity during simple problem-solving. Science **143:** 1190–1192.
JANISSE, M.-P. 1977. Pupillometry: the Psychology of the Pupillary Response. Hemisphere Publishing Co. Washington, D.C.
KAHNEMAN, D. & J. BEATTY. 1966. Pupil diameter and load on memory. Science **154:** 1583–1585.
KAHNEMAN, D. & J. BEATTY. 1967. Pupillary response in a pitch discrimination task. Percept. Psychophys. **2:** 101–105.
KAHNEMAN, D. 1973. Attention and Effort. Prentice-Hall. Englewood Cliffs, NJ.
KOHN, M. & M. CLYNES. 1969. Color dynamics of the pupil. *In* Rein Control, or Unidirectional Rate Sensitivity, a Fundamental Dynamic and Organizing Function in Biology. M. Clynes, Ed. Annals of the New York Academy of Sciences. Vol. 156: 931–950. New York Academy of Sciences. New York.
LEVINE, A. & P. SCHILDER. 1942. The catatonic pupil. J. Nerv. Ment. Dis. **96:** 1–12.
LEVINE, S. 1969. Pupillary Dilation as a Function of Stimulus Uncertainty. Unpublished master's thesis, Queens College of the City University of New York.
LEVIT, R. A., S. SUTTON & J. ZUBIN. 1973. Evoked potential correlates of information processing in psychiatric patients. Psychol. Med. **3:** 487–494.
LIBBY, W. L., JR., B. C. LACEY & J. I. LACEY. 1973. Pupillary and cardiac activity during visual attention. Psychophysiology **73:** 270–294.
LIDSKY, A., G. HAKEREM & S. SUTTON. 1971. Pupillary reactions to single light pulses in psychiatric patients and normals. J. Nerv. Ment. Dis. **153:** 286–291.
LOEWENFELD, I. E. 1958. Mechanisms of reflex dilatation of the pupil. Doc. Ophthalmol. **12:** 185–448.
LOEWENFELD, I. E. 1963. The iris as a pharmacologic indicator. Arch. Ophthalmol. **70:** 42–51.
LOWENSTEIN, O. & I. E. LOEWENFELD. 1952. Disintegration of central autonomic regulation during fatigue and its reintegration by psychosensory controlling mechanisms. I. Disintegration. Pupillographic studies. J. Nerv. Ment. Dis. **115:** 1–21.
LOWENSTEIN, O. & I. E. LOEWENFELD. 1962. The pupil. *In* The Eye. H. Davson, Ed. Vol. 3: 231–267. Academic Press. New York.
LOWENSTEIN, O. & A. WESTPHAL. 1933. Experimentelle und klinische Studien zur Physiologie der pupillenbewegungen. Karger. Berlin.
MAY, P. R. A. 1948. Pupillary abnormalities in schizophrenia during muscular effort. J. Ment. Sci. **94:** 89–98.
METALIS, S. A., B. K. RHOADES, E. H. HESS & S. B. PETROVICH. 1980. Pupillometric assessment of reading using materials in normal and reversed orientation. J. Appl. Psychol. **65:** 359–363.

MORROW, L., P. B. VRTUNSKI, Y. KIM & F. BOLLER. 1981. Arousal responses to emotional stimuli and laterality of lesion. Neuropyschologia **19:** 65–71.

NUNNALLY, J. C., P. D. KNOTT, A. DUCHNOWSKI & R. PARKER. 1967. Pupillary response as a general measure of activation. Percept. Psychophys. **2:** 149–150.

OKADA, F., M. KASE & Y. SHINTOMI. 1978. Pupillary abnormalities in schizophrenic patients during long-term administration of psychotropic drugs: dissociation between light and near reactions. Psychopharmacology **58:** 235–240.

PEAVLER, W. S. 1974. Pupil size, information overload, and performance differences. Psychophysiology **11:** 559–566.

POLICH, J. & T. BURNS. 1987. P300 from identical twins. Neuropsychologia **25:** 299–304.

POOCK, G. K. 1973. Information processing vs. pupil diameter. Percept. Mot. Skills **37:** 1000–1002.

RICHER, F. & J. BEATTY. 1981. Cognitive load affects early brain potential indicators of perceptual processing. Proceedings of the Third Annual Conference of the Cognitive Science Society. 323–325.

RICHER, F., C. SILVERMAN & J. BEATTY. 1983. Response selection and initiation in speeded reactions: a pupillometric analysis. J. Exp. Psychol. Hum. Percept. Perform. **9:** 360–370.

ROGERS, T. D. & I. DEARY. 1991. The P300 component of the auditory event-related potential in monozygotic and dizygotic twins. Acta Psychiatr. Scand. **83:** 412–416.

ROTH, W. T., K. H. DORATO & B. S. KOPELL. 1984. Intensity and task effects on evoked physiological responses to noise bursts. Psychophysiology **21:** 466–481.

ROTH, W. T., J. J. TECCE, A. PFEFFERBAUM, M. ROSENBLOOM & E. CALLAWAY. 1984. ERPs and psychopathology: I. Behavioral process issues. *In* Brain and Information: Event-Related Potentials. R. Karrer, J. Cohen & P. Tueting, Eds. Annals of the New York Academy of Sciences. Vol. 25: 496–522. New York Academy of Sciences. New York.

RUBIN, L. S. 1960. Pupillary reactivity as a measure of autonomic balance in the study of psychotic behavior: a rational approach to chemotherapy. Trans. N. Y. Acad. Sci. **22:** 509–518.

RUBIN, L. S. 1961. Patterns of pupillary dilatation and constriction in psychotic adults and autistic children. J. Nerv. Ment. Dis. **133:** 130–142.

RUBIN, L. S. 1974. The utilization of pupillometry in the differential diagnosis and treatment of psychotic and behavioral disorders. *In* Pupillary Dynamics and Behavior. M.-P. Janisse, Ed. 75–134. Plenum Press. New York.

RUBIN, L. S. & T. J. BARRY. 1968. Autonomic fatigue in psychosis. J. Nerv. Ment. Dis. **147:** 211–222.

RUBIN, L. S. & T. J. BARRY. 1972a. The effect of the cold pressor test on pupillary reactivity of schizophrenics in remission. Biol. Psychiatry **5:** 181–197.

RUBIN, L. S. & T. J. BARRY. 1972b. The reactivity of the iris muscles as an index of autonomic dysfunction in schizophrenic remission. J. Nerv. Ment. Dis. **155:** 265–276.

RUBIN, L. S. & T. J. BARRY. 1976. Amplitude of pupillary contraction as a function of intensity of illumination in schizophrenia. Biol. Psychiatry **11:** 276–282.

RUCHKIN, D. S. & S. SUTTON. 1978. Equivocation and P300. *In* Multidisciplinary Perspectives in Event-Related Brain Potential Research. D. A. Otto, Ed. 175–177. Environmental Protection Agency. Washington, DC.

SHIGA, N. & Y. OHKUBO. 1979. Pupillary responses to auditory stimuli—a study about the change of the pattern of the pupillary reflex dilation. Tohoku Psychol. Folia **38:** 57–65.

SOKOLOV, E. 1963. Perception and the Conditioned Reflex. Macmillan. New York.

SPOHN, H. E. & T. P. PATTERSON. 1979. Recent studies of psychophysiology in schizophrenia. Schizophr. Bull. **5:** 581–611.

STEINHAUER, S. R. 1982. Evoked and Emitted Pupillary Responses and Event-Related Potentials as a Function of Reward and Task Involvement. Unpublished doctoral dissertation, City University of New York.

STEINHAUER, S. R. 1985. Neurophysiological aspects of information processing in schizophrenia. Psychopharmacol. Bull. **21:** 513–517.

STEINHAUER, S. R., F. BOLLER, J. ZUBIN & S. PEARLMAN. 1983. Pupillary dilation to emotional visual stimuli revisited. Psychophysiology **20:** 472.

STEINHAUER, S., G. HAKEREM & B. SPRING. 1979. The pupillary response as a potential indicator of vulnerability to schizophrenia. Psychopharmacol. Bull. **15:** 44–45.

STEINHAUER, S. R., S. Y. HILL & J. ZUBIN. 1986. Event-related potential similarity in adult siblings. Psychophysiology **23:** 464.

STEINHAUER, S. R., J. R. JENNINGS, D. P. VAN KAMMEN & J. ZUBIN. Beat-by-beat cardiac responses in normals and schizophrenics to events varying in conditional probability. Psychophysiology. In press.

STEINHAUER, S. R. & J. ZUBIN. 1982. Vulnerability to schizophrenia: information processing in the pupil and event-related potential. *In* Biological Markers in Psychiatry and Neurology. E. Usdin & I. Hanin, Eds. 371–385. Pergamon Press. Oxford.

STRAUBE, E. R. 1980. Reduced reactivity and psychopathology—examples from research on schizophrenia. *In* Functional States of the Brain: Their Determinants. J. Koukkou, D. Lehmann & J. Angst, Eds. 291–307. Elsevier. Amsterdam.

STRAUBE, E. R. 1982. Pupillometric, cardiac, and electrodermal reactivity of schizophrenic patients under different stimulus conditions. Psychophysiology **19:** 140–141.

SURWILLO, W. W. 1980. Cortical evoked potentials in monozygotic twins and unrelated subjects: comparisons of exogenous and endogenous components. Behav. Genet. **10:** 201–209.

SUTTON, S. 1979. P300—thirteen years later. *In* Evoked Brain Potentials and Behavior. H. Begleiter, Ed. 107–126. Plenum Press. New York.

SUTTON, S., M. BRAREN, J. ZUBIN & E. R. JOHN. 1965. Evoked-potential correlates of stimulus uncertainty. Science **150:** 1187–1188.

SUTTON, S., P. TUETING, J. ZUBIN & E. R. JOHN. 1967. Information delivery and the sensory evoked potential. Science **155:** 1436–1439.

TUETING, P., S. SUTTON & J. ZUBIN. 1970. Quantitative evoked potential correlates of the probability of events. Psychophysiology **7:** 385–394.

ZAHN, T. P. 1986. Psychophysiological approaches to psychopathology. *In* Psychophysiology: Systems, Process, and Applications. M. G. H. Coles, E. Donchin & S. W. Porges, Eds. 508–610. Guilford Press. New York.

ZAHN, T. P., C. D. FRITH & S. R. STEINHAUER. 1991. Autonomic functioning in schizophrenia: electrodermal activity, heart rate, pupillography. *In* Handbook of Schizophrenia. Vol. 5. Neuropsychology, Psychophysiology and Information Processing. S. R. Steinhauer, J. H. Gruzelier & J. Zubin, Eds. 185–224. Elsevier. Amsterdam.

ZUBIN, J., J. MAGAZINER & S. R. STEINHAUER. 1983. The metamorphosis of schizophrenia: from chronicity to vulnerability. Psychol. Med. **13:** 551–571.

ZUBIN, J. & B. SPRING. 1977. Vulnerability: a new view of schizophrenia. J. Abnorm. Psychol. **86:** 103–126.

ZUBIN, J. & S. STEINHAUER. 1981. How to break the logjam in schizophrenia: a look beyond genetics. J. Nerv. Ment. Dis. **169:** 475–491.

ZUBIN, J., S. R. STEINHAUER, R. DAY & D. P. VAN KAMMEN. 1985. Schizophrenia at the crossroads: a blueprint for the 80's. Compr. Psychiatry **26:** 217–240.

ZUBIN, J., S. SUTTON & S. R. STEINHAUER. 1986. Event-related potential methodology in psychiatric research. *In* Brain Electrical Activity and Psychopathology. C. Shagass, R. C. Josiassen & R. A. Roemer, Eds. 1–26. Elsevier. Amsterdam.

P300 Findings for Depressive and Anxiety Disorders[a]

GERARD E. BRUDER

Department of Biopsychology
New York State Psychiatric Institute
722 West 168th Street
New York, New York 10032
and
Department of Psychiatry
Columbia University College of Physicians and Surgeons
New York, New York 10032

INTRODUCTION

The P300 event-related potential (ERP) was discovered over 25 years ago by Samuel Sutton. It was on May 20th, 1964, that Sam and his co-workers first observed P300 (Sutton, Braren, Zubin & John, 1965). They discovered it in a "guessing paradigm," in which either a light flash or click followed a cueing stimulus. There were two different kinds of trials. On the so-called "certain" trials, the cueing stimulus was followed by a test stimulus that was always a light or always a sound. The subject was, therefore, certain as to what the test stimulus would be. On the "uncertain" trials, a different cueing signal was followed by a light on 67% of the trials or a click on 33% of the trials. The subject was then uncertain as to what the stimulus would be and was asked to guess. FIGURE 1 shows the average waveforms for 5 subjects, which were recorded to click stimuli on "certain" and "uncertain" trials. The most striking difference between the waveforms for the "certain" and "uncertain" conditions was the large positive peak that occurred at about 300 msec after click onset. This positive component was much greater for the "uncertain" condition. Here was an ERP component that apparently was associated with the subject's uncertainty as to whether the test stimulus would be a light or sound. Up until this time, cortical-evoked potentials had been used primarily to study shorter latency "exogenous" potentials, such as N100, that vary with the physical properties of the stimulus. The significance of P300 is that it was a function of a psychological or cognitive variable like "uncertainty."

While debate continues on the exact psychological processes associated with P300 and on its origins in the nervous system (Donchin & Coles, 1988; Verleger, 1988), the importance of the P300 component is that it provides a noninvasive means for studying physiological correlates of cognitive processing in humans. Researchers in psychopathology seized upon P300 as a means of studying alterations of cognitive function in psychiatric disorders and the relation of these to possible CNS dysfunction. This paper deals with P300 findings for depressive and anxiety disorders. It presents some new P300 findings for these

[a] This research was supported in part by National Institute of Mental Health Grants MH 36295 and MH 44815.

disorders, and highlights methodological issues relevant for doing research in this area, some of which were initially raised by Sam Sutton over 20 years ago (Sutton, 1969).

Sam was very much aware of the methodological pitfalls of doing P300 research in psychiatric patients. A major problem has to do with lack of control over what

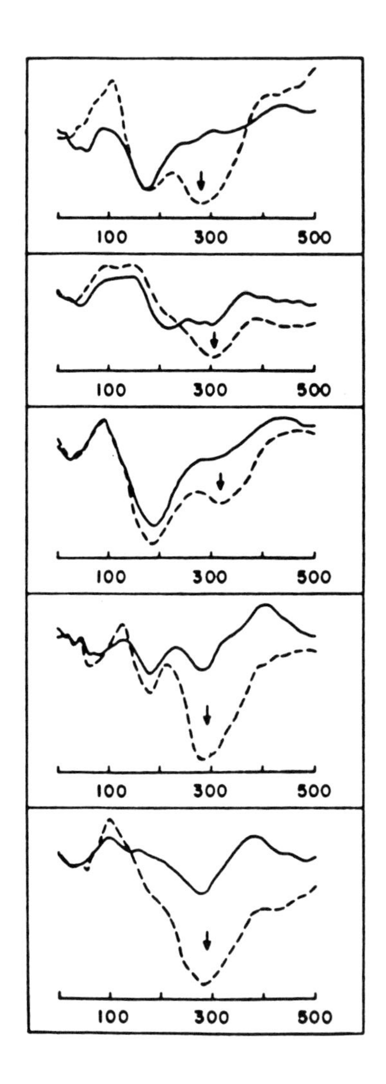

FIGURE 1. Average waveforms in response to clicks for certain (*solid lines*) and uncertain (*dashed lines*) conditions. (From Sutton *et al.*, 1965. Reprinted by permission from *Science*.)

was called "subject option." Do differences in P300 between groups of psychiatric patients and normal controls result from differences in subject cooperation, motivation or involvement in the task? Or, stated simply, "What was the subject really doing when P300 was measured?" If we simply ask the subjects to listen to clicks or tones, we have little control over subject option. However, if we introduce a challenging task that demands active attention, a specific type of cognitive processing, and a response, then we know more about what the subject was actually doing when P300 was measured. One of the last methodological papers that Sam worked on stressed the importance of reducing or controlling "subject option" (Zubin, Sutton & Steinhauer, 1986). As indicated in that paper, the key ingredients for controlling subject option are the use of a challenging task that engages the subject fully and the evaluation of the subject's engagement in the task by using behavioral accuracy measures.

Most P300 studies in depressed patients have used a simple "oddball" task. In this task, a frequent stimulus, let's say a low intensity click, is repeated about once a second, but on a small percentage of trials a more intense click is presented. The subject's task is simply to press a response button upon hearing the infrequent louder click. In this oddball task, you get a large P300 component to the infrequent target. This is illustrated in FIGURE 2, which shows average ERP waveforms for normal adults recorded at midline electrode locations (Fz, Cz, Pz, Oz) to infrequent target and frequent nontarget clicks. These are typical of the ERP waveforms observed for "oddball" tasks, consisting of negative peaks at 100 and 200 msec after click onset and a late positive peak, referred to as P300. Note that the P300 component is evident only for target clicks and has a scalp distribution that peaks at the parietal electrode (Pz). In a simple "oddball" task, the discrimination of the target and nontarget stimuli is usually so easy that subjects are about 100% correct in their responses. The problem here is that the task may not place enough cognitive demands on subjects. This could result in a lack of control of "subject option" because the subject's mind can wander and yet still perform this simple task. Moreover, since accuracy is close to 100% correct, there is no meaningful way of relating accuracy of performance to the ERP measures.

It is not particularly surprising, therefore, that P300 findings for depressed patients in the simple "oddball" task have not been very consistent across studies. In an extensive review of P300 studies in psychiatric patients, Pfefferbaum found that studies were about equally divided as to whether or not depressed patients had reduced P300 amplitude when compared to normal controls (Roth, Duncan, Pfefferbaum & Timsit-Berthier, 1986). Most studies also found no difference in P300 latency between depressed patients and controls, although several studies did find a nonsignificant trend for depressed patients to have longer P300 latencies (Levit, Sutton & Zubin, 1973; Pfefferbaum *et al.*, 1984; Diner, Holcomb & Dykman, 1985). A more recent study by Blackwood *et al.* (1987), in unmedicated patients, did find significantly smaller P300 amplitude in depressed patients than in controls, but the reduction was not as large as that seen for schizophrenic patients and it was normalized following treatment. Again, there was no difference in P300 latency between depressed patients and controls. In summary, findings for studies using the "oddball" task have not revealed anything particularly distinctive in the P300 data for depressed patients.

Given the cognitive deficits that are thought to accompany depressive disorders,the question arises as to why these deficits were not more apparent in their P300 findings? One answer may lie in the use of simple "oddball" tasks that are not challenging enough to reveal the cognitive dysfunctions in depressed patients. Another contributing factor could be the heterogeneity of patient samples. Only a subgroup of depressed patients may have cognitive deficits that give rise to alterations of P300.

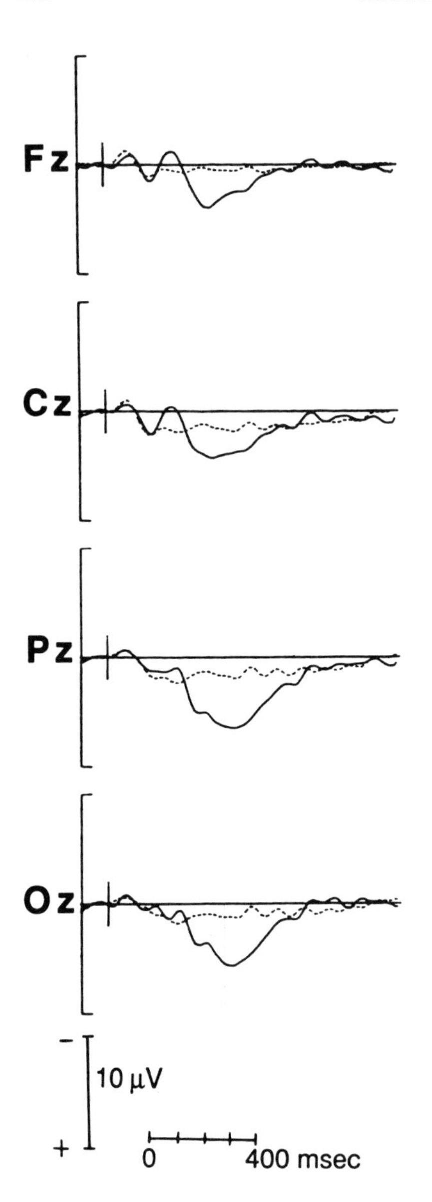

FIGURE 2. Average waveforms for 10 normal adults for targets (*solid lines*) and nontargets (*dashed lines*) in an oddball task.

P300 in Obsessive-Compulsive Disorder

The importance of the cognitive task demand was demonstrated by our recent P300 findings for patients having an obsessive-compulsive disorder (Towey, Bruder, Hollander, Friedman & Erhan, 1990). We used the same type of auditory oddball task as described above, in which the subject was asked to press a response button whenever an infrequent louder click was perceived. The probability of the infrequent target click was 25%. The cognitive demand of the task was manipulated by varying the difficulty of discriminating target and nontarget stimuli. In an Easy condition, the infrequent target click was 12 dB more intense than the nontarget click, and in a Difficult condition, the target click was 8 dB more intense than the

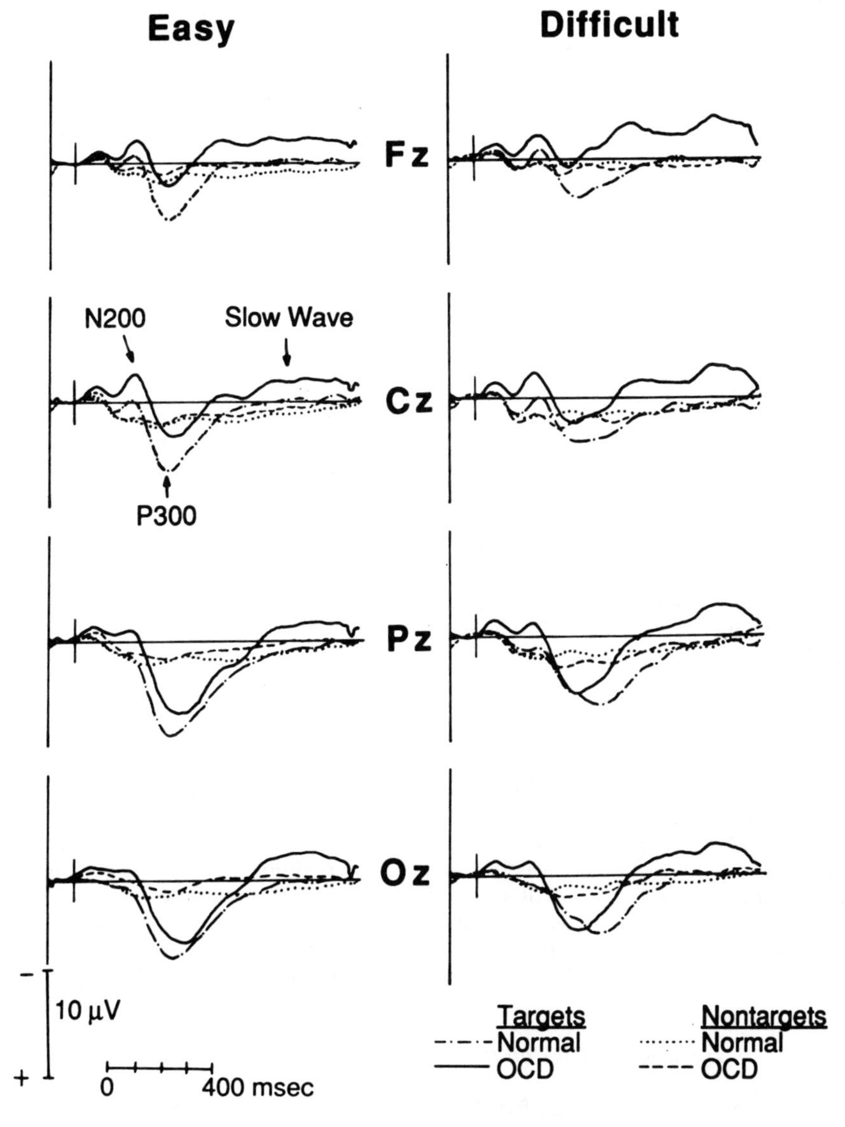

FIGURE 3. Average waveforms for 10 OCD patients and 10 normal controls in easy and difficult oddball tasks. (From Towey *et al.*, 1990. Reprinted by permission from *Biological Psychiatry*.)

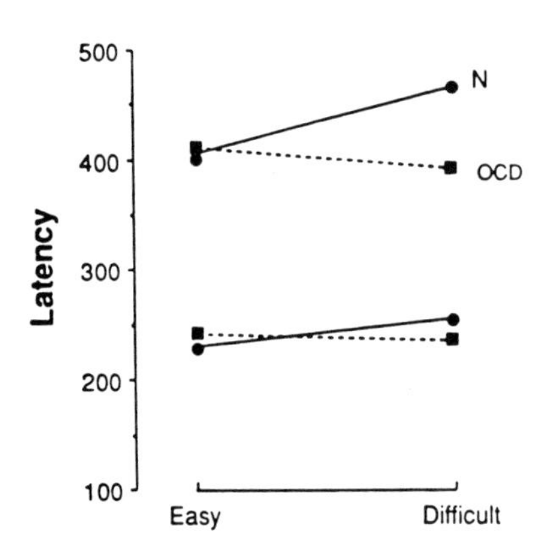

FIGURE 4. Mean latency of N200 and P300 for OCD patients and normal controls in easy and difficult tasks. (Adapted from Towey *et al.*, 1990.)

nontarget click. FIGURE 3 shows the average ERP waveforms for 10 patients who met DSM-III criteria for obsessive-compulsive disorder (OCD) and 10 normal controls. There was no difference between these groups in either P300 amplitude or latency for the Easy oddball task. However, when the discrimination of target and nontarget clicks was more difficult, a clear group difference emerged in P300 latency. P300 latency was shorter in OCD patients than in normal controls! Although there were also significant differences between these groups in negativities in the N200 and later slow wave regions, we shall not concern ourselves with this here.

The important methodological point is that had we used only an "easy" oddball task we might have wrongly concluded that there was no difference in P300 latency between OCD patients and normal controls. The dependence of the latency difference between these groups on task difficulty is evident in FIGURE 4, which shows the mean P300 latency for the OCD and normal groups. Similar findings were previously observed by Beech, Ciesielski & Gordon (1983), who measured P300 during visual discrimination tasks of different complexity.

The use of more difficult oddball tasks also has the advantage of yielding meaningful accuracy data. The OCD patients in our study performed at about the same accuracy level as the normal controls. The percentage of correct target detections for OCD patients was 80% in the Easy condition and 68% in the Difficult condition, while for normals it was 85% in the Easy condition and 64% in the Difficult condition. The similarity of the accuracy data between groups is important because one would be hard-pressed to argue that OCD patients were either more or less attentive to the task when compared to normals.

Influence of Task on P300 Latency in Depressed Patients

In reviewing P300 findings for schizophrenic and depressive patients, Duncan concluded that it would be unwise to continue to restrict P300 research in clinical groups to the oddball paradigm (Roth *et al.*, 1986). By going beyond the simple oddball task, one can test hypotheses about the nature of the cognitive processing deficits in different disorders. For instance, we have been evaluating the hypothesis that some forms of depression involve a disturbance in right hemisphere function (Flor-Henry, 1976; Tucker, Stenslie, Roth & Shearer, 1981; Bruder, Quitkin, Stewart, Martin, Voglmaier & Harrison, 1989). To test this hypothesis, we measured the ERPs of depressed patients while they engaged in cognitive tasks that are thought to involve predominantly left or right hemisphere processing (Bruder, Towey, Stewart, Friedman, Tenke & Quitkin, 1991).

One task was selected based on studies in brain-damaged patients and normal subjects suggesting that the left hemisphere is dominant for fine temporal resolution (Hammond, 1982; Lackner & Teuber, 1973). The general finding in patients with left brain damage, in particular dysphasic patients, is that they perform poorly on tasks that involve the discrimination of differences in duration or temporal order of stimuli. We chose a task that involves discrimination of a small difference in duration of a standard and a test stimulus. Each trial had the following structure: one second after the onset of a fixation light, used to reduce eye movements, a standard train of clicks lasting 240 msec was presented binaurally; two seconds later, a test click train, which was either the same duration or longer than the standard, was presented to the right or left ear. When the fixation light went off, the patient's task was to press a button if the test stimulus was judged to be longer than the standard. Before beginning ERP measurements, a titration procedure was used to select a duration of test stimulus that the patient could discriminate correctly on about 75% of the trials.

The second task was selected on the basis of studies in brain damaged patients indicating that the right hemisphere is dominant for localizing sounds in space (Altman, Balanov & Deglin, 1979; Bisiach, Cornacchia, Sterzi & Vallar, 1984; Ruff, Hersh & Pribram, 1981). These studies have found that damage to the right brain, particularly in posterior regions, interferes with the ability to localize sounds under free-field or dichotic listening conditions. Our audiospatial task used a dichotic paradigm to manipulate the apparent location of click stimuli. When a click is presented by earphones to one ear, followed after a brief delay of 0.1 to 1.0 msec by a click to the other ear, the subject perceives a single "fused" click lateralized toward the ear that receives the lead click. The longer the interaural delay, the further the click is displayed from midline. The audiospatial task involved discriminating a difference in apparent location of standard and test stimuli. The subject first heard a standard stimulus (*i.e.*, a dichotic "fused" click with an interaural delay of 0.3 msec) lateralized toward the right or left hemifield. Two seconds later, a test stimulus was presented at either the same location (*i.e.*, same interaural delay as the standard) or further lateralized in that hemifield (*i.e.*, interaural delay greater than the standard). The subject's task was to press a button if the test stimulus was in a different location than the standard stimulus. As in the temporal task, a titration procedure was used to select an interaural delay for each patient that would yield about 75% correct discrimination.

FIGURE 5 illustrates the average ERP waveforms recorded to test stimuli in the temporal task. At frontal and central sites, the N100 component is most prominant, and at parietal and occipital sites, a late positive complex is evident. The positive complex consists of an early positive peak (200–300 msec), a late positive peak (400–500 msec) and a sustained slow wave potential (600–1,000 msec). The late positive peak appears to correspond to the classical P300 in its scalp distribution. The solid lines in FIGURE 5 show the ERP waveforms for test stimuli that were correctly judged to be "different" from the standard, while the dotted lines show the waveforms for test stimuli that were the "same" as the standard. Since the subject's task was to press a response button when the test stimulus was "differ-

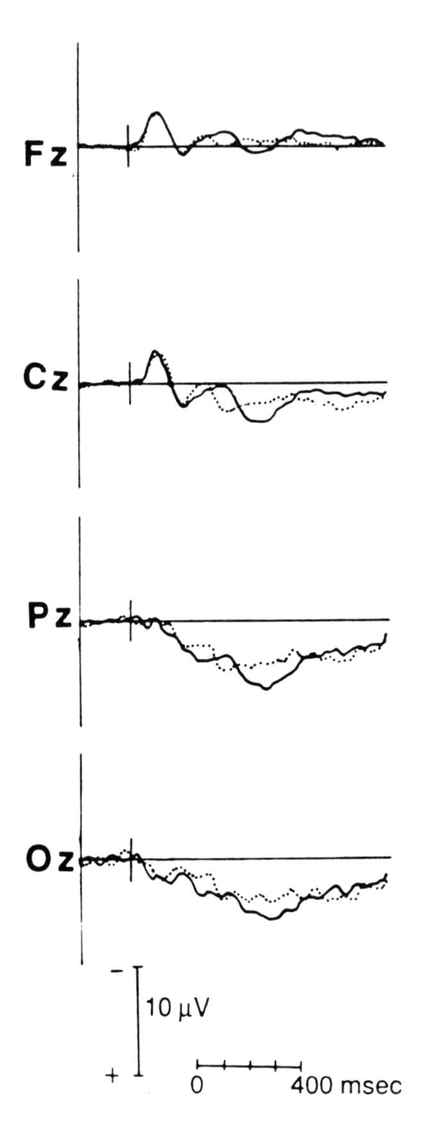

FIGURE 5. Average waveforms for 27 normal subjects to test stimuli in a temporal discrimination task. For correct different (*solid lines*) and same (*dashed lines*) judgments.

ent," the larger amplitude of the late positive peak on these trials could be another example of larger P300 amplitude to target stimuli (Pritchard, 1981). The ERP waveforms for test stimuli in the spatial task show these same components. A separate Principal Components Varimax Analysis (PCVA) was performed on the ERPs obtained for the temporal and spatial tasks so as to help isolate overlapping components in each task (BMDP 4M; Dixon, 1981). FIGURE 6 shows the four components that account for over 90% of the variance of ERP waveforms for each task. The components were highly similar across tasks, consisting of a prominant slow wave, an early positive peak, N100 and a late positive peak. The PCVA components were used in selecting the temporal regions of minimum overlap for measurements of base-to-peak amplitude and latency. P300 was measured at the most positive peak in the region of 276–600 msec after stimulus onset.

The ERPs of 25 unmedicated depressed patients and 27 normal controls were measured during the temporal and spatial tasks. Given the hypothesis of right

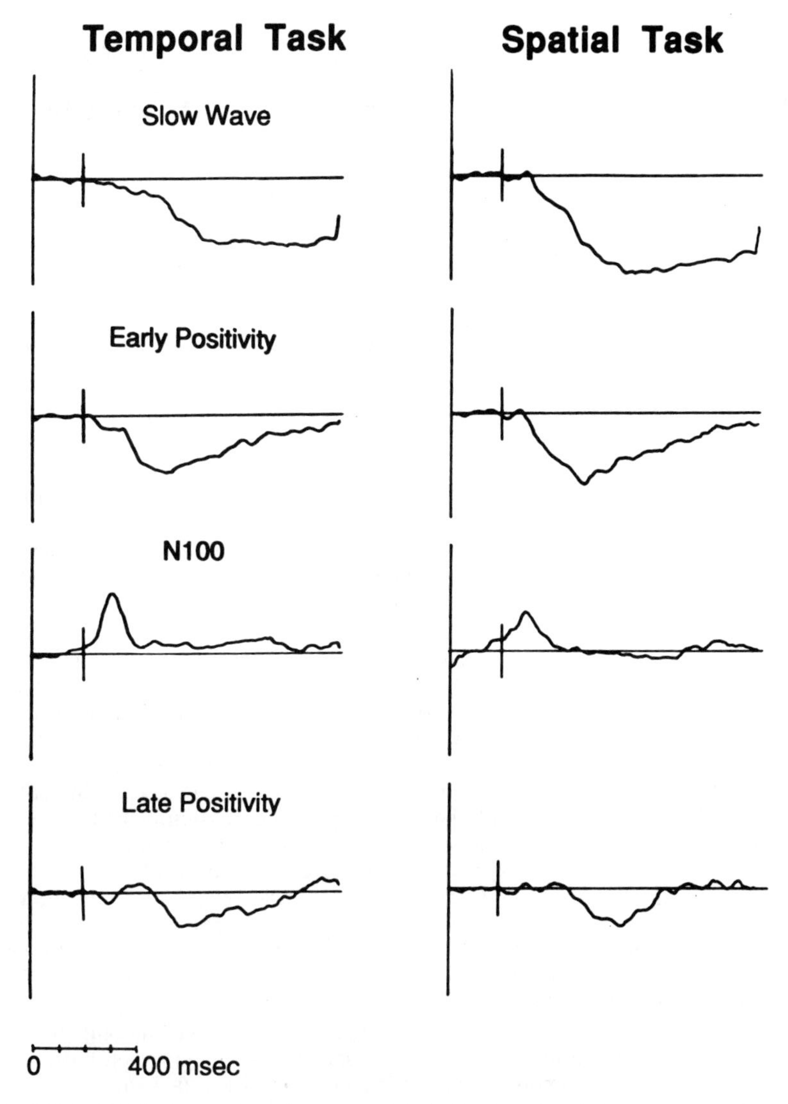

FIGURE 6. PCVA factors for 27 normal subjects for test stimuli in temporal and spatial discrimination tasks.

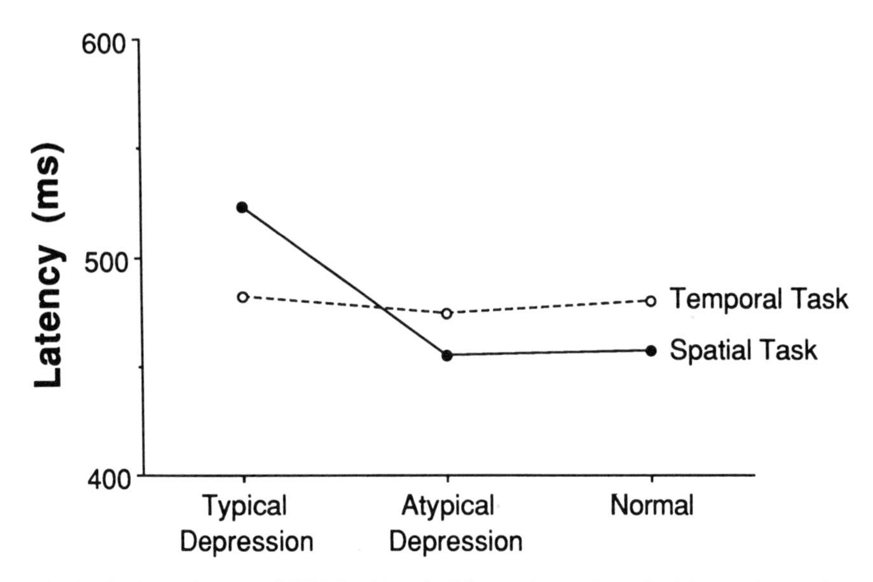

FIGURE 7. Mean latency of P300 for 13 typical depressives, 12 atypical depressives and 27 normal controls averaged over midline electrode sites (Fz, Cz, Pz, Oz).

hemisphere dysfunction in depression (Flor-Henry, 1976; Tucker *et al.*, 1981), we anticipated that depressed patients would be more likely to show abnormal P300 data in the spatial task than the temporal task. We also addressed the issue of diagnostic heterogeneity by dividing the sample of depressed patients into two subgroups, one having a typical major depressive disorder (the majority of whom met DSM-III criteria for melancholia; American Psychiatric Association, 1980) and the other having an atypical depression with symptoms that are in some respects opposite of those for melancholia (Liebowitz *et al.*, 1988). The criteria for atypical depression include the essential feature of reactivity of mood with preserved pleasure capacity, and one or more associated features—hypersomnia, overeating, extreme bodily inertia or rejection sensitivity. Twelve of the depressed patients met criteria for atypical depression and all but 3 of these patients also met research diagnostic criteria for major depressive disorder (Spitzer, Endicott & Robins, 1978). Of the 13 patients having a typical depression, 7 had major depressive disorders with melancholia and 6 had major depressive disorders with mood reactivity but no other symptoms associated with atypical depression (referred to as simple mood reactive depression). There were about an equal number of females and males in the atypical depression (6F/6M), typical depression (7F/6M) and normal control (14F/13M) groups. There was no significant difference in age among groups and all subjects were right handed on the basis of the Edinburgh Inventory (Oldfield, 1971).

Baseline-to-peak measures showed no significant difference among groups in amplitude of P300 in either the temporal or spatial task. There was, however, a task-specific difference among groups in P300 latency. FIGURE 7 shows the mean P300 latency for midline electrode sites for each group. There was no difference in P300 latency among groups for test stimuli in the temporal discrimination task.

Prior studies, in which P300 latency for depressed patients and controls was measured in oddball tasks involving pitch discrimination, have similarly found no difference in P300 latency among groups (Roth *et al.*, 1986; Blackwood *et al.*, 1987). In contrast, there was a marked difference among groups in P300 latency for test stimuli in the spatial discrimination task, with typical depressives having considerably longer latency when compared to either atypical depressives or normal controls.

FIGURE 8 shows the P300 latency values of individuals for the spatial task. The longer mean latency for typical depressives was not due solely to the long latency for patients having a melancholic depression (open circles), but was also present for patients having a simple mood reactive depression. There was no difference in P300 latency between melancholics (M = 526, SD = 54) and simple mood reactive depressives (M = 521, SD = 31), whereas both subgroups had significantly longer latencies than atypical depressives (M = 455, SD = 53) and normal controls (M = 458, SD = 46). Melancholic and mood reactive depressives also share an important clinical feature—*i.e.*, they respond equally well to treatment with either a tricyclic or MAOI antidepressant. In contrast, atypical depressives respond better to treatment with a MAOI than a tricyclic antidepressant, and they may thereby represent a distinct nosologic entity with different pathophysiology (Stewart *et al.*, 1980; McGrath *et al.*, 1986; Liebowitz *et al.*, 1988; Quitkin *et al.*, 1989). There is also evidence that psychomotor retardation in depressed patients is associated with responsiveness to tricyclic antidepressants (Joyce & Paykel, 1989). Since P300 latency is viewed as a measure of stimulus evaluation time (Donchin & Coles, 1988), further study should be given to evaluating the relation between P300 latency and outcome of treatment with a tricyclic antidipressant.

The cognitive task during which P300 was measured was an important factor. Longer P300 latency in typical major depression was observed in an audiospatial

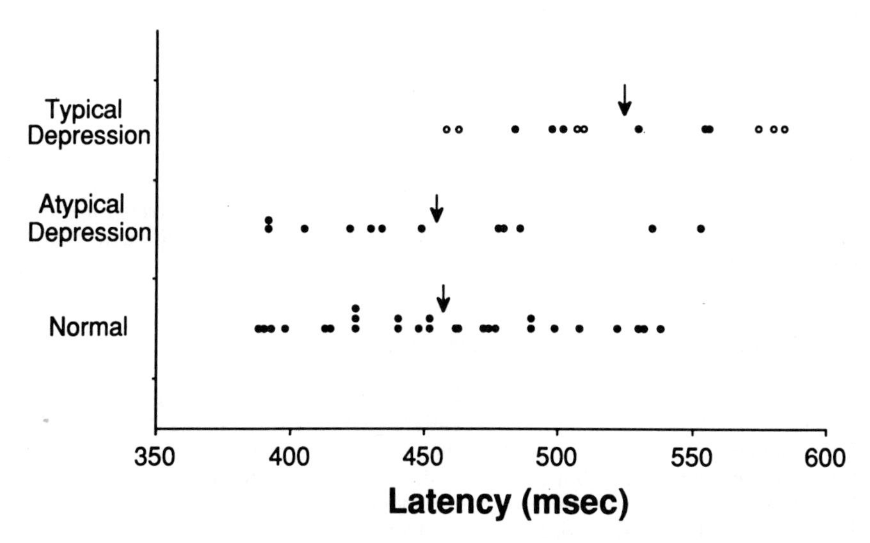

FIGURE 8. Latency of P300 for individual typical depressives (*open circles* = Melancholia), atypical depressives and normal controls.

task but not in a temporal discrimination task. A slowing in processing spatial information could be viewed as consistent with other evidence from dichotic listening measures of perceptual asymmetry (Bruder *et al.*, 1989) and neuropsychological tests of visuospatial function (Flor-Henry, 1976; Fromm & Schopflocher, 1984) that support the hypothesis of right hemisphere dysfunction in depression. However, alternative interpretations need to be ruled out before we can accept this conclusion. The difference in P300 latency between groups for the spatial task was not due to a difference in accuracy of performance between these groups. Recall that we used a titration procedure to adjust stimulus parameters so as to yield about equal accuracy for patients and controls in each task. TABLE 1 shows that there was no difference among groups in d′ measures of accuracy in the temporal or spatial task. All groups did display poorer accuracy in the spatial task than in the temporal task ($F = 22.6$, $p < 0.001$). Both tasks were moderately difficult, and so, it was not merely a difference between Easy vs Difficult tasks, as in the case of P300 latency data for OCD patients. However, we cannot rule out the possibility that the spatial task yielded P300 latency differences between groups because it was more demanding than the temporal task. Further study is needed to clarify the nature of the task demands that contribute to lengthened P300 latency in typical depression.

TABLE 1. Accuracy Measures (d′) for Spatial and Temporal Tasks[a]

		Typical Depression (N = 13)	Atypical Depression (N = 12)	Normal Controls (N = 27)
Temporal	M	1.43	1.65	1.62
task	SD	0.67	0.79	0.70
Spatial	M	0.79	0.87	1.06
task	SD	0.54	0.81	0.77

[a] Task: $F = 22.6$, $p < 0.001$; Group: $F = 0.9$, NS.

It should also be noted that the group differences in P300 latency held equally for midline and for lateral electrode sites over the right and left hemisphere. Also, the spatial and temporal discrimination tasks did not yield significant behavioral asymmetries. The lack of converging behavioral and ERP asymmetries for these tasks limits the conclusions that can be drawn concerning lateralized hemispheric function in depression.

P300 and Behavioral Asymmetries for Complex Pitch Discrimination

We have begun to measure brain ERPs during a dichotic Complex Tone Test that yields a left ear (right hemisphere) advantage in normal adults (Sidtis, 1981). This is a matching task, in which subjects compare the pitch of a binaural complex tone to the pitches of a dichotic pair of complex tones presented 2 seconds earlier. In our initial study of 20 normal right-handed adults (Tenke, Bruder, Towey, Erhan, Leite & Sidtis, in press), ERPs were recorded from homologous sites overlying left and right cerebral hemispheres as a means of examining the relationship between behavioral and ERP asymmetries in the Complex Tone Test. These normal subjects showed a mean left ear advantage of 10%, with 70% of them showing the expected asymmetry. This agrees with prior findings in normal adults and children (Sidtis, 1981; Sidtis, Sadler & Nass, 1987). In order to examine the relationship between individual differences in this asymmetry and ERP measures of hemispheric asymmetry, the subjects were divided at the median behavioral asymmetry into two groups: (1) subjects with a strong left ear advantage (S-LEA) and (2) subjects with little or no left ear advantage (N-LEA). FIGURE 9 shows the average ERP waveforms of the S-LEA and N-LEA groups for correct Same judgments. The waveforms show N1 and P2 components, which are maximum frontocentrally. A late positive complex with parietal maximum is also evident. It consists of a pair of peaks at about 350 and 550 msec, identified with the classical P300 component. Normal adults showed hemispheric asymmetries of these late positive peaks, which were related to the behavioral asymmetry for complex tones. The S-LEA subjects had greater amplitudes over the right hemisphere than the left, whereas the N-LEA subjects showed a nonsignificant trend toward the opposite hemisphere asymmetry (FIG. 10). Across all normal subjects, individual differences in behavioral asymmetry were significantly correlated with hemispheric asymmetry of P300 at parietal and occipital sites ($r = 0.49$ to $r = 0.69$). Greater P300 amplitude over the right hemisphere in normal adults with a strong left ear advantage may reflect the greater involvement of this hemisphere in complex tonal processing. A recent brain imaging study by Coffey *et al.* (1989) reported similar evidence of hemispheric asymmetries for dichotic pitch discrimination in normal adults.

ERPs to Complex Tones in Depressed Patients

We also measured the ERPs of 22 unmedicated depressed pateints during the Complex Tone Test. Twenty of these patients had a major depressive disorder and 2 had a dysthymic disorder. Unlike normal subjects, the depressed patients did not show a significant behavioral asymmetry for complex tones. Only about half of the depressed patients (12/22) had the expected left ear advantage. Most prior studies have similarly found that groups of depressed patients fail to show the normal left ear (right hemisphere) advantage for nonverbal dichotic listening tasks (Bruder, 1988; Bruder *et al.*, 1989). We compared the ERP data for the 22 depressed patients and our 20 normal subjects so as to examine whether or not they also differ in ERP measures of hemispheric asymmetry. In the same manner as for normal subjects, depressed patients were divided into two subgroups—those with a strong left ear advantage (S-LEA) and those with little or no left ear advantage (N-LEA). This allowed us to evaluate the relationship of patient-normal ERP differences to behavioral laterality. There was no difference between patient and normal groups in age, sex or handedness of subjects. The average ERP waveforms for the patient subgroups showed essentially the same late positive components as seen for normal subjects (FIG. 9). Note that the depressed patients had significantly smaller P300 amplitudes for complex tones when compared to normal controls ($p < 0.01$). This is another instance of reduced P300 amplitude for pitch discrimination in depressed patients (Roth *et al.*, 1986). Also, depressed patients failed to show the behavior-related hemispheric asymmetries seen for normal subjects (FIG. 10). Even depressed patients with a strong left ear advantage (S-LEA) did not show greater P300 amplitude over the right hemisphere than the left! It should also be noted that the difference in behavior-related hemispheric asymmetry between depressed patients and normal controls remained the same after vector scaling to remove overall amplitude differences (McCarthy & Wood, 1985).

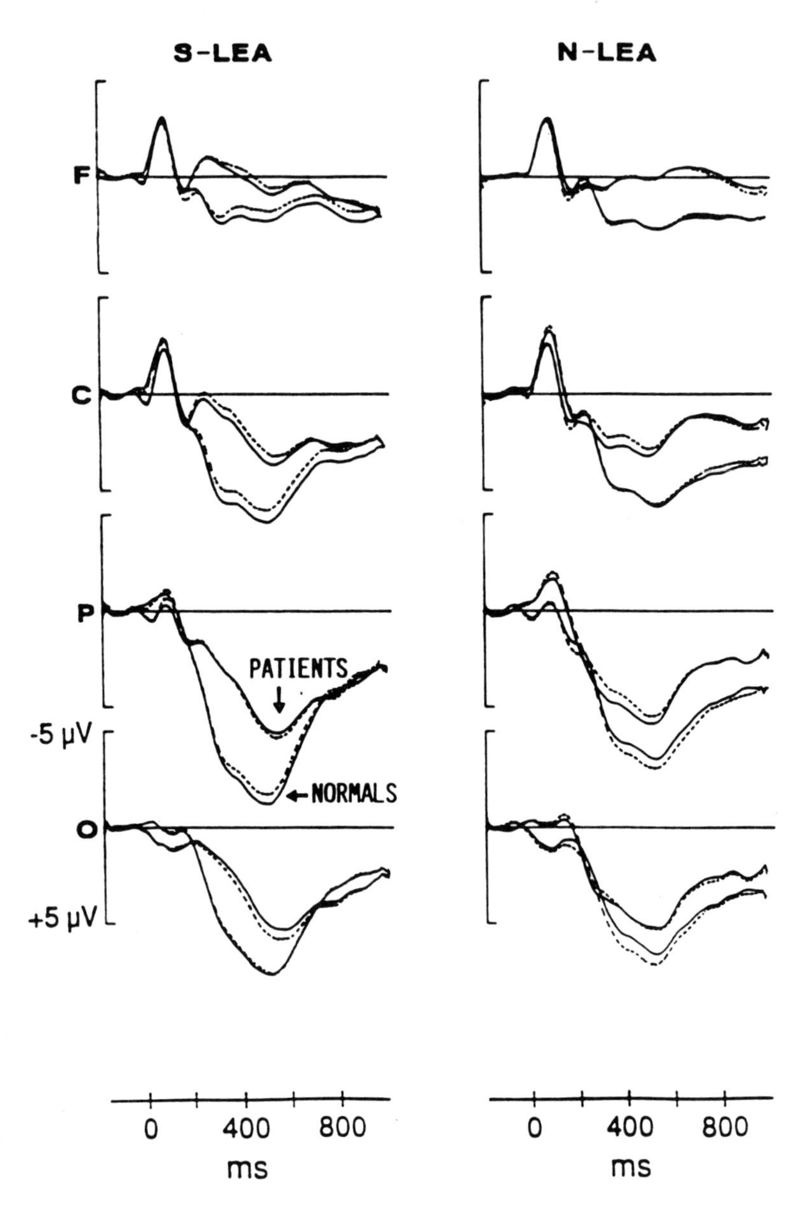

FIGURE 9. Average ERP waveforms over right (*solid lines*) and left (*dashed lines*) hemisphere for depressed patients and normals with a S-LEA or N-LEA.

In summary, depressed patients not only had reduced P300 amplitude, but also failed to show the normal hemispheric asymmetry of P300 for dichotic pitch discrimination. A larger sample of depressed patients should, however, be tested both to replicate these findings and to examine the relationship of these ERP abnormalities to diagnostic subtypes and other clinical features of depression.

CONCLUSIONS

The findings reviewed for depressed patients and patients having obsessive-compulsive disorders illustrate methodological issues that need to be attended to when conducting P300 research with psychiatric groups. Sam Sutton strongly believed that to make sense out of P300 findings for psychiatric patients, one must deal with the psychological or cognitive variables that influence these measures.

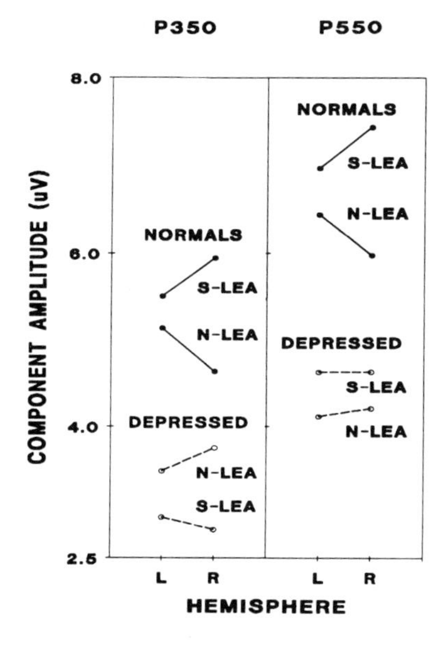

FIGURE 10. Mean P350 and P550 amplitudes over right and left hemisphere for depressed patients and normals with a S-LEA or N-LEA.

The more we know about what patients were doing at the time when P300 was measured, the better we'll be able to understand what patient vs normal differences in P300 actually mean.

It is particularly important in P300 research with patients having depressive or anxiety disorders to use challenging tasks that reduce subject option and tap into cognitive abnormalities in these disorders. The use of cognitively challenging tasks also yields behavioral accuracy scores that provide a measure of the patient's engagement in the task. The relationship between P300 and accuracy measures also serves as an important source of converging evidence for evaluating hypotheses. We have also illustrated the value of manipulating cognitive task demands so as to determine the relation of differences in P300 between patients and normals to difficulty level. Moreover, by going beyond the standard "oddball" task, one can use P300 measurements to test specific hypotheses concerning alterations of cognitive and/or physiological function in depressive and anxiety disorders. For instance, we are finding that depressed patients show alterations of P300 on cognitive tasks for which the right hemisphere is dominant in most normal adults, *i.e.*, audiospatial discrimination and dichotic pitch discrimination. These findings are in accord with the hypothesis that some types of depression involve right hemisphere dysfunction. Finally, there is now evidence for the importance of dealing with the heterogeneity of depressive disorders by comparing P300 for subgroups formed on the basis of diagnostic features (Shagass, 1981; Bruder *et al.*, 1991), or biochemical measures (Verhey & Lemars, 1990).

ACKNOWLEDGMENTS

Findings are presented for two studies begun with Sam Sutton's help. His advice in the design of these studies was of critical importance and we dearly miss his insightful help in the analysis and interpretation of ERP data. These studies have also involved the collaboration and help of many individuals at the New York State Psychiatric Institute. Those who have worked with us in the Psychophysiology Laboratory include: Jim Towey, Craig Tenke, Paul Jasiukaitus, Hulya Erhan, Martina Voglmaier, and Paul Leite. David Friedman and Charles Brown of the Department of Medical Genetics have also been very helpful over the years. Our collaborators at the Depression Evaluation Service, including Jonathan Stewart, Frederic Quitkin, Patrick McGrath, and Wilma Harrison, and at the Anxiety Disorders Clinic, including Eric Hollander and Michael Liebowitz, carried out all aspects of the diagnosis and treatment of patients. They have also helped guide the clinical research aspects of these studies.

REFERENCES

Altman, J. A., L. J. Balanov & V. L. Deglin. 1979. Effects of unilateral disorder of the brain hemisphere function in man on directional hearing. Neuropsychologia **17:** 295–301.

American Psychiatric Association. 1980. Diagnostic and Statistical Manual of Mental Disorders. 3rd edit. (DSM-III). American Psychiatric Press. Washington, DC.

Beech, H. R., K. T. Ciesielski & P. K. Gordon. 1983. Further observations of evoked potential in obsessional patients. Br. J. Psychiatry **142:** 605–609.

Bisiach, E., L. Cornacchia, R. Sterzi & B. Vallar. 1984. Disorders of perceived auditory lateralization after lesions of the right hemisphere. Brain **107:** 37–52.

Blackwood, D. H. R., L. J. Walley, J. E. Christie, I. M. Blackburn, D. M. St. Clair & A. McInnes. 1987. Changes in auditory P3 event-related potential in schizophrenia and depression. Br. J. Psychiatry **150:** 154–160.

Bruder, G. E. 1988. Dichotic listening in psychiatric patients. *In* Handbook of Dichotic Listening: Theory, Methods and Research. K. Hugdahl, Ed. 527–563. Wiley. Chichester, England.

Bruder, G. E., F. M. Quitkin, J. W. Stewart, C. Martin, M. M. Voglmaier & W. M. Harrison. 1989. Cerebral laterality in depression: differences in perceptual asymmetry among diagnostic subtypes. J. Abnorm. Psychol. **98:** 177–186.

Bruder, G. E., J. P. Towey, J. W. Stewart, D. Friedman, C. Tenke & F. M. Quitkin. 1991. Event-related potentials in depression: influence of task, stimulus hemifield and clinical features on P3 latency. Biol. Psychiatry **30:** 233–246.

Coffey, C. E., M. P. Bryden, E. S. Schroering, W. H. Wilson & R. J. Mathew. 1989. Regional cerebral blood flow correlates of a dichotic listening task. J. Neuropsychiatry **1:** 46–52.

Diner, B. C., P. J. Holcomb & R. A. Dykman. 1985. P300 in major depressive disorder. Psychiatry Res. **15:** 175–184.

DIXON, W. T. 1981. The BMDP Biomedical Computer Programs. University of California Press. Los Angeles.

DONCHIN, E. & M. G. H. COLES. 1988. Is the P300 component a manifestation of context updating? Behav. Brain Sci. **11:** 357–374.

FLOR-HENRY, P. 1976. Lateralized temporal-limbic dysfunction and psychopathology. Ann. N. Y. Acad. Sci. **280:** 777–795.

FROMM, D. & D. SCHOPFLOCHER. 1984. Neuropsychological test performance in depressed patients before and after drug therapy. Biol. Psychiatry **19:** 55–71.

HAMMOND, G. R. 1982. Hemispheric differences in temporal resolution. Brain Cognit. **1:** 95–118.

JOYCE, P. R. & E. S. PAYKEL. 1989. Predictors of drug response in depression. Arch. Gen. Psychiatry **46:** 89–99.

LACKNER, J. R. & H. L. TEUBER. 1973. Alterations in auditory fusion thresholds after cerebral injury in man. Neuropsychologia **11:** 409–415.

LEVIT, R. A., S. SUTTON & J. ZUBIN. 1973. Evoked potential correlates of information processing in psychiatric patients. Psychol. Med. **3:** 487–494.

LIEBOWITZ, M. R., F. M. QUITKIN, J. W. STEWART, P. J. MCGRATH, W. HARRISON, J. S. MARKOWITZ, J. RABKIN, E. TRICAMO, D. M. GOETZ & D. F. KLEIN. 1988. Antidepressant specificity in atypical depression. Arch. Gen. Psychiatry **45:** 129–137.

MCCARTHY, G. & C. C. WOOD. 1985. Scalp distributions of event-related potentials: an ambiguity associated with analysis of variance models. Electroencephalogr. Clin. Neurophysiol. **62:** 203–208.

MCGRATH, P. J., J. W. STEWART, W. HARRISON, S. WAGER & F. M. QUITKIN. 1986. Phenelzine treatment of melancholia. J. Clin. Psychiatry **47**(8): 420–422.

OLDFIELD, R. C. 1971. The assessment and analysis of handedness: the Edinburgh Inventory. Neuropsychologia **9:** 97–113.

PFEFFERBAUM, A., W. T. ROTH, B. WENEGRAT, J. M. FORD & B. S. KOPELL. 1984. Clinical applications of the P3 component of the event-related potential. II. Dementia, depression and schizophrenia. Electroencephalogr. Clin. Neurophysiol. **59:** 104–124.

PRITCHARD, W. S. 1981. Psychophysiology of P300. Psychol. Bull. **89:** 506–540.

QUITKIN, F. M., P. J. MCGRATH, J. W. STEWART, W. HARRISON, S. G. WAGER, E. NUNES, J. G. RABKIN, R. N. TRICAMO, P. H. MARKOWITZ & D. F. KLEIN. 1989. Phenelzine and imipramine in mood reactive depressives. Arch. Gen. Psychiatry **46:** 787–793.

ROTH, W. T., C. C. DUNCAN, A. PFEFFERBAUM & M. TIMSIT-BERTHIER. 1986. Applications of cognitive ERPs in psychiatric patients. *In* Cerebral Psychophysiology: Studies in Event-Related Potentials (EEG Suppl. 38). W. C. McCallum, R. Zappoli & F. Denoth, Eds. 419–438. Elsevier. Amsterdam.

RUFF, R. M., N. A. HERSH & K. H. PRIBRAM. 1981. Auditory spatial deficits in the personal and intrapersonal frames of reference due to cortical lesions. Neuropsychologia **19:** 435–443.

SHAGASS, C. 1981. Neurophysiological evidence for different types of depression. J. Behav. Ther. Exp. Psychiatry **12:** 99–111.

SIDTIS, J. J. 1981. The complex tone test: implications for the assessment of auditory laterality effects. Neuropsychologia **19:** 103–112.

SIDTIS, J. J., A. E. SADLER & R. D. NASS. 1987. Dichotic speech and pitch discrimination in children ages seven to twelve. Dev. Neuropsychol. **3:** 227–238.

SPITZER, R. L., J. ENDICOTT & E. ROBINS. 1978. Research diagnostic criteria. Rationale and reliability. Arch. Gen. Psychiatry **35:** 773–782.

STEWART, J. W., F. QUITKIN, A. FYER, A. RIFKIN, P. MCGRATH, M. LIEBOWITZ, L. ROSNICK & D. F. KLEIN. 1980. Efficacy of desipramine in endogenomorphically depressed patients. J. Affective Disord. **2:** 165–176.

SUTTON, S. 1969. The specification of psychological variables in an average evoked potential experiment. *In* Average Evoked Potentials—Methods, Results and Evaluations. E. Donchin & D. B. Lindsley, Eds. 237–262. U.S. Government Printing Office. Washington, DC. (Catalog No. 78-600641.)

SUTTON, S. 1979. P300—thirteen years later. *In* Evoked Brain Potentials and Behavior. H. Begleiter, Ed. 107–126. Plenum. New York.

SUTTON, S., M. BRAREN, J. ZUBIN & E. K. JOHN. 1965. Evoked-potential correlates of stimulus uncertainty. Science **150:** 1187–1188.

TENKE, C. E., G. E. BRUDER, J. P. TOWEY, P. LEITE & J. J. SIDTIS. Correspondence between brain ERP and behavioral asymmetries in a dichotic complex tone test. Psychophysiology. In press.

TOWEY, J., G. BRUDER, E. HOLLANDER, D. FRIEDMAN & H. ERHAN. 1990. Endogenous event-related potentials in obsessive-compulsive disorder. Biol. Psychiatry **28:** 92–98.

TUCKER, D. M., C. E. STENALIE, R. S. ROTH & S. L. SHEARER. 1981. Right frontal lobe activation and right hemisphere performance: decrement during a depressed mood. Arch. Gen. Psychiatry **38:** 169–174.

VERLEGER, R. 1988. Event-related potentials and cognition: a critique of the context updating hypothesis and an alternative interpretation of P3. Behav. Brain. Sci. **11:** 343–356.

VERHEY, F. H. M. & T. H. LAMERS. 1990. P300, depression and dexamethasone suppression test (DST). *In* Psychophysiological Brain Research, C. H. M. Brunia, A. W. K. Gaillard & A. Kok, Eds. Vol. 2: 232–237. Tilburg University Press. Le Tilburg.

ZUBIN, J., S. SUTTON & S. R. STEINHAUER. 1986. Event-related potential and behavioral methodology in psychiatric research. *In* Electric Brain Potentials and Psychopathology. C. Shagass & R. C. Josiassen, Eds. 1–26. Elsevier. Amsterdam.

Pharmacologic Challenge in ERP Research[a]

PATRICIA A. TUETING,[b] JOHN METZ,[b]
BARRY K. RHOADES,[c] AND NASHAAT N. BOUTROS[d]

[b]*Illinois State Psychiatric Institute*
1153 North Lavergne Avenue
Chicago, Illinois 60651

[c]*Center for Network Neuroscience*
University of North Texas
Denton, Texas 76203

[d]*Department of Psychiatry*
Ohio State University
473 West 12th Avenue
Columbus, Ohio 43210-1228

INTRODUCTION

In 1968, Sam Sutton (1969) wrote a critical review on the subject of experimental design in event-related potential (ERP) research and defined the "triangular experimental paradigm." He argued that in the ideal ERP experiment, all three corners of a triad—stimulus events, physiological events, and psychological events—should be studied *simultaneously*. He astutely noted that it was more common, and much easier, to examine only two corners at a time. Unfortunately, what is lost in the two-corner approach, he noted, is the ability to assess critical sources of variance in an ERP experiment. He was pointing to the value of using independent sources of information in construct validation. Concepts emphasized in his review, like the need for experimental control over "subject option" and the need for greater refinement in behavioral assessments during ERP recording were to heavily influence the field for the next two decades as they still do today.

In this paper, we shall review the strategy of examining different classes of variables simultaneously for the purposes of obtaining converging evidence about the nature of ERP response to pharmacologic challenge and understanding the pathophysiology of psychosis. Advantages and problems in extending the triangular research paradigm to include more levels of the physiological corner of the triad will be demonstrated. We shall use our own study of acute challenge with the psychotomimetic N,N-dimethyltryptamine (DMT) to illustrate how information simultaneously derived from multiple levels of independent measurement can potentially provide insight for interpreting behavioral and physiological response to a drug. In this case, the multilevel approach involved combining behavioral and neurotransmitter/neuroendocrine data with ERP and Hoffman-reflex (H-reflex) measures before and after administration of DMT alone or DMT in conjunction

[a] This research was supported in part by United States Public Health Service Grants MH30938, MH29206, and MH137808 (Herbert Y. Meltzer), and by the State of Illinois, Department of Mental Health and Developmental Disabilities.

with serotonin (cyproheptadine) or dopamine (haloperidol) antagonists. Our data support the recent popularity of the temporolimbic model of psychosis, which has been used to interpret information from several areas of investigation including ERP research on late cognitive components (see Tueting, 1991 for a review).

A new model linking late components to the N-methyl-D-aspartate (NMDA) receptor (McCarley, Faux, Shenton, Nestor & McAdams, 1991), and a recent study of auditory evoked potentials in monkeys following challenge with a PCP-like agent (Javitt, Schroeder, Arezzo & Vaughan, in press) would appear to make our study of timely importance, even though it was designed and completed some time ago. In addition, Strassman, Uhlenhuth, Kelner, and Qualls (1991) have recently replicated some aspects of our study, involving the psychological and endocrine findings, with low-dose DMT in human subjects.

In order to place acute pharmacologic challenge in perspective, we shall briefly review chronic pharmacologic challenge in ERP research, emphasizing the literature on stimulant and major tranquilizing drugs. These are the pharmacological agents most relevant to standard theorizing about psychosis in terms of dopaminergic, serotoninergic, and noradrenergic neurotransmitter systems.

Chronic Drug Challenge Studies

ERP studies can be "piggybacked" onto clinical treatments or onto ongoing clinical investigations. While still problematic for ERP research, a clinical investigation is preferable to treatment at therapists' discretion, in that a drug treatment protocol is standardized and planned in advance. Objective diagnostic and clinical outcome measures, as well as assessments of drug metabolism and effects on neurotransmitter and receptor systems, are frequently included as part of the study, which allows the ERP researcher to be "blind" as to these clinical measures. The ERP investigator is often not a principal investigator of the clinical study, but it is worth noting that the FDA does not require the sponsor of an investigational new drug exemption (IND) to be a physician (Tueting, 1991b). A physician must, however, prescribe the drug.

Chronic challenge studies are not easy, and are almost without exception extremely expensive in time and effort. It may take weeks, months, or sometimes years rather than hours or days to complete data collection on a single subject. More significantly, the patient's clinical treatment and well-being understandably take precedence over ideal research design. Once the study is ongoing, a dropout rate of 20–30% or higher is usual with dropouts occurring for a wide variety of reasons, many of which can be related to the length of the study. Adequate control groups are rarely used because ill patients may be in immediate need of treatment and often cannot be assigned to conditions, like a no-treatment group, because of clinical concerns. The solution of using a standard drug treatment group as a control is problematic too in that such a design reverses the usual statistical logic. Finding no difference—to find that a drug is as good as a currently used standard treatment rather than to find a difference between drug conditions—is the salient result. Finally, the distance between clinical and scientific perspective and career reward systems can potentially hinder successful collaborative research efforts.

On the positive side, as investigators realize that standard techniques do not apply in longitudinal research because of individual differences in response, missing data, and correlated repeated measures, new scientific and statistical methods aimed specifically at these issues with clinical investigations are being developed. Gibbons and Hedeker (Gibbons, Hedeker, Waternaux & Davis, 1988; Hedeker,

Gibbons & Davis, 1991) have developed a general statistical model for the analysis of within-subject psychiatric data. The model is based on the concept of "random regression" and permits missing data, irregularly spaced observations, correlated errors, time varying covariates, and estimates of the rate of change for each individual subject.

Major Tranquilizers and Stimulants

There is a large literature on the effect of major tranquilizers on ERPs. Results have depended upon stimulus type, ERP measure (component, electrode location, etc.), dose and type of drug, length of treatment, and characteristics of the subject sample. What is remarkable is that reports of dramatic and stable effects of drugs on ERPs remain few. What has emerged instead is a rather complex picture of subtle effects, as should have been predicted in the early 1970s when the first data on this issue were acquired. There was initial puzzlement in Sutton's laboratory when the P300 component appeared to reflect the drug status of the patient to only a minor degree (Levit, Sutton & Zubin, 1973), although the finding was reassuring because of the difficulty of taking state hospital patients off medication. After more than 20 years of work, current conclusions are similar overall with respect to the influence of antipsychotic drug treatment on the P300 component (see McCarley *et al.*, 1991; Pfefferbaum, Ford, White & Roth, 1989). There is space to provide only a limited overview of the complex findings from the chronic pharmacological challenge literature highlighting issues related to these findings.

Tolerance, Withdrawal, Sensitization

Most chronic drug challenge experiments have only before and after treatment as time points of measurement. Thus, the standard drug treatment study typically involves a single comparison of ERP data obtained during a drug-free period with data obtained after a period of drug treatment. The length of the withdrawal period is usually short for most chronic treatment studies if chronic psychosis and major tranquilizing medication are involved because it is difficult to maintain these patients without medication. A major determinant of the length of a drug trial is whether short-term treatment of an episode or maintenance treatment is the focus. In the former, clinical outcome measures are the major dependent variables, while in the latter it is usually relapse. The standard drug treatment research paradigm is often inadequate for resolving issues of concern to researchers in neurophysiology.

Lack of basic knowledge about the time course of drug effects can make precise research design and interpretation of data difficult. Changes in dependent measures early in the first few weeks of treatment can differ substantially from changes after chronic administration. Many drug effects are long lasting and there may be tolerance, withdrawal, and re-treatment effects related to prior sensitization to the drug. To maintain homeostasis, receptor and neuronal feedback systems require finely tuned adjustments and these processes may have an unknown time course. The development of tardive dyskinesia (TD) with extended neuroleptic treatment is one of the most dramatic examples of a longer latency drug effect. Lack of knowledge concerning drug-induced changes over time is exemplified by the fact that we still do not have a complete picture of the etiology and treatment of TD.

The expected timing of critical events, such as the time window during which positive clinical response should be expected, when and how receptors change, and the time courses of phenomena like tolerance, sensitization, kindling, etc. in an individual, turn out to be extremely difficult research questions. A large time window is sometimes selected in drug treatment studies apparently on the assumption that the timing of these events is modulated by unknown factors. One could argue, however, that a phenomenon which occurs at 1 week post drug may not be the same as a qualitatively similar phenomenon which occurs at 3 weeks.

On the other hand, careful study of the time course of events can yield important information relevant to the pathophysiology and prognosis of disordered function. Despite some design issues that can be dealt with related to habituation and repeated measures, the use of ERPs to probe the time course of drug effects should be explored further. From a practical standpoint, repeated measurements of ERPs can be made with relative safety and ease in patients. Conceivably, measures could be developed to predict the time course of therapeutic responses and side effects to psychotropic drugs, sparing psychiatric patients from the trial-and-error approach (Tueting, 1991a). Ultimately, baseline ERPs could prove useful in assessing individual differences in response to drug treatments for specific psychiatric disorders.

A case study by Buchsbaum, Post, and Bunney (1977) can serve as an example of using ERPs as probes to study basic questions about the timing of drug effects. The authors used a change in the slope of the amplitude/intensity curve of an early visual evoked potential component (P100) to predict a bipolar patient's switch from depression into mania 8–10 days in advance of the switch. Changes in a metabolite in body fluids of the neurotransmitter norepinephrine (MHPG) paralleled ERP findings. This study, even though it involved only one subject, demonstrates how an ERP component can be used as an independent measure useful in marshalling important information on an individual's drug response during treatment. At the same time, careful attention to the time course of events may also lead to more powerful conclusions about the processes underlying ERP changes and psychopathology.

The issue of long lasting effects of drug treatment has led to an interest in drug-naive subjects, if they can be found. In a recent study, Radwan, Hermesh, Mintz & Munitz (1991) reported that late components were smaller in drug-naive schizophrenic patients than in controls for both a task and a no-task condition. Smaller amplitude late components is a common finding in schizophrenic patients whether on drug treatment or not. However, Radwan *et al.* (1991) failed to find the smaller task-related increase in late component amplitude commonly reported in schizophrenic patients compared to normal subjects. The authors concluded that the reported lack of responsivity of P300 in schizophrenia could be associated with chronic neuroleptic treatment rather than with the disorder itself, since they failed to find reduced responsivity in their drug-naive patients. However, the population of drug naive-patients from which Radwan *et al.* (1991) selected their sample could have been less seriously ill and therefore more responsive in terms of both performance and underlying neurophysiology. More studies like Radwan *et al.*'s (1991) are needed to tease apart sources of variance in ERP studies of drug treatment effects.

Neuronal Excitability and Sensory Gating

There is a tendency to interpret decreases in latency, especially of early ERP components, following neuroleptic treatment in terms of increased "neuronal excitability," a concept which can be viewed as either useful or evidence of "fuzzy" thinking depending on one's point of view. A common variation of this construct implicates dopamine. The hypothesis is that short latency early ERP components in unmedicated schizophrenic and manic patients may be related to excess levels of the neurotransmitter dopamine, hence to excessive neuronal excitability and overly fast information processing. Excessive latency variability, and consequently reduced amplitude of several ERPs, is sometimes implied in these hypotheses as well.

An example in the auditory modality of use of a model of this type, was recently reviewed by Freedman, Waldo, Bickford-Wimer, and Nagamoto (1991). They reported that auditory P50 amplitude is increased with neuroleptic treatment, presumably due to more precise timing of neuronal events underlying this component as excitability of the system is reduced by the major tranquilizing drug. Similarly, the latency of the visual evoked potential has also been reported to be short in schizophrenic patients and latency has been reported to increase with neuroleptic treatment (Bodis-Wollner, Yaar, Mylin & Thornton, 1982). Further construct validation is provided by the finding of an increased latency of the early visual ERP in Parkinson's disease, a condition associated with low dopaminergic function, and a normalization in ERP latency with L-dopa treatment. Neuroleptic treatment, however, did not change the P50 ratio of conditioning (S1) amplitude to test stimulus (S2) amplitude. The authors related the P50 amplitude ratio at 500 msec interstimulus interval to sensory gating, and specifically to failure to gate early sensory information in schizophrenia. Information overload due to faulty gating has been linked to cognitive dysfunction in schizophrenia and to low amplitude of the late components of the ERP as a concomitant of this dysfunction. The issue of the extent to which the P50 amplitude ratio is altered by neuroleptic treatment in different subject groups and in animals has been the topic of a recent debate (Braff & Geyer, 1990; Cohen, 1991).

Models like the above which involve considerable theorizing about neurotransmitter functioning may be useful aids in thinking about phenomena. Such models, however, are often limited in terms of their predictive power in designing experiments. It is unfortunately the case that there is such an abundance of rapidly accumulating information on neurotransmitter systems that almost any ERP result can be explained in multiple, often opposite ways.

Hemispheric Asymmetries and Change with Drug Treatment

Much of the large literature on asymmetries of the brain in schizophrenia has focused on a presumed dysfunction in left hemisphere processing. Amplitude of the P300 component over the left hemisphere has been reported to be lower than over the right hemisphere presumably reflecting this dysfunction, but it is unclear how much of this effect is attributable to drug treatment (McCarley *et al.*, 1991; Morstyn, Duffy & McCarley, 1983; Radwan *et al.*, 1991) and how much to the disorder. Neuroleptic treatment itself appears to be related to changes in asymmetry (Mintz, Tomer & Myslobodsky, 1982). Roemer and Shagass (1990), using a waveshape stability measure, reported lower left-than-right hemisphere visual and auditory evoked potential stability in chronic schizophrenia which is also suggestive of left hemisphere dysfunction. Antipsychotic medication was found to alter the asymmetry of waveshape stability.

Thus, it is unclear at the present time whether neuroleptic treatment normalizes asymmetries found in schizophrenia or whether the drug treatment itself induces the asymmetry which has been reported in schizophrenia. The usefulness of simultaneous measurement of ERPs together with more molecular neurochemical measures in drug challenge studies needs to be assessed in regard to this question. In an isolated study, Gottfries, Perris, and Roos (1974) reported a higher correlation between CSF homovanillic acid (HVA) and ERP amplitude for the left hemisphere than for the right hemisphere. In contrast, CSF 5-hydroxyindoleacetic acid (5-HIAA) correlated higher with ERP amplitude on the right. Since HVA is a metabolite of dopamine and 5-HT is a metabolite of serotonin, the finding suggests a relationship between ERPs and dopamine in the left hemisphere and serotonin in the right hemisphere. The finding of Gottfries *et al.* corresponds to known distributions of serotonin and dopamine in cortex and would suggest that different types of antipsychotic drugs should be studied separately in relation to asymmetries in ERP topography. Different antipsychotic drugs are known to have differential effects on dopamine and serotonin. Roemer and Shagass (1990) did analyze results separately for different types of antipsychotic drugs in the study noted above and they found differences in their asymmetry measure as a function of class of antipsychotic drug.

Fewer ERP studies have examined the effects of chronic stimulant use on the ERP than have examined the effects of major tranquilizers. The literature, however, is intriguing, in that measurement of stimulants and major tranquilizers may provide a more complete picture of the status of neurophysiological systems reflected by ERPs. Animal models used in the development of new antipsychotic drugs characteristically involve titration of the new antipsychotic against the known effects of stimulants such as amphetamine. Research psychiatrists have also been interested in stimulants for their potential usefulness in the treatment of some depressive conditions and for the treatment of hyperactivity in children. Stimulants are thought to raise the level of catecholamine activity, such as dopamine and norepinephrine, which are presumed to be deficient in terms of depression. The theoretical foundation for use of methylphenidate in children includes receptor and neuronal network feedback systems.

Chronic administration of methylphenidate in children for the treatment of hyperactivity or attention deficit disorder has been studied with ERPs (Shagass, Ornitz, Sutton & Tueting, 1978). Some studies have shown that children who respond to methylphenidate, as compared to nonresponders, have ERPs that show pre- to posttreatment differences, or that "normalize" with a positive response to the drug (*e.g.*, Prichep, Sutton & Hakerem, 1976; Satterfield, Cantwell, Lesser & Podosin, 1972). Early ERP components appear to decrease in latency as a function of stimulant challenge, which is consistent with increased "neuronal excitability" and "enhanced dopamine activity." The effect on late cognitive components, however, seems to go in the same direction as the major tranquilizer results. Children who improve cognitively on methylphenidate show enhanced P300 activity just as the psychotic patients who improved on neuroleptic treatment (Peloquin & Klorman, 1986; Prichep *et al.*, 1976). Prichep and colleagues were more prepared for this finding because of earlier similar findings with major tranquilizers in psychotic patients from Sutton's laboratory.

Acute Drug Challenge

Statisticians have pointed to intrinsic problems with long-duration experiments as we noted in the Introduction. One such problem is that random error increases as a function of the length of the time interval between measures. The magnitude of such a problem is easy to intuitively understand in regard to psychiatric patients.

In addition to drug treatment effects, there are many uncontrolled sources of variation that may affect a patient between pre- and posttreatment measurement times. A patient may improve with a change in the weather, a better diet, a new situation, a change in living situation or social milieu, and so forth. Of course, besides patient variation, errors of measurement due to changes in equipment, procedures, and research personnel tend to increase with time. Acute pharmacologic challenge has the statistical advantage of a compressed time frame. Repeated measurements taken at intervals of only minutes or hours reduce the error variance associated with more extended time intervals between measurements. Interrelationships among variables at different levels can also be examined with greater reliability and validity. Placebo control sessions and control groups can be used with the different ethical issues involved and without excessive expense.

The acute pharmacologic challenge paradigm can also be used to address questions related to changes occurring over long periods of time by repeating acute challenge tests at long intervals; for example, before and after treatment, or by using a baseline acute challenge "test" to predict clinical and physiological response to long-term treatment response. However, the same issues outlined above for the chronic pharmacologic challenge situation will then, of course, apply to the long-term aspect of the study.

Acute pharmacologic challenge designs are by no means problem free. Tight planning is necessary to insure that everything during the session goes smoothly and according to plan. There is little room for error and data are sometimes lost. Problems with repeating some measures, such as occur when the same rater repeatedly rates a subject on the same behavioral variables, are magnified rather than reduced by the short time frame. Challenge sessions are stressful for some subjects, and the risk/benefit ratio of undertaking the study in psychiatric patients needs to be carefully considered. Patient subjects may have little to gain directly from participating, other than contribution to generalizable knowledge, unless a clinical challenge test is being developed as a product of the research. Moreover, stress itself may influence measures, and collection of data for one measure may inadvertently affect data on the other measures. Some of these problems can be solved by designing studies to be as low risk and as comfortable as possible, *e.g.*, by taking medical precautions in case of emergency and by using drug dosages that are as low as theoretically practical.

Some stimulants appear to mimic aspects of psychosis in normal persons or to exacerbate psychosis in the mentally ill. There has been considerable interest in acute stimulant challenge in psychosis because the same neurotransmitter pathways have been theorized to underly psychosis and response to stimulants. The role of stimulant abuse in the etiology of at least some schizophrenic illnesses is suspected on clinical and theoretical grounds. Moreover, Lieberman, Kinon, and Loebel (1990) and others have distinguished subgroups of schizophrenic patients on the basis of their behavioral response to a test dose of a stimulant drug like methylphenidate. Research indicates that whether or not patients become more psychotic in response to stimulants may predict relapse when they are withdrawn from neuroleptic treatment. However, a stimulant challenge test has not yet been developed for widespread use and is still pending further research.

The extreme sensitivity of ERPs to cognitive and motivational variables makes ERPs an ideal tool for the study of individual differences in normal brain function and in pathologic brain function, where there is obvious disruption of information processing and response selection. Normal individuals vary greatly with respect to changes in ERPs induced by drugs (*e.g.*, Callaway & Naghdi, 1981). For instance, dextroamphetamine may either increase alertness or create paradoxical drowsiness and dysphoria in normal subjects. Contingent negative variation (CNV) has been shown to mirror these differences in behavioral response to dextroamphetamine (Tecce, Savingnano-Bowman & Cole,1982). "Paradoxical" behavioral responses to drugs have often been reported experimentally and in folklore. The "coffee nightcap" is a case in point. It is significant that the statistical result would be no significant mean effect if subjects show different, or even opposite behavioral effects to the administration of an acute drug challenge. If different subjects show radically different, or even opposite, real behavioral effects to an acute drug challenge, group statistics fail to reflect treatment effects. The problem for researchers is to somehow establish covergent validity for real individual differences in response to acute stimulant challenge. Can ERPs be useful in this regard?

There is evidence that scalp recorded electrophysiological activity is sensitive to differences in the biological substrates mediating acute methylphenidate effects. The EEG itself changes in response to methylphenidate: beta activity has been reported to be reduced by methylphenidate (Hynek, Faber, Tosovsky & Cerny, 1974), whereas relative power in the alpha frequency has been reported to increase (Berchou, Chayasirisobhon, Green & Mason, 1986; Craggs, Wright & Werny, 1980).

The influence of methylphenidate, however, appears selective, as shown by ERP experiments in which behavior and ERPs have been simultaneously examined (Berchou *et al.*, 1986; Brumaghim, Klorman, Strauss, Lewine & Goldstein, 1987; Naylor, Halliday & Callaway, 1985). It is well known that stimulants speed reaction time (RT), and ERP/RT studies have been designed to assess whether perceptual processing or motor processing or both are speeded. Reaction times to stimuli of varying complexity usually decrease with acute administration of methylphenidate. The speed/accuracy tradeoff does not appear to be a major consideration as normal subjects do not usually make significantly more errors while on methylphenidate than on placebo. Naylor *et al.* (1985) reported that in their sample of young adults the ability of methylphenidate to speed RT was even greater when the stimuli were more complex. Some evidence in support of an individual difference factor in RT is provided by Halliday, Callaway, Naylor, Gratziner, and Prael (1986). They reported that stimulants speeded response processing in the young but not in elderly subjects.

ERP results of the above studies indicate that methylphenidate primarily affects motor processing. The amplitude of ERP components related to stimulus information processing, *e.g.*, the P300 component, does not appear to be as affected by acute methylphenidate challenge, at least not by oral administration in normal adults. The latency of P300 was also not affected by methylphenidate in these studies, although there were exceptions depending on the nature of the experimental situation. Similar conclusions were made by Herning, Jones, Hooker, and Tulanay (1985) based on the results of their study of auditory ERPs in adults who received intravenous doses of cocaine, a drug with many effects similar to methylphenidate.

Naylor *et al.* (1985) suggested that their P300 data showed evidence of curvilinearity reminiscent of the Yerkes-Dodson law, which posits an inverted-U relationship between arousal and performance. Perloquin and Klorman (1986) argue against the dramatic individual differences in response to methylphenidate that would be predicted by a model hypothesizing differences in arousal level. The authors investigated the effect of acute and oral administration of methylphenidate on normal children. RT and P300 findings indicated that methylphenidate enhanced the quality of stimulus evaluation in a memory scanning task, and the speed of response processing in a vigilance task. Since these effects for normal children were the same as for children with attention deficit disorders and for normal adults, Peloquin and Klorman argued against the existence of a paradoxical response to methylphenidate in different subject groups. The addition of measures designed to reflect the effect of methlyphenidate on dopaminergic systems might prove more successful in isolating sources of subject variance in future ERP studies of acute stimulant challenge.

We turn now to a similar model of psychosis based on hallucinogenic drugs. Human studies of hallucinogens were conducted primarily in the late 1950s to early 1970s. The research grew out of renewed cultural interest in these mind-altering drugs, but initially tended to be anecdotal and uncontrolled. The research focused on the observation that the behavior of persons under the influence of hallucinogenic drugs tended to mimic temporarily the behavior of psychotically ill persons. This fact led to the development of the psychotomimetic model of schizophrenia, or of psychosis in general, as a tool for understanding mental disorder. While models based on stimulant drugs tend to have focused on dopamine, psychotomimetic models have traditionally focused more on serotonin pathways. This always was an oversimplification, of course, as many brain pathways are probably involved. Recently, a psychotomimetic model hypothesizing a dysfunction of the NMDA receptor has been introduced (Javitt & Zukin, 1991). The model has been linked to ERP data by Javitt *et al.* (in press) and by McCarley *et al.* (1991).

N,N-Dimethyltryptamine (DMT) Study

Several years ago we conducted ERP and H-reflex studies "piggybacked" on an endocrine study designed to explore hypothalmic-pituitary responses to acute challenge with the hallucinogen DMT. The neuroendocrine findings have been published (Meltzer, Boutros, Simonovic, Gudelsky & Fang, 1983). The H-reflex findings were presented at a Society for Neurosciences meeting (Metz, Boutros, Tueting, Grimm & Meltzer, 1982). The ERP and behavioral findings were presented at the Seventh International Conference on Event-Related Potentials of the Brain (EPIC VII) (Tueting & Metz, 1983). A more detailed description of the results of this study, especially with respect to psychological effects and effects on the H-reflex and ERP, seems appropriate now in light of recent theorizing about the possible temporolimbic sources of late components of the ERP (Javitt *et al.*, 1990; McCarley *et al.*, 1991; Tueting, 1991a). Moreover, Strassman *et al.* (1991) recently reported data obtained from 12 experienced hallucinogen users in a study of DMT similar in many respects to ours.

DMT is an indole hallucinogen having properties of an endogenous psychotogen (Gillin, Kaplan, Stillman & Wyatt, 1976; Rodnight, Murray, Don, Brockington, Nicholls & Berley, 1976). The drug produces a variety of cognitive and affective changes in normal volunteers. Szara (1956) reported that intramuscular (i.m.) DMT produced visual hallucinations and illusions, distortion of spatial perception and body image, distortion of thoughts, and euphoria in normal subjects one to five minutes after injection. Turner and Merlis (1959) reported that subjects became restless, actively withdrew from social contact, and experienced fear, visual hallucinations, irrational thoughts, and dizziness following i.m. injection of DMT. Rosenberg, Isbell, and Miner (1963) reported that DMT produced anxiety, hallucinations (usually visual), and perceptual distortion in normal volunteers. Gillin, Kaplan, Stillman, and Wyatt (1976) also found that i.m. administration of DMT to normal volunteers produced an acute psychotic reaction, including visual illusions and hallucinations, excitation, and faster thinking. They alluded to the usefulness of DMT as a model for schizophrenia. However, they also noted that symptoms which are frequently seen in endogenous psychoses such as auditory hallucinations and paranoid delusions usually fail to develop, at least not in the same character as in psychosis.

It has been proposed, largely on the basis of animal studies, that DMT produces its psychological effects primarily through its influence on serotonergic systems, although dopamine, acetylcholine, and other neurotransmitters have also been implicated (Commissaris, Lyness, Moore & Rech, 1981; Glennon & Rosencrans, 1981; Meltzer, Simonovic, Fang & Goode, 1981; Sloviter, Drust, Damiano & Connor, 1980).

Pilot data for these studies provided preliminary evidence that DMT is a serotonin (5-HT) agonist in man. DMT (0.7 mg/kg i.m.) administered to three experienced drug users produced increases in serum prolactin, growth hormone and cortisol, consistent with the 5-HT agonist hypothesis (Meltzer, Wiita, Tricou, Simonovic, Fang & Manov, 1982). However, cyproheptadine, a serotonin antagonist, appeared to block the growth hormone increase induced by DMT, but had no significant effect on prolactin and cortisol secretion. Contrary to a serotonin-based hypothesis, the overall DMT experience was reported as more pleasurable when DMT was combined with cyproheptadine. (Meltzer *et al.*, 1982). Pretreatment of one subject with the neuroleptic drug, haloperidol (2 mg), a dopamine receptor antagonist, also appeared to intensify the psychotomimetic effect.

A larger controlled study was designed to explore further the apparent blocking effect of cyproheptadine on growth hormone and to quantify the psychological effects of DMT and the effect of cyproheptadine and haloperidol pretreatment on DMT's psychotomimetic action. A rating scale was developed and the study was conducted double-blind. In the main study (Experiment 1), DMT was administered following pretreatment with the serotonin antagonist cyproheptadine, the dopamine antagonist haloperidol, or a placebo-control. The effects of DMT on behavioral ratings, prolactin, growth hormone and cortisol levels in serum (Meltzer *et al.*, 1983), and on the somatosensory evoked potential and the H-reflex recovery curve (HRRC), were also measured. ERP and H-reflex were "piggybacked" onto the study to explore the extent to which DMT effects paralleled previous ERP and H-reflex findings with psychotic subjects, specifically the decrease in late component amplitude and the increase in secondary facilitation of the Hoffman reflex (H-reflex) recovery curve in schizophrenia (Metz, Goode & Meltzer, 1980a). It was hoped that this information would clarify the role of serotonin in psychosis and provide insight on the role of the ERP and H-reflex measures in the psychotomimetic model of psychosis.

To our knowledge, there have been no studies in man involving the H-reflex and hallucinogenic drugs. However, EEG and ERP studies in human subjects were conducted mainly during the popular surge in psychedelic drug use in this country. The EEG response to the hallucinogen phencyclidine (PCP) involves generalized theta, bilaterally synchronous high voltage low paroxyms suggesting deafferentation of cortical neurons (Fiariello & Black, 1978). Shagass, Schwartz, and Marrazzi (1962) reported that the hallucinogen lysergic acid diethylamide (LSD) failed to alter the amplitude of the major component of the somatosensory evoked potential elicited by ulnar nerve stimulation, even though the dose of LSD was large enough to change affect, emotion and thinking. Brown (1968; 1969) reported different effects of LSD on visual evoked potential color specificity in subject groups differentiated on the basis of the ability to visualize.

Chapman and Walter (1965) reported different effects of small vs large doses of LSD on the amplitude of visual evoked potentials, and latency of late components increased following LSD administration. However, PCP and LSD have a different biochemical structure than DMT. Therefore, the most interesting ERP data obtained during this period with respect to our study were reported by Arnold (1975). He directly compared auditory evoked potential response to DMT and LSD, and reported that DMT, but not LSD, markedly depressed auditory evoked response amplitude. The fact that evoked potential findings in normal subjects following DMT administration mimicked findings in schizophrenic patients supports a psychotomimetic model of schizophrenia that incorporates the drug DMT.

METHOD

Subjects

Five experienced male drug users participated in the main study, Experiment 1, although H-reflex data were obtained from four of these subjects only. One female subject completed two out of the four sessions of the main study; results for the two sessions were consistent with the more complete data set, but will not be reported here. Three subjects returned for a follow-up study of haloperidol in combination with DMT in Experiment 2. Finally, two of the original subjects and a new male subject participated in a follow-up study of the time course of DMT challenge to resolve issues regarding the time course of DMT's effect. Subjects' ages ranged from 20 to 32 with a mean age of 27. All had extensive previous experience with a variety of hallucinogenic drugs, and were recruited through a newspaper advertisement. They were prescreened to exclude physical or psychiatric problems or a family history of psychiatric illness other than drug abuse. They were instructed to refrain from consuming alcohol, psychotomimetic drugs or other drugs for one week prior to each study session. The studies were approved by an Institutional Review Board and an Investigational New Drug Exemption (IND) was obtained from the FDA.

PROCEDURE

Experiment 1: Main Study

Each of the five subjects in the main study was tested under four conditions. Each condition paired a drug pretreatment with a drug injection at the time of the experimental session. The four conditions were: 1) placebo-placebo or pl-pl, 2) cyproheptadine-DMT or cyp-DMT, 3) cyproheptadine-placebo or cyp-pl, 4) placebo-DMT or pl-DMT. The four sessions took place on separate days, with at least one week between sessions. The order of the four conditions was randomly determined for each subject. Both the subjects and the experimenters were blind as to the order of drug treatment conditions.

Each subject was given three capsules containing either 4 mg cyproheptadine or an inactive substance. The subject was instructed to take one capsule per day for three days prior to the experimental session and to have a light breakfast on the study day. Sessions were conducted between 12 noon and 5 p.m. During the session, the subject received an i.m. injection of either DMT or saline. DMT was administered at a dosage of 0.7 mg/kg of body weight up to a maximum of 50 mg.

The Brief Quantified Mental Status (BQMS) is a 25-item scale developed by Meltzer *et al.* (1983) to measure psychological variables that seemed important in the DMT experience (see TABLE 1). Each variable was rated by the experimenter on a scale of 0 (not present) to 3 (severe). The rating scale was administered by the same experimenter (NB) throughout the study. During each session, it was

TABLE 1. Items on Brief Quantified Mental Status

Items
Perceptual Effects
1. Depersonalization
2. Auditory "hallucinations"
3. Visual hallucinations
4. Sense of unreality
5. Illusions
Cognitive Effects
6. Confusion
7. Disorientation
8. Blocking
9. Concreteness
10. Loose association
11. Delusions
12. Poor concentration
Emotional Effects
13. Sadness
14. Giddiness
15. Anger
16. Anxiety
17. Lability
18. Agitation
19. Fluctuation
20. Elation
Side Effects
21. Nausea
22. Vomiting
Miscellaneous Effects
23. Inappropriateness
24. Negative towards test
25. Lack of involvement with examiner

administered 30 minutes prior to drug injection, immediately following the injection, and at 15, 30, 60, and 90 minutes thereafter. At the end of the session, a 30-item DMT Experience Questionnaire for obtaining additional phenomenological information about the intensity and nature of the experience was also administered to the subject.

In addition to these psychological/experiential measures, endocrine and psychophysiological data were acquired throughout the session. An indwelling catheter was placed in an arm vein 60 minutes prior to drug injection. Periodic blood samples were taken via the catheter for endocrine studies. Electrodes for the ERP and H-reflex recordings were also attached during the period before injection and recordings made with the subject seated in a chair in a soundproof chamber during the intervals between repeated administrations of the BQMS rating scale.

H-Reflex Procedures

H-reflex measures were obtained preinjection and at approximately 20 min after injection of DMT. The H-reflex is the electrically evoked monosynaptic spinal reflex. The H-reflex recovery curve (HRRC) is produced by varying the interval between paired stimuli. The ratio of the reflex produced by the second stimulus relative to that produced by the first is an indication of the excitability of the alpha motor neuron pool. H-reflex testing procedures have been fully described previously (Goode, Crayton & Meltzer, 1977; Hugon, 1973). Briefly, recording electrodes were taped to the skin over the gastrocnemius-soleus muscles. Pulse-pairs were presented every 10–15 seconds. The posterior tibial nerve was stimulated at the popliteal fossa (1-msec pulses) to provide maximum amplitude of an unconditioned monosynaptic reflex (H1). The stimulus for a conditioned H-reflex (H2) was the same intensity, with the interstimulus interval varying in 10 standard steps between 50 and 300 msec. Intervals in this range (the secondary facilitation phase of the recovery curve or SF) are most sensitive to pharmacological actions (Metz, Busch & Meltzer, 1981; Metz, Goode & Meltzer, 1980a; Metz, Holcomb & Meltzer, 1980b) as well as to psychiatric illnesses (Metz *et al.*, 1980a), even though complete recovery of the conditioned H-reflex does not occur until the interstimulus interval is 2–4 seconds. The average of the 10 intervals provides a single number which summarizes the subject's performance on a given test. Between 1 and 5 trials were presented at each interval, depending on the stability of the responses. Electrical stimuli were presented to the subjects via a Grass S48 stimulator, through stimulus isolation and constant current units. Stimulus presentations followed a preset sequence controlled by a PDP 11/03 laboratory computer, which also digitized and recorded the muscle responses and measured peak-to-peak amplitudes. The experimenter also monitored the EMG on an oscilloscope screen.

ERP Procedures

EEG electrodes were applied by a standard technique at Cz, Pz, C3x (1 cm posterior and 2 cm lateral to C3), and at C4x with linked ear reference, and below and at the outer canthus of the eye for EOG recording. ERP stimuli were low-voltage electrical pulses to the median nerve of the right wrist. Stimuli were presented in a target detection paradigm with the target (a slightly higher intensity stimulus) probability set at 20%. The average interstimulus interval was 10 sec. There were 320 stimuli (64 targets and 256 nontargets) presented preinjection, 320 after injection, and 320 postinjection. Subjects were instructed to move their finger from a key as quickly as possible when a target stimulus was presented. The response latency criterion was set at 1000 ms poststimulus. Stimulus presentations were controlled by the computer, which also digitized and recorded ERP epochs for off-line analysis. Individual trials with eye movement contamination were removed from the average by visual analysis of single scalp and EOG epochs. Trials were separated into hits, misses, correct rejections, and false alarms. Only trials associated with hits and correct rejections were averaged. ERP measures were visually identified and measured with the aid of the computer. Component identification was more difficult under drug conditions. In some cases, an amplitude measurement was arbitrarily made at the component's expected latency based on pre- and postdrug baseline conditions where the component was more easily identified.

Experiment 2: Haloperidol-DMT (Hal-DMT)

After completion of these four experimental days of the main study, three of the subjects returned for an additional test using a similar paradigm involving pretreatment with the dopamine antagonist haloperidol. This study involved a single nonblind session. Each subject received 0.5 mg haloperidol i.m. 60 minutes prior to DMT injection. The DMT dosage, psychological measures, and physiological tests were administered as in Experiment 1 above. The H-reflex was tested before the haloperidol, before the DMT, and 20 minutes after the DMT. Complete H-reflex data were recorded for only two of the three subjects.

Experiment 3: Time course of DMT Effect

Three of the original five subjects and an additional subject participated in another session designed to more accurately establish the time course of the DMT effect, particularly with respect to the H-reflex. No ERP or blood samples were collected. This study also involved a single nonblind session. Baseline data were obtained at about 1:00 p.m. A single DMT injection was given as in the main study, and there was no drug pretreatment. Behavioral ratings were made 15 minutes prior to injection, at the time of injection, and at 5, 10, 15, 20, 30, and 45 minutes thereafter. H-reflex recordings were made during the intervals between behavioral ratings at approximately 20 seconds after the injection and then every 5 minutes for 30 minutes, then again at 40 and 60 minutes. Each test of the H-reflex recovery curve required approximately 3 minutes. Subjects were allowed to engage in conversation or listen to the radio during H-reflex recording periods, and the situation was generally more relaxed than in earlier sessions with these same subjects.

RESULTS

Psychological Changes

For all subjects, administration of DMT resulted in clear "DMT experiences," as indicated by the DMT Experiences Questionnaire. Subjects reported that these experiences were of comparable intensity to their previous experiences with street hallucinogens, and rated the DMT conditions as "moderate" to "strong" drug experiences. Their DMT experience involved perceptual changes (*e.g.*, odors were more intense, objects looked distorted, and things seemed unreal), cognitive distortions (*e.g.*, strange thoughts, racing thoughts, inability to concentrate), and emotional changes, notably elation. Both visual and auditory hallucinations were also reported. Subjects reported that they saw things with their eyes closed, but that these things were unreal. They heard sounds that were not present in the room, but not specific words. Thus, these sensory changes did not appear to have the same quality as hallucinations in schizophrenia where distinct words convey distinct messages, but were more like pseudohallucinations. The intensity and pleasure of the experience was, if anything, rated greater when DMT was administered jointly with cyproheptadine or haloperidol.

The BQMS items were grouped into the following categories: perceptual, cognitive, emotional, and side effects (see TABLE 1). BQMS results were evaluated in terms of these four factors as well as an overall composite score. There were no psychological effects apparent on the BQMS for either the pl-pl or cyp-pl conditions. By contrast, both pl-DMT and cyp-DMT treatments resulted in marked effects in all subjects for composite, perceptual, cognitive, and emotional scores (FIG. 1). Very little effect of pretreatment with cyproheptadine is apparent in the ratings. The peak effect across subjects for each treatment was at 15 minutes after the injection (T15). Cognitive and emotional effects appeared to have a slightly longer duration than perceptual effects under both DMT conditions. Perceptual effects were largely over by T30 whereas cognitive and emotional effects were still pronounced at T30. Subjects experienced visual "hallucinations" when DMT was administered, whether or not cyproheptadine was also administered. Auditory "hallucinations" were experienced at T15 by three subjects under pl-DMT and by four subjects under cyp-DMT. Side effects were largely absent, except for one subject who experienced mild nausea under both DMT conditions.

It should be mentioned that for all subjects, administration of DMT resulted in psychological and behavioral effects that were obvious to both the subjects and the experimenters. Some subjects commented on noticeable effects of the cyproheptadine pretreatment, such as dryness of the mouth and drowsiness (presumably due to the drug's anticholinergic properties). Thus, although the study was ostensibly double-blind, both subjects and experimenters could make a fair appraisal of the actual drug treatment condition for a given session, before some of the specific psychological measures were taken.

Two of the three subjects given haloperidol in a nonblind manner prior to DMT showed an intensified, more pleasant drug experience (as indicated by the DMT Experience Questionnaire) relative to the pl-DMT condition. The third subject showed a slightly attenuated effect. Haloperidol pretreatment tended to prolong the DMT effects on the BQMS (see FIG. 2). This prolongation was most pronounced and most consistent across subjects for the emotional category. Auditory "hallucinations" were again present in two of the three subjects.

A two-way-repeated-measures ANOVA was run on average composite scores for T0, T15, and T30 crossed by the five drug treatment conditions in Experiments 1 and 2. BQMS composite scores for the three subjects who participated in both studies were included. Both the time effect ($F(2,4) = 164.19$, $p < 0.01$) and the drug effect ($F(4,8) = 41.33$; $p < 0.01$) were significant. We used the Scheffe method for post-hoc comparisons. The only significant contrast was that between the two pretreatment-saline conditions vs the three pretreatment-DMT conditions. This supports the BQMS as a measure of the psychological effects of DMT and suggests that both cyproheptadine and haloperidol had no significant effect on the psychological effects of DMT.

BQMS ratings tended to be substantially smaller for Experiment 3 (time course experiment) than for the earlier DMT conditions. Three of the four subjects failed to report auditory "hallucinations" at any time period. The peak DMT effect, in terms of the composite, perceptual, and emotional BQMS scores was at T5.

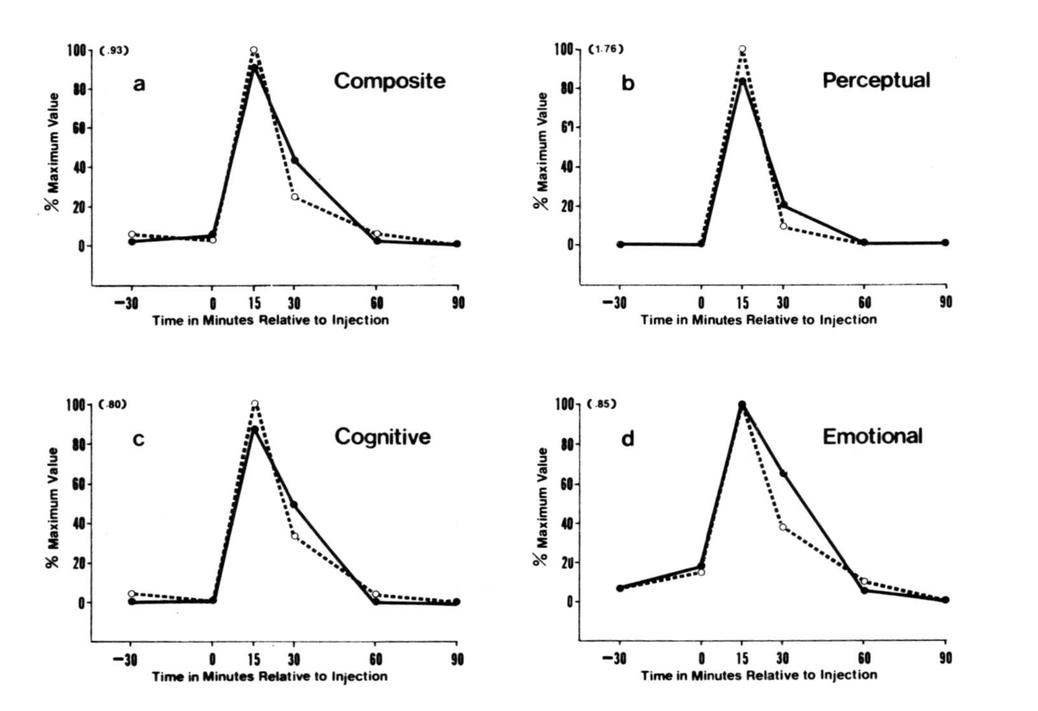

FIGURE 1. Time course of the mean values for each BQMS category, averaged across the five subjects in the Main Experiment, Experiment 1. Values are plotted as percentages of the maximum value across conditions for each BQMS variable set. The maximum value to which each graph is scaled appears in parentheses at the 100% mark. Only two DMT conditions are represented.

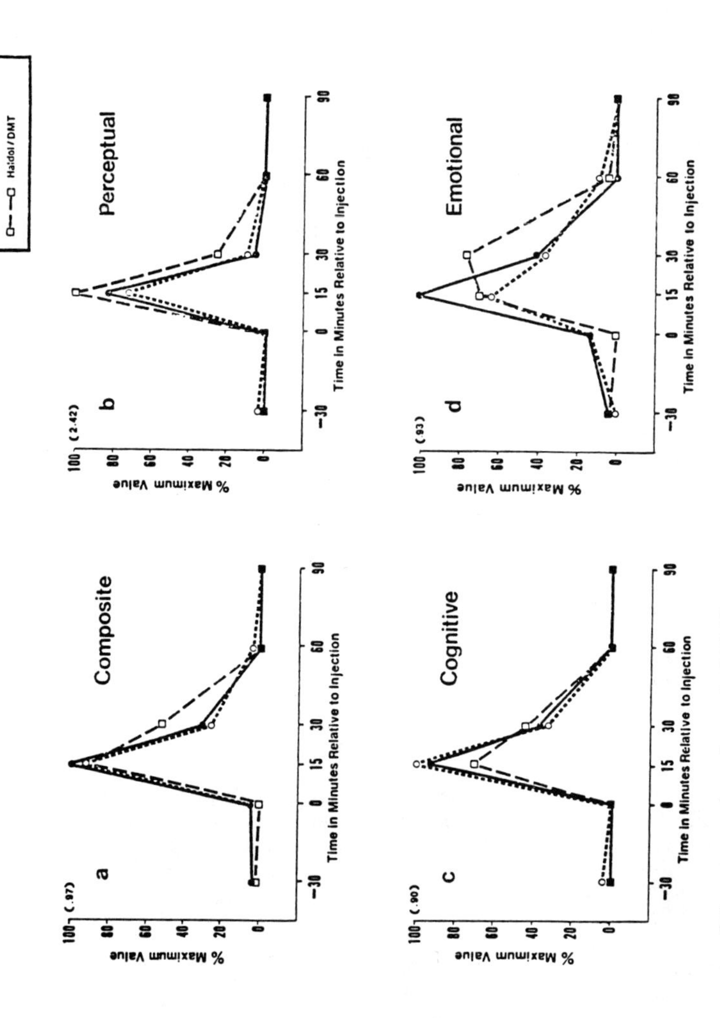

FIGURE 2. Time course of the mean values for each BQMS category, averaged across the three subjects who participated in both Experiment 1 and 2. Values are plotted as percentages of the maximum value across conditions for each BQMS variable set. The maximum value to which each graph is scaled appears in parentheses by the 100% mark. Only the three DMT conditions are represented.

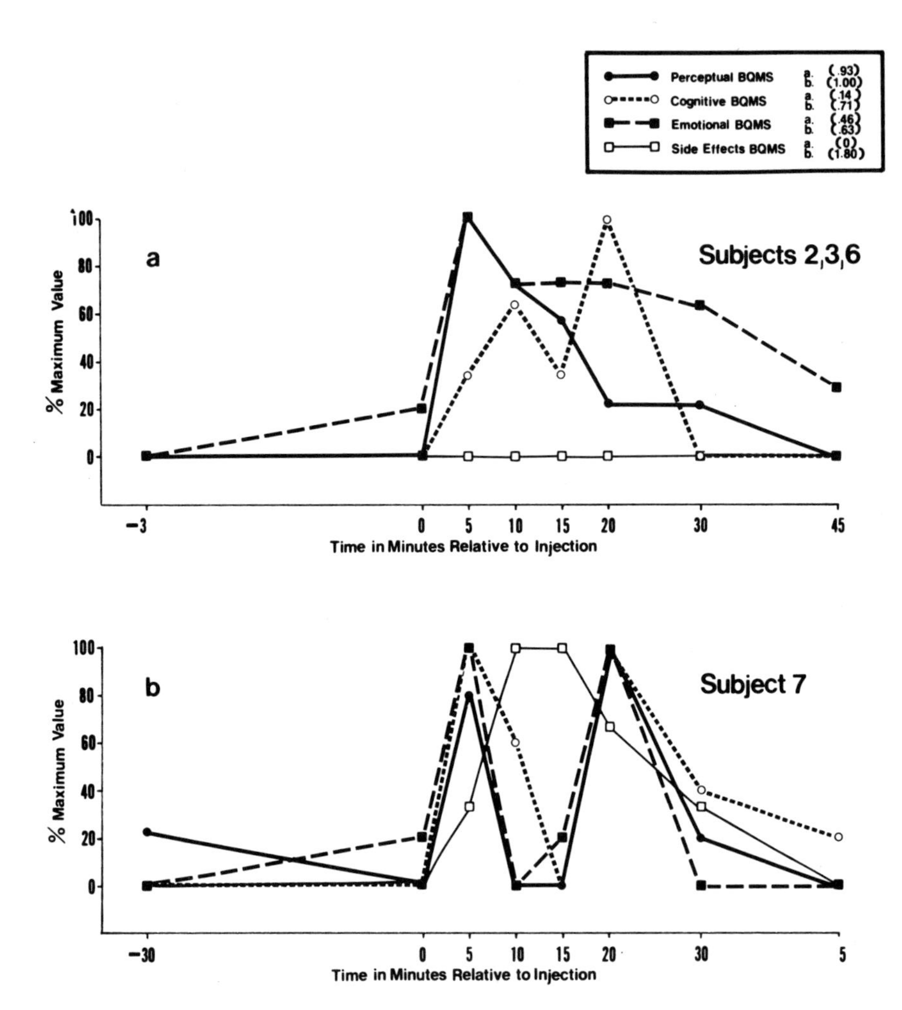

FIGURE 3. Time course of the mean values for each BQMS variable set for Experiment 3. Values are plotted as percentages of the maximum value for each BQMS category. The maximum value to which each trace is scaled appears in parentheses in the key. **(a)** Values averaged across subjects 2, 3, and 6; **(b)** values for subject 7.

Cognitive effects were very small and peaked at T20. As in Experiments 1 and 2, cognitive and emotional effects lasted longer than perceptual effects. There was a trend across subjects for DMT effects to be biphasic, with peaks at T5 and T20 as shown in FIGURE 3. However, this trend was primarily due to one subject who experienced nausea, which apparently temporarily interfered with expression of the other DMT effects. It is interesting to note that for this subject, there was a rebound of typical DMT experiences as the nausea subsided. This rebound occurred at a time (T20) at which the DMT effect had largely dissipated for the other three subjects.

H-Reflex

TABLE 2 shows the values of secondary facilitation (SF) and unconditioned monosynaptic reflex (H1) averaged across intervals for each of the subjects in the different conditions. H1 showed small and nonsystematic changes in the different baseline conditions. Relative to baseline, H1 increased in all subjects (increased very dramatically in subjects 2 and 4) in the postinjection test on the P1-DMT day. This effect of DMT on H1 is apparently blocked by cyproheptadine. H1 also was slightly reduced in both subjects after haloperidol; this is in contrast to our findings with up to 25 mg of chlorpromazine, which has no effect on H1 (Metz *et al.*, 1980b). Otherwise, H1 did not change in a consistent manner between baseline and postinjection tests in the different conditions. Both H1 and SF were less stable in baseline conditions than we had found in previous studies (Metz *et al.*, 1980b), which could be an effect of cyproheptadine. However, since even the placebo tests were not as stable as in earlier studies, it may be that the frequent drug usage of these subjects also contributed to instability in these measures.

FIGURE 4 summarizes the changes in the H-reflex recovery curve (HRCC) in each of the five conditions for the subjects. Because of the variability of baseline secondary facilitation (SF) from condition to condition and because of individual differences in subjects (Crayton & King, 1981; Metz *et al.*, 1980a,b), points in this graph represent SF after the injection divided by SF before the injection (before the first injection in the case of Hal-DMT). Thus, if a subject did not change after the injection this ratio would be 1.0; if SF increased after the injection the ratio would be greater than 1.0; if SF decreased after the injection the ratio would be less than 1.0.

The main findings of this study are clear from the figure; SF was dramatically increased following DMT, and this effect was antagonized by pretreatment with cyproheptadine. Even though there were few subjects, the treatment effect in the

TABLE 2. Values of Unconditioned Monosynaptic Reflex (H1) and Secondary Facilitation (SF) for Each Subject in Different Conditions

	Pl-Pl		Cyp-Pl		Pl-DMT		Cyp-DMT		Hal-DMT	
Subject	H1	SF	H1	SF	H1	SF	H1	SF	H1	SF
S1										
Base	5.2	51	4.2	34	4.0	47	5.0	28	3.7	47
Inj 1	4.6	50	4.4	40	4.5	59	4.1	27	1.8	28
Inj 2									9.8	58
S2										
Base	10.2	96	8.5	85	7.9	73				
Inj 1	9.8	83	7.9	83	15.3	95				
S3										
Base	8.0	29	6.6	18	7.2	12	3.4	18	8.0	21
Inj 1	7.5	25	6.6	8	7.5	50	4.7	32	6.7	23
Inj 2									8.1	34
S4										
Base	4.9	31	2.9	37	7.2	21	2.2	29		
Inj 1	6.4	28	2.6	40	12.3	45	2.9	28		
S6										
Base	7.6	17	4.5	8	3.3	13	9.1	16		
Inj 1	5.8	12	4.5	8	6.6	22	12.3	14	6.3	17
Inj 2									6.5	36

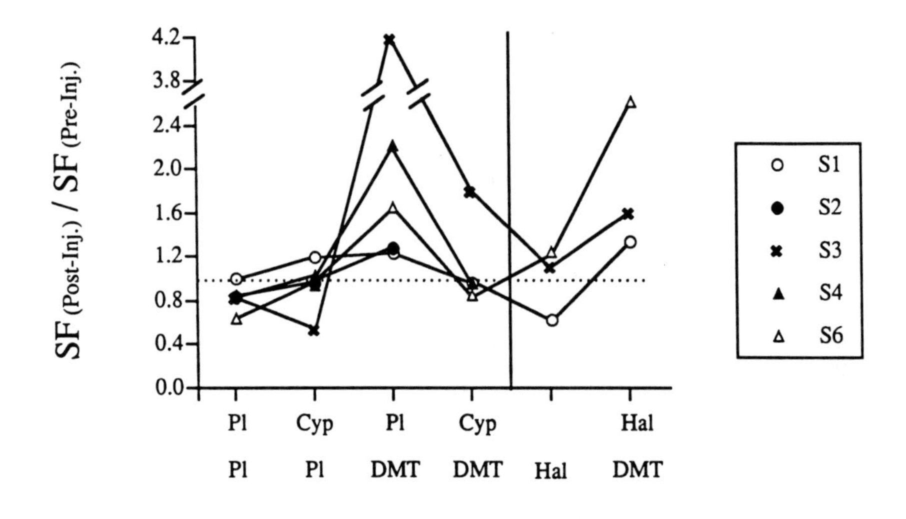

FIGURE 4. Effect of DMT and pretreatments on the H-reflex recovery curve (HRRC) and secondary facilitation (SF).

three conditions for which we have data from all subjects was statistically significant (Friedman 2-way analysis of variance statistic = 6.50, $p < 0.05$. The increased variability among subjects in the cyp-pl condition compared to pl-pl confirms that cyproheptadine may induce some instability in the HRRC. Haloperidol does not appear to consistently either block or enhance the DMT effect on H-reflex.

The main experiment and the haloperidol follow-up study established that DMT increases both the unconditioned monosynaptic H-reflex (H1) and secondary facilitation (SF). FIGURES 5 and 6 summarize the effects of DMT on SF over time

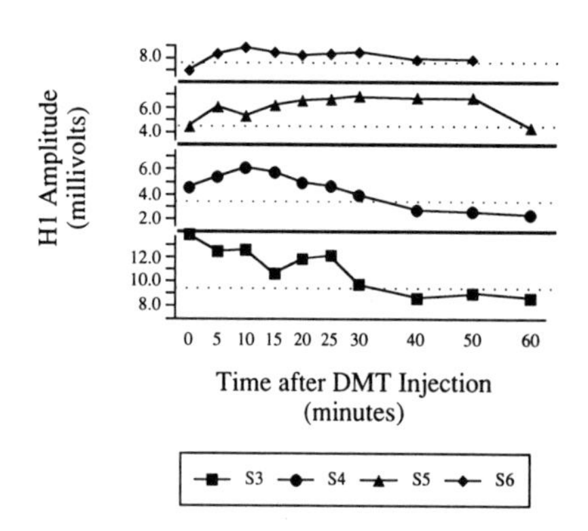

FIGURE 5. Effect of DMT on the unconditioned monosynaptic reflex (H1) as a function of time after injection in Experiment 3. *Broken horizontal line* indicates level of preinjection H1.

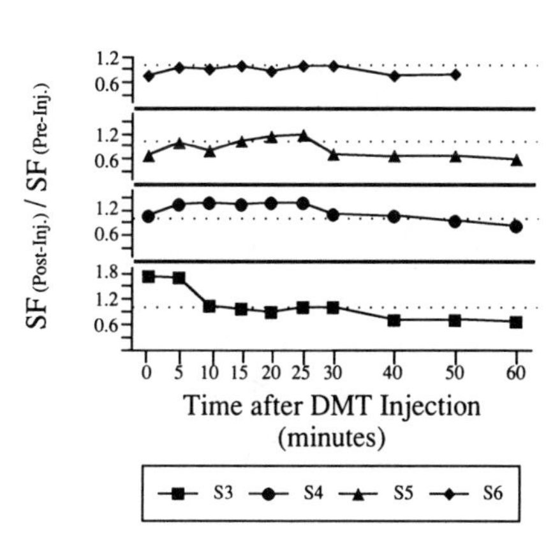

FIGURE 6. Effect of DMT on secondary facilitation (SF) as a function of time after injection in Experiment 3. *Broken horizontal line* indicates level of preinjection SF.

in the time course follow-up study. In all subjects, the unconditioned monosynaptic reflex (H1) clearly increased within 10 minutes of the injection and remained elevated until at least 25 minutes after the injection. Three of the subjects showed increases in secondary facilitation (SF), although the timing of the effect was different in each. In the case of one subject, the increase was somewhat earlier and much smaller than we had observed in this same subject in the main study (Experiment 1). Subject 6 showed a slight decrease in SF following DMT. All subjects had moderate decreases in SF after 30–60 minutes.

ERP Changes

TABLE 3 provides information on task performance during ERP recording. These data reflect accuracy within the reaction time (RT) criterion window, *i.e.*, a hit outside of the RT criterion of 1000 ms would be automatically scored by the program as a miss. Four of the five subjects did well overall, but one subject with long RTs did rather poorly overall. The important observation is the difference in hit rate for the different conditions. Surprisingly, subjects continued to perform fairly well after the administration of DMT and recovered quickly to the baseline level in the third set of trials.

Changes in ERP waveshape with DMT were dramatic whether or not DMT was

TABLE 3. Mean Performance of the Five Subjects in Behavioral Task in Percentage of Total Trials

	Preinjection				Injection				Postinjection			
Condition	Miss	Hit	CR	F+	Miss	Hit	CR	F+	Miss	Hit	CR	F+
pl-pl	4	16	80	0	4	16	80	0	2	18	80	1
Cyp-pl	3	17	80	0	3	17	80	0	4	16	80	0
Pl-DMT	3	17	80	0	6	14	79	1	3	17	80	0
Cyp-DMT	2	18	80	0	7	13	77	3	2	18	80	0
Hal-DMT[a]	1	19	80	0	6	14	75	5	2	18	80	0

[a] Based on 3 out of the 5 subjects in single blind session.

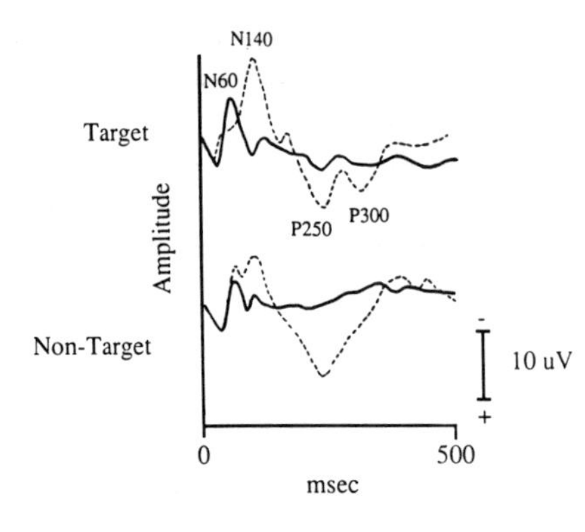

FIGURE 7. ERPs elicited by somatosensory target and nontarget stimuli for one subject at the C3x electrode. The *dotted line* represents a baseline condition (cyproheptadine pretreatment) and the *solid line* represents the drug combination of DMT + cyproheptadine.

given in conjunction with cyproheptadine or haloperidol. Large late components, a large negative component with a mean latency of about 135–140 ms and large late positive components in the P200 and P300 latency range were dramatically reduced and in some cases obliterated completely in four of the five subjects who participated in Experiment 1 and in the three subjects who participated in Experiment 2 (see FIG. 7).

The decrease in late components following DMT administration was accompanied by an increase in early components. P45 and N60 were identified over contralateral cortex (C3x). P90 was also present at C3x but had a bilateral distribution

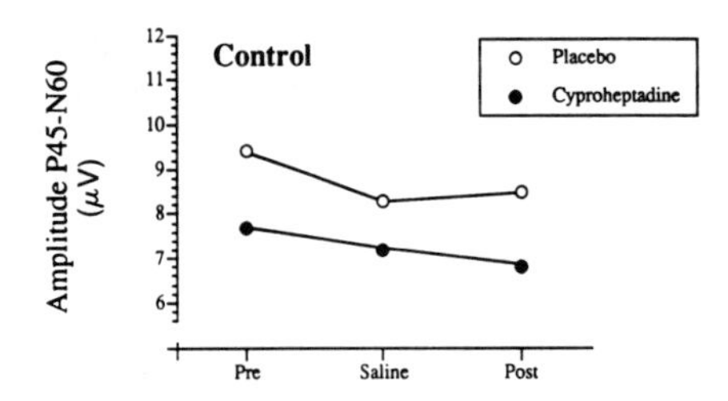

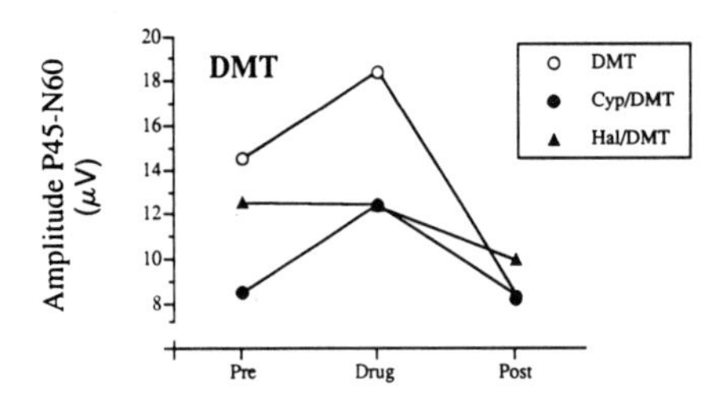

FIGURE 8. Mean amplitude of the P45-N60 somatosensory component recorded from C3x over somatosensory cortex contralateral to the stimulated wrist for control and drug conditions.

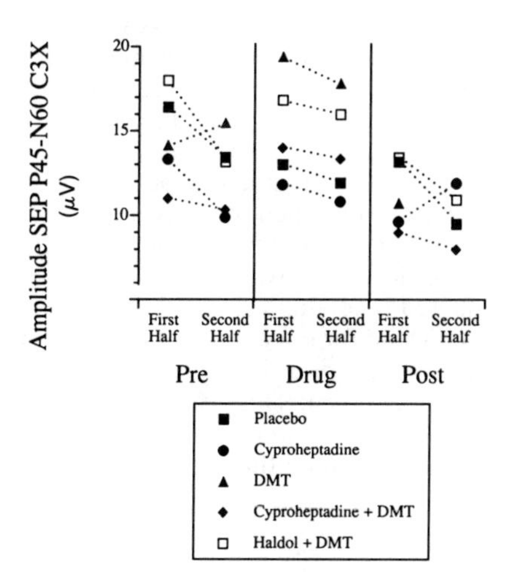

FIGURE 9. Same data as for FIGURE 8 except that data are plotted differently to show the split-half replicability feature and habituation.

with a Cz maximum. Peak-to-peak measurements of these components were made and mean amplitude of P45-N60 is presented in FIGURE 8. Cyproheptadine is associated with a decrease in the amplitude of P45-N60 for all conditions, pre- and posttreatment as well as during saline or DMT. The effect of DMT is large and interrupts the habituation curve. ERP data for the preinjection, injection, and postinjection trial runs were split in half for purposes of examining habituation and reliability in greater detail. These data for the amplitude of P45-N60 and for N60-P90 at the C3x electrode location are presented in FIGURES 9 and 10. Cyprohepta-

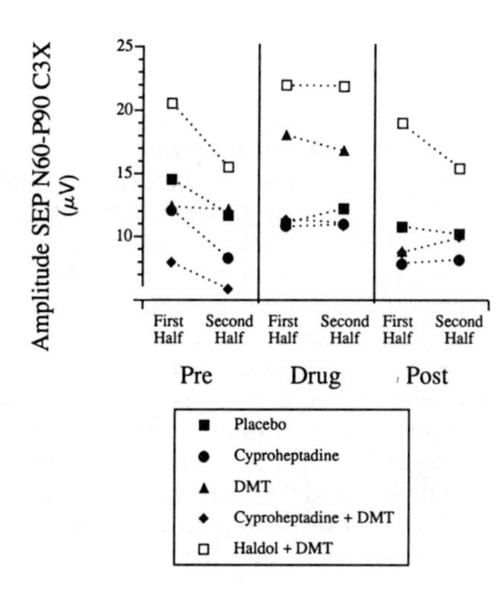

FIGURE 10. Data plotted in the same way as for FIGURE 9 but for a later component, N60-P90, at the C3x electrode.

dine, in addition to being associated with a reduction in amplitude of early components, was associated with an increase in latency. Measurements of the latency of the N90 component are presented in TABLE 4. Both cyproheptadine and DMT, but not haloperidol/DMT, appear to be associated with small increases in the latency of this component.

Data for the late components were not averaged because of the one atypical subject who failed to show a decrease in these components, and in fact, showed a slight increase. This subject's ERPs were extremely low amplitude for baseline conditions and waveshape was atypical. Normal recording sessions were conducted before and after this subject's sessions and the results were similar for the DMT and for the cyp-DMT session. He was the oldest subject and had the longest and most serious history of multiple drug abuse, having actually been hospitalized for alcoholism. This subject was also the only one who failed to show an increase in prolactin response after DMT. The subject's BQMS data were, however, similar to those of the other subjects.

TABLE 4. Mean Latency (ms) of Somatosensory Component P90 at Cz

Condition	Pl-pl	Cyp-Pl	Cyp-DMT	Pl-DMT	Hal-DMT
Target					
Pre	87	90	89	90	90
Inj	91	93	103	90	88
Post	88	93	95	87	83
Nontarget					
Pre	94	95	95	102	93
Inj	94	94	102	106	92
Post	94	96	96	107	92

DISCUSSION

Phenomenological Effects

DMT produced moderate to strong, but brief, psychotomimetic responses in all subjects tested. The nature and time course of the response was comparable to that previously described by other investigators (Kaplan, Mandel, Walker, Vandenheuvel, Stillman, Gillin & Wyatt, 1974; Szara, 1956; Turner & Merlis, 1959). The response was largely pleasurable to these experienced users, which is what we had anticipated. No serious side effects, mental or physical, were noted other than transient nausea.

Our data are consistent with a general failure to find blocking of the effects of indole hallucinogens by serotonin or dopamine antagonists in man (Isbell, Logan & Miner, 1959) or in other primates (Schlemmer & Davis, 1983). Individual differences and subtle changes in the quality of behavior, however, have been a consistent finding and these changes may be associated with serotonergic or dopaminergic pathways. For example, the results suggest that adjunctive cyproheptadine and haloperidol may alter the time course of the effect of DMT on human experience.

Our data indicate that the cognitive and emotional effects induced by DMT tend to last longer than the perceptual alterations. In contrast to previous studies which reported only visual hallucinations (*e.g.*, Gillin *et al.*, 1976), DMT also produced auditory "pseudo" hallucinations in our subjects. Auditory hallucinations were more prominent in Experiment 1, in which subjects were new to the experiment and relatively more isolated in the soundproof chamber. Failure to report auditory hallucinations in Experiment 3, the time course study, could be attributed to subjects conversing or listening to the radio during the H-reflex tests, or being generally more relaxed after becoming accustomed to the experimental situation after four or five previous sessions. The situation in Experiment 3 was more comparable to previous anecdotal studies of DMT experiences where conversation was permitted throughout.

Recently, Strassman *et al.* (1991) reported findings similar to ours in a double-blind placebo controlled dose/response study (0.05, 0.1, 0.2, and 0.4 mg/kg DMT) of 12 experienced hallucinogen users. Psychological effects were found to be dose dependent as was the effect of DMT on blood level increases of beta-endorphin, ACTH, cortisol, and prolactin, as well as blood pressure, heart rate, rectal temperature, and pupil diameter. Their findings in regard to cortisol and prolactin were similar to neuroendocrine responses obtained in the present study (Meltzer *et al.*, 1983). Growth hormone responses to DMT obtained by Strassman *et al.* (1991) were inconsistent which also could be related to the delay in growth hormone response to DMT found by Meltzer *et al.* (1983).

H-Reflex Findings

The H-reflex findings, namely the increase in the unconditioned monosynaptic H-reflex (H1) and in secondary facilitation (SF) with DMT were similar to those reported for psychosis (Metz *et al.*, 1980a). The serotonergic aspects of the secondary facilitation effect are supported by the fact that cyproheptadine blocked the effect of DMT on H1 and SF, while haloperidol failed to so, at least not to the same extent as cyproheptadine.

The differences in time course, the different response of one subject in Experiments 1 and 3, and the failure of one subject to show a SF increase in Experiment 3 demonstrate that the physiological effects of DMT are variable and may be influenced by factors in the testing situation. In Experiment 1, in addition to the HRRC test at 20 minutes after the injection, subjects were given somatosensory stimulations (shock to median nerve) and were required to make a stimulus discrimination for purposes of providing evoked potential data. In Experiment 3, subjects were electrically stimulated (much more intensely) every 10–15 seconds through most of an hour, but were not required to attend to the stimulations. It is possible that our findings reflect the fact that demands of the task situation or associated changes in stress, can modulate the physiological effect of DMT. Systematic study of the effect of varying task demands during drug challenge, is needed in future studies.

ERP Findings

DMT, alone, or in combination with cyproheptadine or haloperidol decreased late component amplitude (N100, P200, P300) from Cz, C3x, C4x, and Pz for all but one subject. These findings with the somatosensory evoked potential correspond to data on the auditory evoked potential following DMT in human subjects reported previously by Arnold (1975). Late components to target stimuli were virtually eliminated in some cases, even though the subject continued to respond adequately to target stimuli. Reaction times were, however, considerably delayed by DMT. The one subject who showed an increase in amplitude of late components with DMT also had extremely small baseline evoked potentials, had the longest history of drug abuse, and was the only subject who failed to show an elevation of prolactin with DMT. For this subject, the evoked potential had a distribution that was maximum over the contralateral somatosensory area following the DMT injection, and cyproheptadine blocked the increase in the evoked potential over the somatosensory cortex. Thus, behavioral, ERP, and neuroendocrine data all show that this subject was atypical.

The finding that DMT decreased late ERP components in the majority of subjects is also consistent with the numerous reports of decreased amplitude of late components in schizophrenia and other psychotic conditions and supports the psychotomimetic model of schizophrenia. This decrease in late components could be consistent with the inhibitory effects of amines such as serotonin and dopamine on brain electrical activity, but given the lack of effect of cyproheptadine and haloperidol in this study, is perhaps more consistent with an effect on limbic NMDA receptors.

Since our study was conducted, the neurotoxic properties of the neurotransmitter glutamate and related excitatory amino acids has led to intense research on their role in neurodegenerative disorders. One excitatory amino acid receptor, the N-methyl-D-aspartate (NMDA) receptor is also being studied for possible relevance to memory, epilepsy, and psychosis (Olney, 1989). It is known that the deep temporal lobe and limbic areas of the brain, assumed to be involved in memory, epilepsy, and psychosis, are especially rich in NMDA receptors. Recent theorizing implicates the NMDA receptor in P300 component generation (McCarley *et al.*, 1991). Although there is no evidence that DMT affects the NMDA receptor directly, it is known that another hallucinogen, phencyclidine (PCP) does directly block the ion channel in the NMDA receptor via its own receptor. However, PCP and DMT differ in chemical structure and in psychotomimetic properties. PCP is known to actually induce a psychotic state lasting a week or more, while DMT's psychotogenic effects are thought to be over within hours.

The same pattern of findings, a dramatic reduction in late component amplitude, without a reduction in early components of the auditory evoked response was recently reported by Javitt *et al.* (in press) following challenge with a PCP-like drug, MK801. Their findings with a psychotogen closely linked to PCP, and therefore to the NMDA receptor, in monkeys were strikingly similar to our findings with DMT in humans. It is not known for certain whether DMT, like PCP, is closely related to the NMDA receptor, but this is a theoretical possibility. In any case, both sets of data are consistent with the psychotomimetic model of psychosis with respect to late components of the ERP and temporolimbic function. McCarley *et al.* (1990) have also incorporated the concept of the NMDA receptor into their model of late component generation and of abnormality in psychiatric patients.

The early components (P45-N60, N60-P90) recorded contralaterally to the stimulated hand (C3x) increased in amplitude on DMT, which also parallels the Javitt *et al.* (in press) findings in monkeys as well as a set of somatosensory evoked potential findings in chronic schizophrenia patients. Early somatosensory evoked potential components including N60 have been reported to be larger in amplitude and more stable in chronic schizophrenic patients (Shagass *et al.*, 1978a, 1981).

The increased somatosensory evoked potential amplitude associated with both schizophrenia and DMT also supports the incorporation of DMT findings into a psychotomimetic model for schizophrenia.

Cyproheptadine Effects

A major finding of this study was that cyproheptadine had little effect on DMT-induced changes in cognition, affect, and perception and on late components of the ERP. Cyproheptadine, however, was associated with a decrease in the amplitude of the P45-N60 component both pre- and post-DMT and during the drug's peak effect (FIGS. 8 & 9). The failure of cyproheptadine to block the behavioral effect of DMT contrasts with the fact that cyproheptadine tended to block the growth hormone response (Meltzer *et al.*, 1982) and the H-reflex response (both the increase in H1 and the increase in secondary facilitation induced by DMT) and decreased the amplitude of the early ERP and increased latency. However, small qualitative differences in psychological response to DMT, which could be attributed to cyproheptadine, were present in some subjects. There was some indication that cyproheptadine shortened the duration of the DMT effect. In addition, cyproheptadine appeared to modulate the intensity of the DMT experience in a subject-specific manner, with some subjects actually reporting enhanced experiences.

There have been no previous attempts, to our knowledge, to study the effect of cyproheptadine on the psychotomimetic effects of DMT in man. However, Sai-Halasz (1962) reported that methysergide, another 5-HT antagonist, intensified the psychotomimetic effects of DMT in man. Isbell *et al.* (1959) also found that 1-benzyl-2-methyl-5-methoxytryptamine (BAS), a serotonin antagonist, failed to block the effect of LSD-25, an indole hallucinogen also believed to act through a serotonergic mechanism (Aghajanian & Haigler, 1976). Our results are in agreement with the results of these studies and indicate that classic 5-HT antagonists fail to block sensory experiences elicited by indole hallucinogens in human subjects and might even intensify the experience to a slight extent in some subjects.

Failure to find marked blocking of the DMT experience by cyproheptadine does not rule out involvement of serotonin in the effect of DMT. The fact that cyproheptadine had differential effects on individual subjects in our study points to the possibility that cyproheptadine may be an effective antagonist in some individuals at a lower or higher dose than that given in the present study. Moreover, it is now clear that there are multiple types of serotonin receptors. The ability of some 5-HT antagonists to block and others to potentiate the effects of indole hallucinogens in rodents and man, is consistent with *in vitro* binding data. Serotonin antagonists vary in their affinity for different receptors (Whitaker-Azmitia & Peroutka, 1990) and various types of 5-HT receptors appear to mediate complex responses to indole hallucinogens. The fact that cyproheptadine has significant effects on the histamine and cholinergic receptors also complicates interpretation of its effects on 5-HT receptors, and especially on cognition and H-reflex measures which could be affected via DMT's anticholinergic effects. The involvement of the NMDA receptor as noted above may also be pertinent to this set of findings.

Further, our data and the findings noted above are in general agreement with serotonergic theories of indole hallucinogen action at different levels of the nervous system (Agahajanian & Haigler, 1976; Davis, Kehne, Commissaris & Geyer, 1984). Davis *et al.* (1984) noted that inhibitory 5-HT receptors (presynaptic, autoreceptors on 5-HT cell bodies and at synapses in the forebrain target area) are not blocked by classic antagonists. However, excitatory 5-HT receptors located primarily on motor neurons are blocked by classic antagonists. Thus, sensory experiences elicited by DMT, which were not blocked by cyproheptadine in our subjects, could be mediated by inhibitory 5-HT receptors at higher brain centers and modulated by NMDA receptors. Correspondingly, the effect of DMT on the H-reflex recovery curve, which was substantially blocked by cyproheptadine, may involve excitatory 5-HT receptors on motor neurons. The effect of cyproheptadine on the H-reflex is also in agreement with the data and conclusions of Heyme, Rasmussen, and Jacobs (1984) and Trulson, Heyms, and Jacobs, (1981) concerning the possible role of 5-HT postsynaptic supersensitivity in mediating the behavioral response to psychotomimetics. Typical behavioral measures in animals, *e.g.*, the limb flick response in cats, may be more comparable to the H-reflex measures than to measures of human sensory experience. Both the H-reflex in humans and several of the behavioral measures used in animal studies may be mediated by excitatory postsynaptic 5-HT receptors rather than inhibitory 5-HT receptors.

Haloperidol Effects

In this study, haloperidol pretreatment also failed to block the effect of DMT. In addition, the effect of haloperidol on the H-reflex and on the ERP response to DMT were minimal. Two effects of haloperidol in the present study are intriguing: 1) haloperidol tended to lengthen the time course of DMT, and 2) haloperidol tended to have more of an effect on cognitive and emotional aspects of the DMT experience than on perceptual aspects. These changes could be mediated by an effect of DMT on dopaminergic neurons. Altogether, these results suggest increased dopaminergic activity is not essential for the hallucinogenic effects of DMT in man, but dopamine may nevertheless be involved in modulating the effect. DMT has been reported to increase dopamine synthesis (Smith, 1977), to lower dopamine concentration by enhancing dopamine release (Haubrich & Wang, 1977) and to enhance dopamine turnover in the striatum (Waldemeier & Maitre, 1977) and the hypothalamus (Simonovic, Gudelsky & Meltzer, 1981). These results have led to the hypothesis that the psychotomimetic effects of DMT are due to direct or indirect effects on the dopaminergic system. The lack of effect of haloperidol reported here is contrary to this hypothesis, although it is possible to argue that minimal and subject-specific effects are important or that a different dose of haloperidol might have been effective. However, the dose of haloperidol used was sufficient to produce maximal prolactin stimulation.

Previous reports of the effect of dopamine blockers on the response to indole hallucinogens have also been mixed and inconclusive in regard to the dopamine hypothesis of the action of indole hallucinogens (Blackburn Cox, Heapy & Lee, 1982; Domino, 1975; Jenner, Marsden & Thanki, 1980; Schlemmer & Davis, 1983; Silverman & Ho, 1980; Snyder, 1972). Moore, Hatada, and Domino (1976) reported that low doses of a variety of neuroleptics potentiate the ability of DMT to decrease the amplitude of evoked potentials in the visual cortex and lateral geniculate following stimulation of the optic chiasm. Higher doses inhibited this effect of DMT, indicating that the dose of dopamine antagonist may indeed be a very critical factor.

Interpretation of the haloperidol/DMT results is dependent upon the low dose of haloperidol administered, although the dose was sufficient to produce maximal prolactin responses. However, it is quite possible that higher doses of haloperidol could block the behavioral effects of DMT. These findings reinforce the general conclusion that antipsychotic drugs are poor antagonists of indole hallucinogens and that neuroleptics may even intensify their effects on consciousness under some circumstances (Domino, 1975).

CONCLUSION

The multilevel approach—in this case, combining behavioral and neurotransmitter/neuroendocrine data with evoked potentials—was moderately successful in our study of DMT challenge. How much this study contributes to the psychotomimetic model of psychosis in relation to underlying molecular biology remains to be seen, although results seem consistent with the literature. Regardless, the effects of DMT and cyproheptadine were clearly and reliably different depending on the measure, with H-reflex and even early and late somatosensory evoked potentials differing with respect to response to DMT and the effect of the antagonist. The large reduction in late-component amplitude reflected the cognitive and perceptual distortions experienced by the subjects even though they were still able to perform reasonably well on a well-practiced simple vigilance reaction time task.

The ability of an extension of Sutton's triangular research paradigm to handle the issue of individual differences in drug response was only moderately encouraging. An atypical subject was identified, both on the basis of late component amplitude and neuroendocrine response, but not on the basis of other measures, such as psychological response to DMT and H-reflex. We are unable to explain this pattern of findings further at the present time other than in terms of the subject's considerable drug-abuse history. It is possible that the functional dose of DMT could have been minimal for this subject, and our ability to interpret our findings was clearly hampered by the fact that single doses were selected without careful prior dose-response studies. Strassman *et al.* (1991) have now completed a dose-response study which may help in interpreting the psychological and neuroendocrine data from our study when the results are published. However, the doses used by Strassman *et al.* (1991) were considerably lower than the dose selected for our study. Another issue regarding within-subject studies of this nature, in which subjects participate in many sessions and in which multiple measures are obtained, is that the N may not be large enough to examine the issue of individual differences in depth unless the converging evidence from the multiple levels of measurement is convincing in terms of differences in response pattern.

One issue in acute pharmacological challenge studies of this type is the fact that while ostensibly double-blind, the study may not be completely so. This "pseudo" blindness may be less of a problem with lower drug dosages, which should be kept in mind in designing future studies.

Another important observation, which Sutton might have predicted, is that acute pharmacological challenge is very sensitive to changes in the experimental situation and not only in terms of what the subject is asked to do specifically. The experimental situation, and consequently subject performance and stress on the subject, can change depending upon what the experimenter is doing as well as what the subject is supposed to do. The addition or subtraction of dependent variable measures in the experimental situation is likely to influence the response to drugs, and this could be responsible for differences in psychological and H-reflex response to DMT in the same subject participating in the time course follow-up experiment in which the evoked potential vigilance task was eliminated from the design. Therefore, it may be prudent to add variables cautiously and only after considerable prior research in which task demands are systematically varied, preferably in relation to drug dose.

Our efforts at studying the time course of drug effects in some depth led to several interesting findings and examination of time course should be incorporated in challenge studies whenever possible. Study of time course is essential for evoked potential measures as shown in this study where habituation and drug effects were compounded in the somatosensory P45-N60 component. An added feature of studying time course is that it can provide an estimate of the reliability and stability of the measure.

ACKNOWLEDGMENTS

The authors are grateful for the contributions of Herbert Y. Meltzer and Charles Grimm.

REFERENCES

AGHAJANIAN, G. K. & H. Y. HAIGLER. 1976. Hallucinogenic indoleamines: preferential action upon presynaptic serotonin receptors. Psychopharmacol. Commun. **1:** 619–629.

ARNOLD, VON O. H. 1975. N,N-dimethyltrpytamine: Einige erste Vergleichsergebnisse. Drug Res. **25:** 972–974.

BERCHOU, R., S. CHAYASIRISOBHON, V. GREEN & K. MASON. 1986. The pharmacodynamic properties of lorazepam and methylphendiate drugs on event-related potentials and power spectral analysis in normal subjects. Clin. Electroencephalogr. **17**(4), 176–180.

BLACKBURN, T. P., B. COX, C. G. HEAPY & T. F. LEE. 1982. Possible mechanism of 5-methoxy-N,N dimethyltryptamine-induced turning behavior in DRN lesioned rats. Pharmacol. Biochem. Behav. **16:** 7–11.

BODIS-WOLNER, I., M. D. YAAR, L. MYLIN & J. THORNTON. 1982. Dopaminergic deficiency and delayed visual evoked potentials in humans. Ann. Neurol. **11:** 478–483.

BROWN, B. B. 1968. Subjective and EEG responses to LSD in visualizer and non-visualizer subjects. Electroencephalogr. Clin. Neurophysiol. **25:** 372–379.

BROWN, B. B. 1969. Effect of LSD on visually evoked responses to color in visualizer and non-visualizer subjects. Electroencephalogr. Clin. Neurophysiol. **27:** 356–363.

BRAFF, D. L. & M. A. GEYER. 1990. Sensorimotor gating and schizophrenia: human and animal studies. Arch. Gen. Psychiatry **47:** 181–188.

BRUMAGHIM, J. T., R. KLORMAN, J. STRAUSS, J. D. LEWINE & M. G. GOLDSTEIN. 1987. What aspects of information processing are affected by methylphenidate? Findings on performance and P3b latency from two studies. Psychophysiology **24:** 361–373.

BUCHSBAUM, M. S., R. M. POST & W. E. BUNNEY, JR. 1977. AER in a rapidly cycling manic-depressive patient. Biol. Psychiatry **12:** 83–99.

CHAPMAN, L. F. & R. D. WALTER. 1965. Actions of lysergic acid diethylamide on averaged human cortical evoked responses to light flash. *In* Recent Advances in Biological Psychiatry. J. Wortis, Ed. Vol. 7: 23–36. Plenum Press. New York.

COHEN, M. R. 1991. Pitfalls in animal models (Letter to the editor and response by Braff and Geyer). Arch. Gen. Psychiatry **48:** 379–380.

COMMISSARIS, R. L., W. H. LYNESS, K. E. MOORE & R. H. RECH. 1981. Central 5-hydroxytryptamine and the effects of hallucinogens and phenobarbital on operant responding in rats. Pharmacol. Biochem. Behav. **15:** 595–601.

CRAGGS, M. D., J. J. WRIGHT & J. S. WERRY. 1980. A pilot study of the effects of methlyphenidate on the vigilance-related EEG in hyperactivity. Electroencephalogr. Clin. Neurophysiol. **48:** 34–42.

CRAYTON, J. W. & S. KING. 1981. Inter-individual variability of the H-reflex in normal subjects. Electromyogr. Clin. Neurophysiol. **21:** 183–200.

DAVIS, M., J. H. KEHNE, R. L. COMMISSARIS & M. A. GEYER. 1984. Effects of hallucinogens on unconditioned behaviors in animals. *In* Hallucinogens: Neurochemical, Behavioral, and Clinical Perspectives. B. L. Jacobs, Ed. 35–75. Raven Press. New York.

DOMINO, E. F. 1975. Indole alkyl amines as psychotogen precursors—possible neurotransmitter imbalance. *In* Neurotransmitter Balance Regulating Behavior. E. F. Domino & J. M. Davis, Eds. 185–224. U. Michigan. Ann Arbor.

FIARIELLO, R. G. & J. A. BLACK. 1978. Pseudoperiodic bilateral EEG paroxysms in a case of phencylidine intoxication. J. Clin. Psychiatry **39:** 579–581.

FREEDMAN, R., M. WALDO, P. BICKFORD-WIMER & H. NAGAMOTO. 1991. Elementary neuronal dysfunctions in schizophrenia. Schizophr. Res. **4:** 233–243.

GIBBONS, R. D., D. HEDEKER, C. WATERNAUX & J. M. DAVIS. 1988. Random regression models: a comprehensive approach to the analysis of longitudinal psychiatric data. Psychopharmacol. Bull. **24:** 438–443.

GILLIN, J. C., J. KAPLAN, R. STILLMAN & R. J. WYATT. 1976. The psychedelic model of schizophrenia: the case of N,N-dimethyltryptamine. Am. J. Psychiatry **133:** 203–207.

GLENNON, R. A. & J. A. ROSECRANS. 1981. Speculations on the mechanism of action of hallucinogenic indolealkylamines. Neurosci. Behav. Rev. **5:** 197–207.

GOODE, D. J., H. Y. MELTZER, J. W. CRAYTON & T. A. MAZURA. 1977. Physiological abnormalities of the neuromuscular systems in schizophrenia. Schizophr. Bull. **3:** 121–138.

GOTTFRIES, C. G., C. PERRIS & B. E. ROOS. 1974. Visual averaged evoked responses (AER) and monoamine metabolites in cerebrospinal fluid (CSF). Acta Psychiatr. Scand. Suppl. **255:** 135–142.

HALLIDAY, R., E. CALLAWAY, H. NAYLOR, P. GRATZINER & R. PRAEL. 1986. The effects of stimulant drugs on information processing in elderly adults. J. Gerontol. **41:** 748–757.

HAUBRICH, D. R. & P. F. L. WANG. 1977. N,N-dimethyltryptamine lowers rat acetylcholine and dopamine. Brain Res. **131:** 158–161.

HEDEKER, D., R. D. GIBBONS & J. M. DAVIS. Random regression models for multicenter clinical trials data (1991). Psychopharmacol. Bull. **27**(1), 73–78.

HEYM, J., K. RASMUSSEN & B. L. JACOBS. 1984. Some behavioral effects of hallucinogens are mediated by a postsynaptic serotonergic action: evidence from single unit studies in freely moving cats. Eur. J. Pharmacol. **101:** 57–68.

HUGON, M. 1973. Methodology of the Hoffmann reflex in man. *In* New Developments in Electromyography and Clinical Neurophysiology. J. Desmedt, Ed. Vol. 3: 277–293. S. Karger. Basel.

HYNEK, K., J. FABER, J. TOSOVSKY & M. CERNY. 1974. The effect of combination of methylphendiate and diazepam on the EEG and its evaluation by means of discriminant analysis. Act. Nerv. Super. (Praha) **16:** 257–258.

ISBELL, H., C. R. LOGAN & E. J. MINER. 1959. Studies of lysergic acid diethylamide (LSD-25). Arch. Neurol. Psychiatry **81:** 20–27.

JAVITT, D. C., C. E. SCHROEDER, J. C. AREZZO & H. G. VAUGHAN, JR. 1992. Selective inhibition of processing-contingent auditory event-related potential components by the PCP-like agent MK-801. Electroencephalogr. Clin. Neurophysiol. In press.

JAVITT, D. C. & S. R. ZUKIN. 1991. Recent advances in the phencyclidine model of schizophrenia. Am. J. Psychiatry **148**(10), 1301–1307.

JENNER, P., C. D. MARSDEN & C. M. THANKI. 1980. Behavioral changes induced by N,N-dimethyltryptamine in rats. Br. J. Pharmacol. **69:** 69–80.

KAPLAN, J., L. R. MANDEL, R. W. WALKER, W. I. A. VANDENHEUVEL, R. STILLMAN, J. C. GILLIN & R. J. WYATT. 1974. Blood and urine levels of N,N-dimethyltryptamine following administration of psychoactive dosages to human subjects, Psychopharmacologia **38:** 239–245.

LEVIT, R. A., S. SUTTON & J. ZUBIN. 1973. Evoked potential correlates of information processing in psychiatric patients. Psychol. Med. **3:** 487–494.

LIEBERMAN, J. A., B. J. KIHON & A. D. LOEBEL. 1990. Dopaminergic mechanisms in idiopathic and drug-induced psychoses. Schizophr. Bull. **16:** 97–110.

MCCARLEY, R. W., S. F. FAUX, M. E. SHENTON, P. G. NESTOR & J. ADAMS. 1991. Event-related potentials in schizophrenia: their biological and clinical correlates and a new model of schizophrenic pathophysiology. Schizophr. Res. **4:** 209–231.

MELTZER, H. Y., N. N. BOUTROS, M. SIMONOVIC, G. A. GUDELSKY & V. S. FANG.

1983. Hallucinogenic drugs and neuroendocrine secretion. *In* Integrative Neurohumoral Mechanisms. E. Endroczi, Ed. 463–477. Elsevier. The Netherlands.

Meltzer, H. Y., M. Simonovic, V. S. Fang & D. J. Goode. 1981. Neuroendocrine effects of psychotomimetic drugs. J. McLean Hosp. **6:** 115–138.

Meltzer, H. Y., B. Wiita, B. J. Tricou, M. Simonovic, V. Fang & G. Manov. 1982. Effect of serotonin precursors and serotonin agonists on plasma hormone levels. *In* Serotonin in Biological Psychiatry. B. T. Ho, C. Schoolar & E. Usdin, Eds. 117–139. Raven Press. New York.

Metz, J., N. Boutros, P. Tueting, C. Grimm & H. Y. Meltzer. 1982. Effect of N,N-dimethyltrptamine on H-reflex recovery curve. Soc. Neurosc. Abstr. **8:** 363.

Metz, J., D. A. Busch & H. Y. Meltzer. 1981. Des-tyrosine-endorphin: H-reflex response similar to neuroleptics. Life Sci. **28:** 2003–2008.

Metz, J., D. J. Goode & H. Y. Meltzer. 1980a. Descriptive studies of H-reflex recovery curves in psychiatric patients. Psychol. Med. **10:** 541–548.

Metz, J., H. H. Holcomb & H. Y. Meltzer. 1980b. Effect of chlorpromazine on H-reflex recovery curve. Soc. Neurosci. Abstr. **6:** 694.

Mintz, M., R. Tomer & S. Myslobodsky. 1982. Neuroleptic-induced lateral asymmetry of visual evoked potentials in schizophrenia. Biol. Psychiatry **17:** 815–828.

Moore, R. H., K. Hatada & E. F. Domino. 1976. The effects of N,N-dimethyltryptamine on electrically evoked responses in the cat visual system and modification by neuroleptic agents. Neuropharmacology **15:** 535–539.

Morstyn, R., F. H. Duffy & R. W. McCarley. 1983. Altered P300 topography in schizophrenia. Arch. Gen. Psychiatry **40:** 729–734.

Naylor, H., R. Halliday & E. Callaway. 1985. The effect of methylphenidate on information processing. Psychopharmacology **86:** 90–95.

Olney, J. W. 1989. Excitatory amino acids and neuropsychiatric disorders. Biol. Psychiatry **26:** 505–525.

Peloquin, L. J. & R. Klorman. 1986. Effects of methlyphenidate on normal children's mood, event-related potentials and performance in memory scanning and vigilance. J. Abnorm. Psychol. **95:** 88–98.

Pfefferbaum, A., J. M. Ford, P. M. White & W. T. Roth. 1989. P_3 in schizophrenia is affected by stimulus modality, response requirements, medication status, and negative symptoms. Arch. Gen. Psychiatry **46:** 1035–1044.

Prichep, L. S., S. Sutton & G. Hakerem. 1976. Evoked potentials in hyperkinetic and normal children under certainty and uncertainty: a placebo and methylphenidate study. Psychophysiology **13:** 419–428.

Radwan, M., H. Hermesh, M. Mintz & H. Munitz. 1991. Event-related potentials in drug-naive schizophrenic patients. Biol. Psychiatry **29:** 265–272.

Rodnight, R., R. M. Murray, M. C. H. Oon, J. F. Brockington, P. N. Nicholls & J. L. T. Berley. 1976. Urinary dimethlytryptamines and psychiatric symptomatology and classification. Psychol. Med. **6:** 649.

Roemer, R. A. & C. Shagass. 1990. Replication of an evoked potential study of lateralized hemispheric dysfunction in schizophrenics. Biol. Psychiatry **28:** 275–291.

Rosenberg, D. E., H. Isbell & E. J. Miner. 1963. Comparison of a placebo, N,N-dimethyltryptamine and 6-hydroxy-N-dimethyltryptamine in man. Psychopharmacologia **4:** 39–42.

Sai-Halasz, A. 1962. The effect of antiserotonin on the experimental psychosis induced by dimethyltryptamine. Experientia **18:** 137.

Satterfield, J. H., D. P. Cantwell, L. I. Lesser & R. L. Podosin. 1972. Physiological studies of the hyperactive child. Am. J. Psychiatry **128:** 102–108.

Schlemmer, R. F. & J. M. Davis. 1983. A comparison of three psychotomimetic-induced models of psychosis in nonhuman primate social colonies. *In* Ethopharmacology: Primate Models of Neuropsychiatric Disorders. K. A. Miczck, Ed. 33–78. Alan R. Liss, Inc. New York.

Shagass, C., E. M. Ornitz, S. Sutton & P. Tueting. 1978. Event-related potentials and psychopathology. *In* Event Related Brain Potentials in Man. E. Callaway, P. Tueting & S. Koslow, Eds. 443–510. Academic Press. New York.

Shagass, C., M. Schwartz & A. Marrazzi. 1962. Some drug effects on evoked cerebral potentials in man. J. Neuropsychiatry **3:** 49.

SILVERMAN, P. B. & B. T. HO. 1980. The discriminative stimulus properties of 2,5-dimethoxy-4-methylamphatamine (DOM): differentiation from amphetamine. Psychopharmacology **68:** 209.

SIMONOVIC, M., G. A. GUDELSKY & H. Y. MELTZER. 1981. Role of tuberinfundibular dopamine neurons in the biphasic effect of 5-methoxy-N,N-dimethyltryptamine in rat prolactin secretion. Soc. Neurosci. Abstr. **7:** 22.

SLOVITER, R. S., E. G. DRUST, B. P. DAMIANO & J. D. CONNOR. 1980. A common mechanism for lysergic acid, indolealkylamine and phenethylamine hallucinogens: serotonergic mediation of behavioral effects in rats. J. Pharmacol. Exp. Ther. **214:** 231–238.

SMITH, T. L. 1977. Increased synthesis of striatal dopamine by N,N-dimethyltryptamine. Life Sci. **21:** 1597–1602.

SNYDER, S. H. 1972. CNS stimulants and hallucinogens. *In* Chemical and Biological Aspects of Drug Dependence. S. J. Muie & C. H. Brill, Eds. 55–63. CRC Press. Cleveland.

STRASSMAN, R. J., E. H. UHLENHUTH, R. KELLNER & C. QUALLS. 1991, December. Psychopharmacology and neuroendocrinology of DMT. Proc. Am. Coll. Neuropsychopharmacol. (San Juan, Puerto Rico). Abstr., p. 196.

SUTTON, S. 1969. The specification of psychological variables in an average evoked potential experiment. *In* Average Evoked Potentials: Methods, Results, and Evaluations. E. Donchin & D. B. Lindsley, Eds. 237–298. NASA SP-191. Washington, D.C.

SZARA, S. 1956. Dimethyltryptamine: its metabolism in man: the relation of its psychotic effect to serotonin metabolism. Experientia **12:** 441–442.

TECCE, J. J., J. SAVINGNANO-BOWMAN & J. O. COLE. 1981. Drug effects on contingent negative variation and eyeblinks: the distraction-arousal hypothesis. *In* Psychopharmacology: a Generation of Progress. M. A. Lipton, A. D. Masero & K. F. Killin Ban, Eds. 745–758. Raven Press. New York.

TRULSON, M. E., J. HEYMS & B. C. JACOBS. 1981. Dissociation between the effects of hallucinogenic drugs on behavior and raphe unit activity in freely moving cats. Brain Res. **215:** 275–293.

TUETING, P. 1991a. Electrophysiologicy of schizophrenia: EEG and ERPs. Curr. Opinion Psychiatry **4**(1), 7–11.

TUETING, P. 1991b. Investigational Drugs and Research. *In* A Selective Update on Psychopharmacologic and Somatic Therapies in Psychiatry—II. Psychiatric Medicine. R. C. W. Hall, Ed. Vol. 9(2): 333–347. Ryandic Publishing. Longwood, FL.

TUETING, P. 1991c. Psychophysiological predictors of drug treatment response. *In* A Selective Update on Psychopharmacologic and Somatic Therapies in Psychiatry—I. Psychiatric Medicine. R. C. W. Hall, Ed. Vol. 9(1): 145–162. Ryandic Publishing. Longwood, FL.

TUETING, P. & J. METZ. 1983, September. Evoked potential changes associated with N,N-dimethyltryptamine (DMT). Paper presented at the Seventh International Congress on Event Related Potentials of the Brain, Florence, Italy.

TURNER, W. J. & S. MERLIS. 1959. Neurochemical investigations of the interaction of N,N-dimethyltryptamine with the dopaminergic system in rat brain. Psychopharmacology **52:** 137–144.

WALDEMEIER, P. C. & L. MAITRE. 1977. Neurochemical investigations of the interaction of N,N-dimethyltryptamine with the dopaminergic system in rat brain. Psychopharmacology **52:** 137–144.

WHITAKER-AZMITIA, P. M. & S. J. PEROUTKA. EDS. 1990. The Neuropharmacology of Serotonin. Annals of the New York Academy of Sciences. Vol. 600. New York Academy of Sciences. New York.

Event-Related Potentials and Factor Z-Score Descriptors of P3 in Psychiatric Patients[a,b]

E. R. JOHN[c,d] AND L. S. PRICHEP[c,d]

[c]*Department of Psychiatry*
New York University Medical Center
550 First Avenue
New York, New York 10016
and
[d]*Nathan S. Kline Institute for Psychiatric Research*
New York State Department of Mental Health
Orangeburg, New York 10962

Since the discovery of P3 in the early 1960's (Sutton, Braren, John & Zubin, 1965), there have been numerous demonstrations of its abnormalities in psychiatric disorders. Diminished P3 in the auditory, visual and somatosensory modalities has been reported in schizophrenics (Baribeau-Braun, Picton & Gosselin, 1983; Buchsbaum, Awsare, Holcomb, Delisi, Hazlett, Carpenter, Pickar & Morihisa, 1986; Galderisi, Maj, Mucci, Monteleone & Kemali, 1988; Levit, Sutton & Zubin, 1973; Pfefferbaum, Wenegrat, Ford, Roth & Kopell, 1984b; Pritchard, 1986; Romani, Zerbi, Mariotti, Callieco & Cosi, 1986; Shagass, Josiassen, Roemer, Straumanis & Slepner, 1983; Steinhauser & Zubin, 1982). Lack of change in P3 abnormalities, in some modalities, with clinical improvement has been interpreted by some as suggestive of a "trait" marker for schizophrenia (Duncan, 1988; Kutcher, Blackwood, Clair, Gaskell & Muir, 1987). The topography of reduced auditory P3 amplitude in chronic schizophrenics suggests the greatest abnormality in the left temporal region (Faux, Torello, McCarley, Shenton & Duffy, 1988; Morstyn, Duffy & McCarley, 1983). Diminished frontal distribution of P3 has been observed in the hebephrenic, or disorganized, subtype of schizophrenics (Maurer, Kierks, Ihl & Laux, 1989).

A large number of studies report increased P3 latencies in Alzheimer's (SDAT) patients relative to normals (Goodin & Aminoff, 1986; Goodin, Squires & Starr, 1978; Gordon, Kraiuhin, Harris, Meares & Howson, 1986; Pfefferbaum *et al.*, 1984b; Patterson, Michalewski & Starr, 1988; Polich, Ehlers, Otis, Mandell & Bloom, 1986; Clair, Blackburn, Blackwood & Tyrer, 1988; Surwillo and Iyer, 1989; Visser, Tilburg, Hooifer, Jonker & Rijke, 1985), but this did not differentiate between SDAT patients and those with other dementing disorders (Goodin *et al.*, 1978; Gordon *et al.*, 1986; Pierrot-Deseilligny, Turell, Penet, Legrigand, Pillon, Chain & Agid, 1989; Polich *et al.*, 1986) nor between "cortical" or "subcortical" dementias (Goodin & Aminoff, 1986). Auditory P3 has been reported to be generally reduced in depression (Blackburn, Clair & McInnes, 1987; Josiassen, Roemer,

[a] Dedicated to our dear friend Sam Sutton, whose collaboration and thoughtful judgment would surely have helped illuminate these complex findings.

[b] Supported in part by Cadwell Laboratories, Kennewick, WA.

Shagass & Straumanis, 1986; Pfefferbaum *et al.*, 1984b; Shagass, 1983). Increased P3 latency and reduced amplitude have been correlated with cumulative alcohol intake (Polich & Bloom, 1986), and reduced P3 amplitude has been reported in alcohol-naive sons of alcoholics (Begleiter, Porjesz, Bihari & Kissin, 1984) and in subjects with a positive family history (Elmasian, Neville, Woods, Schuckit & Bloom). Shorter P3 latencies were found in patients with obsessive-compulsive disorders (OCD) compared to normal during a difficult P3 condition (Beech, Ciesielski & Gordon, 1983; Towey, Bruder, Hollander, Friedman, Erhan, Liebowitz & Sutton, 1990).

While these numerous studies suggest P3 abnormalities as a common and sensitive correlate of psychopathology, very few have investigated the specificity of these findings by evaluating several disorders in the same study. In the present study, the topography of P3 was compared in normals and patients with senile dementia, schizophrenia or obsessive-compulsive disorder, in order to estimate the specificity of P300 abnormalities in cognitive disturbances. A passive oddball paradigm was used, to minimize the requirement for active engagement of the patient in a mental activity. In addition, in view of the current extensive clinical utilization of P3, it seemed desirable to explore a standardized procedure for its quantitative evaluation. Principal component analysis with Varimax rotation was carried out on a sample of normal individuals. Using the resulting factor waveshapes as standardized descriptors, a quantitative statistical evaluation of P3 waveshapes was obtained by Z-transformation of individual factor scores relative to the distribution of factor scores in the normative group (John, Prichep, Friedman & Easton, 1989; Pfefferbaum, Ford, Wenegrat, Roth & Kopell, 1984a).

METHODS

Subjects

The inclusion and exclusion criteria for "normal" subjects have been given in detail elsewhere (John, Prichep, Fridman & Easton, 1988). All patients were DSM III or DSM III-R diagnosed by multiple raters, had no evidence of neurological disease, no history of head injury, no history of drug or alcohol abuse and all patients, except the schizophrenics, were drug-free for at least 7 days prior to evaluation. Since there is much controversy related to considering as "drug-free" chronic schizophrenics who were off medication for 7 days, and since longer drug-free periods were impossible to obtain, we decided to test the schizophrenics while on treatment doses of neuroleptics (most commonly haloperidol). The degree of age-associated cognitive deterioration was measured in the dementia group using the Global Deterioration Scale (GDS) (Reisberg, Ferris, de Leon & Crook, 1982). Patients with GDS scores ranging from mild (GDS = 3) to moderately severe (GDS = 6) were included for study.

Using the above criteria, the following groups of subjects were included in this study:

(1) Normally functioning adults, n = 19, (10 females and 9 males with a mean age of 30.7, and age range 17.5–54.1 years);
(2) Patients with chronic schizophrenia, n = 37, (7 females and 30 males with a mean age of 36.1 and an age range 19.9–70.5 years);
(3) Patients with primary degenerative dementia, n = 26, (14 females and 12 males with a mean age of 69.7 and age range 54.9–81.5 years);

(4) Patients with obsessive-compulsive disorder, n = 14, (1 female and 13 males with a mean age of 37.6 and age range 22.6–59.5 years).

ERP Task

Due to our belief that a useful ERP condition for the evaluation of psychiatric patients should be applicable regardless of age or degree of cognitive impairment, a "passive" paradigm was adopted, in which the subject was instructed to listen to tones of two different frequencies (both frequencies having been previously demonstrated to them). The decision to use a "passive" paradigm was also supported by a recent study comparing an active and a "passive sequence paradigm." The passive paradigm yielded P300 waveforms, scalp distributions and peak latencies remarkably similar to a "simple oddball" paradigm (Polich, 1989).

Tones with two different frequencies (1000 Hz = "common," and 3000 Hz = "rare") were used in a 2:1 ratio. The interstimulus interval was 0.97 sec. Only artifact-free trials were averaged; no mental task related to either stimulus was requested. A total of 300 artifact-free trials were collected, 200 with "common" tones and 100 with "rare" tones. The choice of a 2:1 ratio was made in an attempt to keep the signal-to-noise-ratio in the common and rare averages more comparable without unduly prolonging the procedure.

Data Acquisition

Data were collected from the 19 channels of the International 10/20 system, referenced to linked ears, using the Cadwell Spectrum 32. Transorbital eye leads were used to record eye movement. Electrode impedances were below 5000 Ωs. As on-line artifact rejection algorithm was used, based upon voltage windows. Patients were seated comfortably in a sound and light attenuated recording chamber. Amplifier gains were set at 10,000, filter settings were 1.0–70.0 Hz, and the sampling rate was 1600 Hz. A 500 ms analysis epoch was used. Stimuli were delivered binaurally via head phones, at 80 dB SPL. Patients were requested to keep their eyes closed during the recording session.

Data Analyses

Group grand average ERPs were computed for the *common* and *rare* stimuli and difference waves for *rare* minus *common* ERPs were constructed. ERPs elicited by the common tone in the normal subjects were also subjected to principal component analysis followed by Varimax rotation (PCVA). Factors were extracted which accounted for the greatest proportion of the variance. For each lead, the mean and standard deviation were computed for each of the factor scores required to optimally reconstruct the common ERP. Using these statistical parameters, the factor scores for the rare stimuli were Z-transformed relative to the distribution of factor scores for the common stimuli. Thus, all factor scores were expressed in probabilistic terms and significant deviations from expected normal values ($p > 0.05$) were those with a Z value ≥ 1.96 (John, Easton, Prichep & Friedman, 1991b; John, Easton, Prichep & Friedman, 1991a; John *et al.*, 1989; Pfefferbaum *et al.*, 1984a).

RESULTS

Normal Subjects

The group grand average ERP to the common tone (1000 Hz) is shown in the top panel of FIGURE 1 and to the rare tone (3000 Hz) in the bottom panel. In response to the rare tone, P3 appears in the group average as a late positive complex with peaks at 255–265 ms and 320 ms, most probably corresponding to P3a and P3b (Squires, Squires & Hillyard, 1975). For purposes of simplicity we shall refer to these components as "early" and "late" P3. In response to the common tone, note the small complex component with peaks at about 250 ms and 350 ms seen in the frontal and central regions and at 315 ms and 400 ms in the posterior regions.

Other differences between the rare and common ERPs can also be discerned. The most salient of these are that in C_z and other central and parietal regions, N1 and P2 are *smaller* to the rare than to the common event. In contrast, N1 in F_z is slightly larger but P2 slightly smaller to the rare than to the common event. These differences can perhaps be better visualized by considering FIGURE 2A, which shows the "difference waves" obtained by subtracting common from rare ERPs.

In the frontal poles, the rare event elicits a very large early negativity (N1) which is absent to the common event. P2 is essentially the same to both events. In the frontal regions, (F_7, F_8, F_3, FP_1, FP_2), N1 is markedly more negative in response to rare than to common events, while P2 is slightly smaller. In central regions, both N1 and P2 are smaller in response to the rare event. In parietal and other posterior regions, N1 is smaller but P2 essentially unchanged to the rare event. This suggests two different anatomically localized processes related to selective attention. In the frontal regions, the data suggest greater attention to the rare than the common event. In the more posterior regions, the data suggest greater attention to the common than to the rare event. This may indicate that the common event elicits greater vigilance. Perhaps vigilance for an unpredictable event is more difficult to construct neurophysiologically than for a well-defined event. Even with the decrease in vigilance, the rare event elicits a greater P3 than does the common, reflecting the separation of the brain processes associated with uncertainty from those associated with selective attention and vigilance. These data further imply that anterior regions may be more involved in attention. Posterior brain regions may be more involved than anterior in maintaining vigilance. Both regions are involved in responding to uncertainty of those stimuli which are perceived.

Three factors derived from the principal component analysis with Varimax rotation accounted for 87.6% of the variance of the full set of waveshapes elicited by the common stimulus, and are shown in FIGURE 3. The peak latencies for Factors 1–3 are 98 ms, 166 ms and 323 ms, respectively. It is noted that expanding these results to a four-factor solution separates the "early" and "late" components with Factor 3 at 235 ms and Factor 4 at 323 ms. These results will be discussed elsewhere.

The "common" ERPs of the normal group were individually well reconstructed by these "common factors." The group grand average reconstructed ERPs are shown in FIGURE 4, and are almost identical with the original waveforms shown in FIGURE 1, top.

When a PCVA was similarly carried out on the "rare" ERPs from the normal group, we found it was not possible to obtain a set of factors which could reconstruct all of the "rare" waveshapes with an acceptably small variance of the factor scores. This variability of factor scores required to reconstruct the rare ERP may be

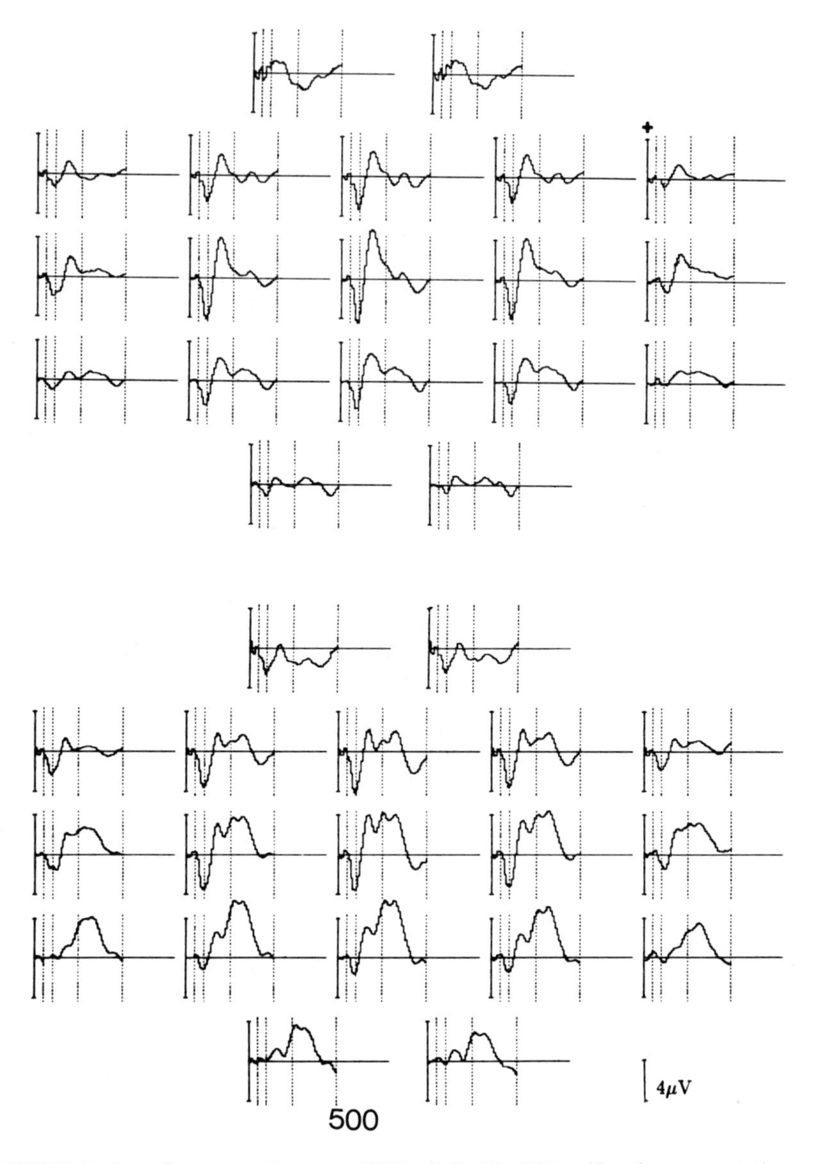

FIGURE 1. *Top*. Group grand average ERPs elicited by 200 artifact-free presentations of the common stimulus (1000 Hz) in a group of 19 normal subjects. The waveshapes are arrayed topographically as if one were looking down on the 10/20 System with face upward. The *vertical dotted lines* correspond to 50, 100, 250 and 500 milliseconds. *Bottom*. Group grand average ERPs elicited by 100 artifact-free presentations of the rare stimulus (3000 Hz) in the same normal subjects.

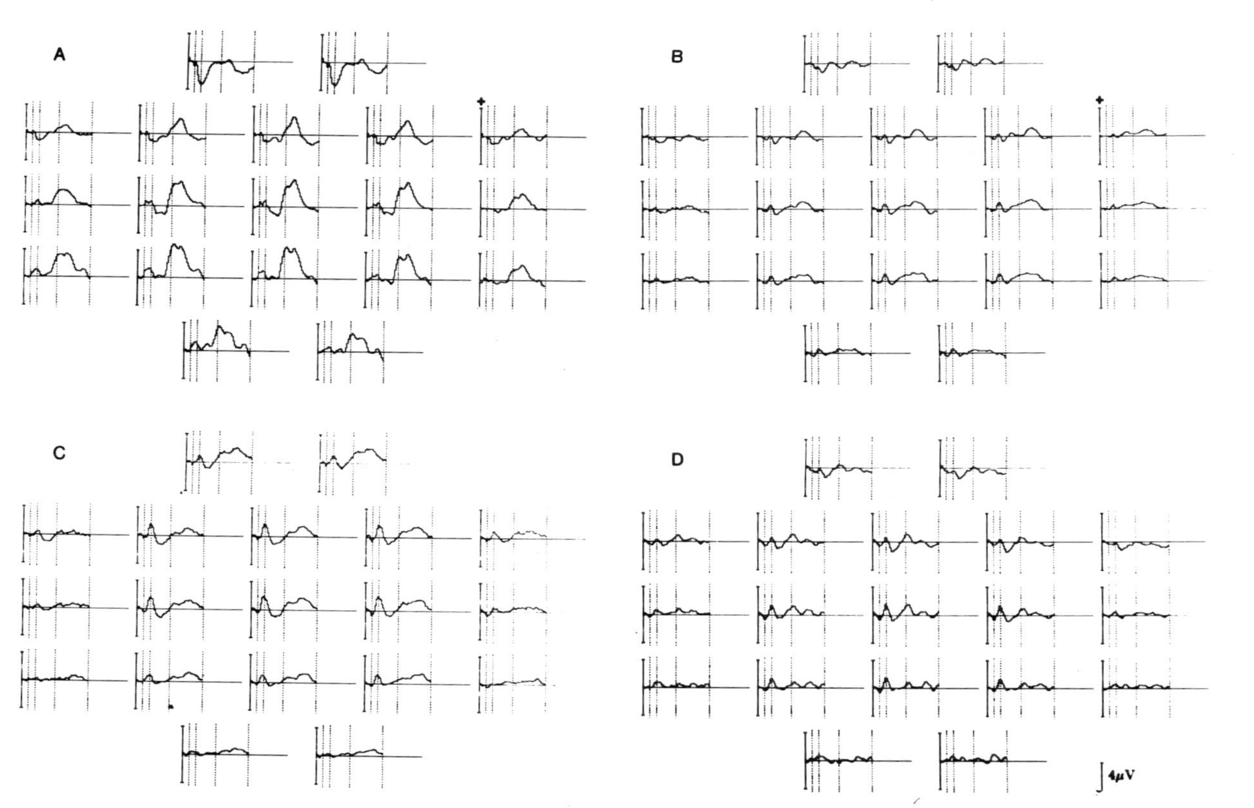

FIGURE 2. "Rare minus common" ERP difference waveshapes for **(A)** normals, **(B)** schizophrenic patients, **(C)** dementia patients, and **(D)** obsessive-compulsive patients.

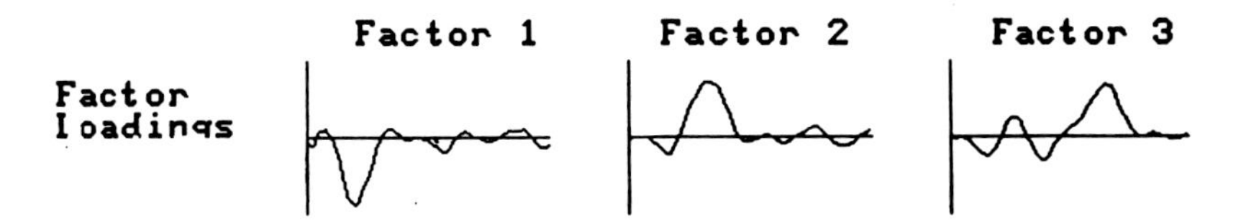

FIGURE 3. Waveforms (factor loadings) of the three factors obtained by Varimaax rotation after a principal component analysis of the average ERPs elicited in the 19 electrode positions of the 19 normal subjects in response to the common stimuli. Analysis epoch was 490 ms. The peak latency for Factor 1 was 98 ms, for Factor 2 was 166 ms and for Factor 3 was 323 ms.

due in part to the low ratio of common to rare events, which increased intersubject variability. Consequently, we decided to describe the "rare" ERPs as deviations from the factor structure of the "common" ERPs. The "rare" ERPs of the normal group were then individually reconstructed using the 3 factors obtained from PCVA of the common ERPs. As expected, large positive factor Z-scores were obtained for Factor 3, in most leads of most subjects. The normal group grand average Factor 3 Z-score map is shown in Figure 5A. Inspection of FIGURE 5A shows that, in the normals, the rare ERP contains an excess of the process described by Factor 3 over the whole head, with maximum values in temporal, central, parietal and occipital regions.

The percentage of significant Factor 3 Z-scores ($\geq$1.96) for each electrode site

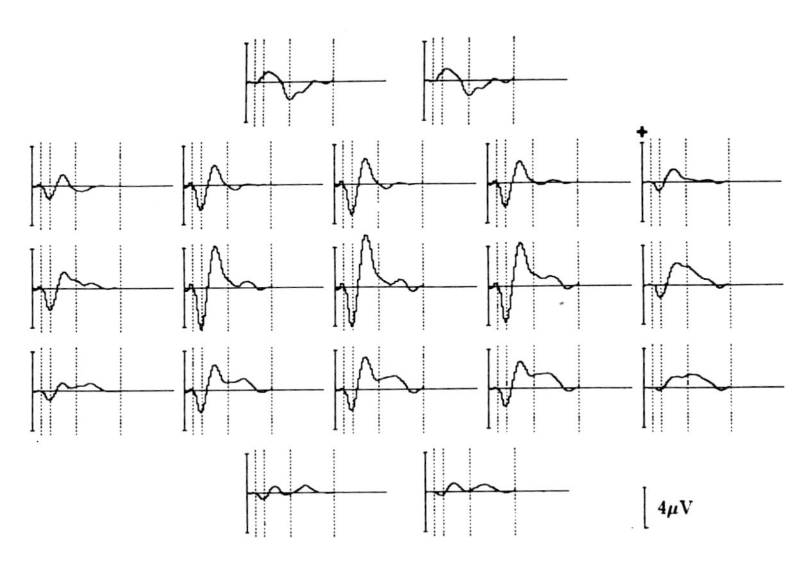

FIGURE 4. Reconstructed waveshapes of the grand average of the normal subjects as a linear combination of the three Varimax factor loadings. *Vertical dotted lines* correspond to 50, 100, 250 and 500 ms.

is shown in TABLE 1 for the normal (line 1) and abnormal groups (lines 2–4). That is, the percent of individuals in each group in which the P300 factor Z-score for the rare ERP was statistically significant compared with that for the common ERP. Note that the highest entry on line 1 is 58%. The *total %* of normals showing a significant P3 was higher, but the anatomical locus varied from person to person.

Schizophrenic Patients[1]

The schizophrenic group grand average *common* and *rare* ERPs are shown in the top and bottom of FIGURE 6 averaged across the 37 patients. P50 is essentially normal. In frontal regions, N1 and P2 are slightly smaller in schizophrenics than normals in response to the common event. In central regions and anterior temporals, N1 and P2 are much smaller in the schizophrenics and there is a late negativity peaking at about 415 ms, lacking the positivity at P250 and P350 ms seen in the normal case. In parietal, posterior temporal and occipital regions, there is little difference between normals and schizophrenics in response to the common event. The rare event elicits a slightly smaller N1 with little change in P2 in frontal and central regions. In parietal regions, N1 disappears in response to the random event. In response to the rare event, P3 is present in the centrals at 320 ms although small, but delayed to about 340 ms in the frontals and to 385 ms in the parietal region, possibly due to medication (Radwan, Hermesh & Munitz, 1991). Although there is a clear difference between the schizophrenics response to common and rare (see FIG. 2B), the schizophrenic rare response appears to most resemble the normal common response shown in FIGURE 1.

The schizophrenic group grand average Factor 3 Z-score map is shown in FIGURE 5B. Some positive values are seen, reflecting the small late positivities in frontal leads evident when the bottom panel of FIGURE 6 is compared with the top panel of FIGURE 1. No significant differences between the schizophrenic response to rare and the normal response to common are seen posterior to the frontal regions. Thus, the rare ERP in schizophrenics seems to resemble the common ERP in normals. However, the difference waveshapes between the schizophrenic response to rare versus common, shown in FIGURE 2B, indicate a diminished early negativity (N1) and an enhanced early positivity (P2) and late (P3) positivity in the rare but not the common ERP, comparable in all but the occipital leads, and clearest in C_z. The percentage of the individuals in the schizophrenic group showing a significant Factor 3 Z-score to rare (Z value $\geq$1.96), when reconstructed using the common factors, was calculated and is shown by lead in line 2 of TABLE 1. A modest percentage of the schizophrenics (11–22%) have significant Factor 3 Z-scores in frontal regions, slightly more on the left than the right, especially comparing F_7 and F_8. More posterior regions do not significantly deviate from the normal response.

These data suggest that, with respect to selective attention, there is no difference in the ERPs elicited in frontal regions of schizophrenic patients by the rare and common stimuli. In central and more posterior regions, the common event elicits a greater N1P2 than the rare event, as in the normals. In the normals, however, the later positive event has two peaks which are anatomically diffuse

[1] This population was studied in collaboration with K. Alper, M.D., Dept. of Psychiatry, New York University Medical Center and Bellevue Hospital Center, New York, NY.

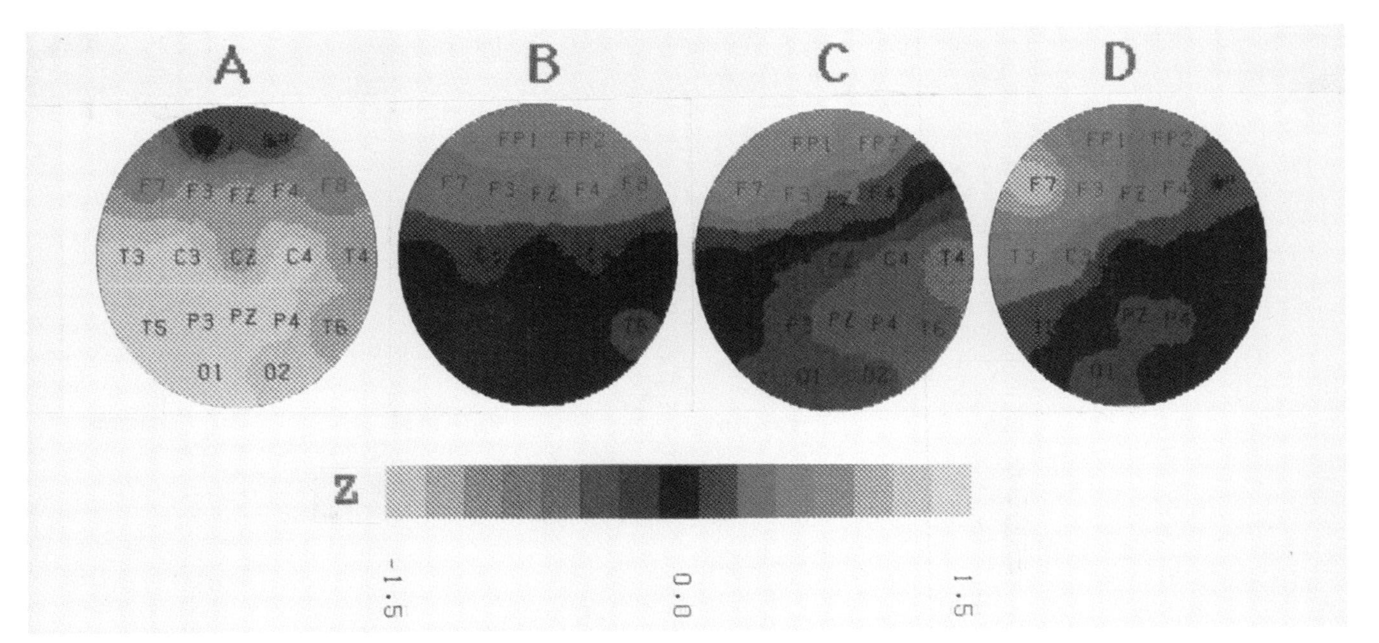

FIGURE 5. Normal group grand average Factor 3 Z-score topographic maps for **(A)** normals, **(B)** schizophrenics, **(C)** demented, and **(D)** obsessive-compulsives. These maps were constructed by a) reconstructing the rare ERP at every electrode in each individual as a linear combination of the three factor loadings, each multiplied by the factor score optimal for best fit; b) the factor scores were converted to standard deviation of the corresponding factor score distributions in the normal group; c) the group average Z-score was computed for each factor at each lead; finally, d) the resulting mean value for each factor at each electrode was color coded using the standard deviation scale at the bottom of the figure, in which increasing excessive values are encoded from red to yellow and increasing deficient values are encoded from blue to green. The significance of Z-scale values for group data can be estimated by multiplying the square root of sample size and standard deviation of each group.

TABLE 1. % Significant[a] Z-Scores for Factor 3 ($p \leq 0.05$)

Group	N	FP_1	FP_2	F_3	F_4	C_3	C_4	P_3	P_4	O_1	O_2	F_7	F_8	T_3	T_4	T_5	T_6	F_z	C_z	P_z	All
Normal	19	16	21	37	42	42	42	53	47	37	32	21	26	42	32	58	32	37	42	53	47
Schizophrenia	37	14	11	19	16	8	11	5	3	3	3	22	14	8	0	3	0	16	8	5	11
Dementia	26	15	12	15	4	4	0	0	0	0	0	23	8	4	0	0	0	4	0	0	8
OCD	14	0	0	29	14	0	0	7	7	0	0	50	14	0	7	0	0	14	0	7	7

[a] P300 of rare ERP was statistically significant ($p \leq 0.05$) compared with common ERP.

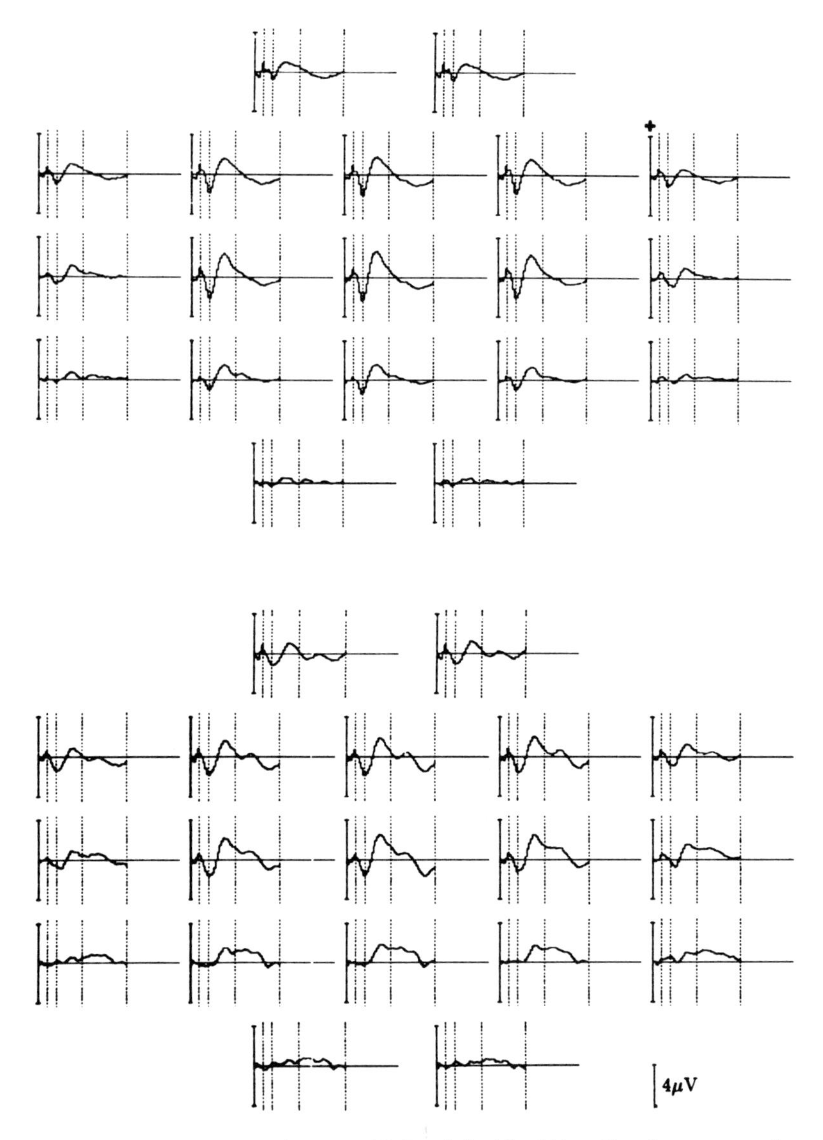

FIGURE 6. *Top*. Group grand average ERPs elicited by 200 artifact-free presentations of the common stimulus (1000 Hz) in a group of 37 medicated schizophrenic patients. The waveshapes are arrayed topographically as if one were looking down on the 10/20 System with face upward. The *vertical dotted lines* correspond to 50, 100, 250 and 500 milliseconds. *Bottom*. Group grand average ERPs elicited by 100 artifact-free presentations of the rare stimulus (3000 Hz) in the same group of schizophrenic patients.

and greatest in central and posterior regions, with a smaller frontal representation. In the schizophrenic, a late positivity at P350 ms is seen to the rare event in most leads, albeit smaller than in normals, but the earlier peak of the late positive complex at P250 ms is absent from all but posterior leads where it is much diminished.

Dementia Patients[2]

The dementia group grand average *common* and *rare* ERPs are shown in FIGURE 7, averaged across 26 patients. P50 is widespread and larger than normal and is the same for both stimuli. In response to the common stimulus, N1 is normal in central, smaller in posterior regions but markedly larger in the frontal regions, especially the frontopolar region. A widespread N1 response is seen to both the common and rare stimulus, but much larger to the common, whereas in the normals N1 is larger to the rare, and absent in the frontopolar regions to the common stimulus. N1P2 is substantially smaller for the rare than the common stimulus in both the frontal regions and the central and more posterior regions in the dementias, whereas this was only true in the central regions for the normal and schizophrenic patients. P2 in response to the common stimulus is the same amplitude as the normals in frontal regions, but it is prolonged. However, P2 appears further prolonged in response to the rare event, apparently due to the appearance in most regions of the first peak of the late positive complex at P250 ms. P2 is smaller in the centrals and more posterior regions. In response to the rare stimulus, P2 is smaller in all regions. The later peak, usually at P350, appears to be substantially delayed to about P420 ms. These differences may be seen more clearly in the "rare minus common" difference waves of the dementia patients, shown in FIGURE 2C.

These data suggest that, with respect to selective attention, there is a far greater decrement in the frontal as well as the central regions of the dementia patients in response to the rare stimulus. However, N1P2 to the common stimulus is similar to the normals and schizophrenics. In the response to the common stimulus, there is a late negativity peaking at 365 ms, largest in the centrals, rather than the slight positive peaks at 250 and 350 ms seen in the normals. In the response to rare, the P3 complex has one peak at 255–265 ms in the centrals and frontals, as in normals, and a second peak at 395 ms. These are differentially delayed in more posterior regions. The earlier late positivity at P250 ms is relatively unchanged in central and anterior leads but delayed to P300 ms in parietal and posterior leads. The later positive peak is delayed to P425 in parietal and occipital regions. This is appreciably later than would be expected from normal aging (Pfefferbaum *et al.*, 1984a).

Thus, in this group of demented patients, the early negative processes related to vigilance appear to be in similar functional condition in both the frontal and more posterior regions. Both regions appear to function comparably to the normal and schizophrenic groups in response to the common stimulus, while both regions show the decrement to the rare stimulus seen only in the posterior regions of the other two groups. The systems responsible for response to the "uncertainty" aspect of the stimulus appear to be impaired in such a way as to produce an anatomically diffuse delay of about 100 milliseconds in all regions.

[2] This population was studied in collaboration with S. Ferris and B. Reisberg, Millhauser Laboratories, Dept. of Psychiatry, New York University Medical Center. Supported in part by National Institute on Aging Grants #AG03051 and #MH32577.

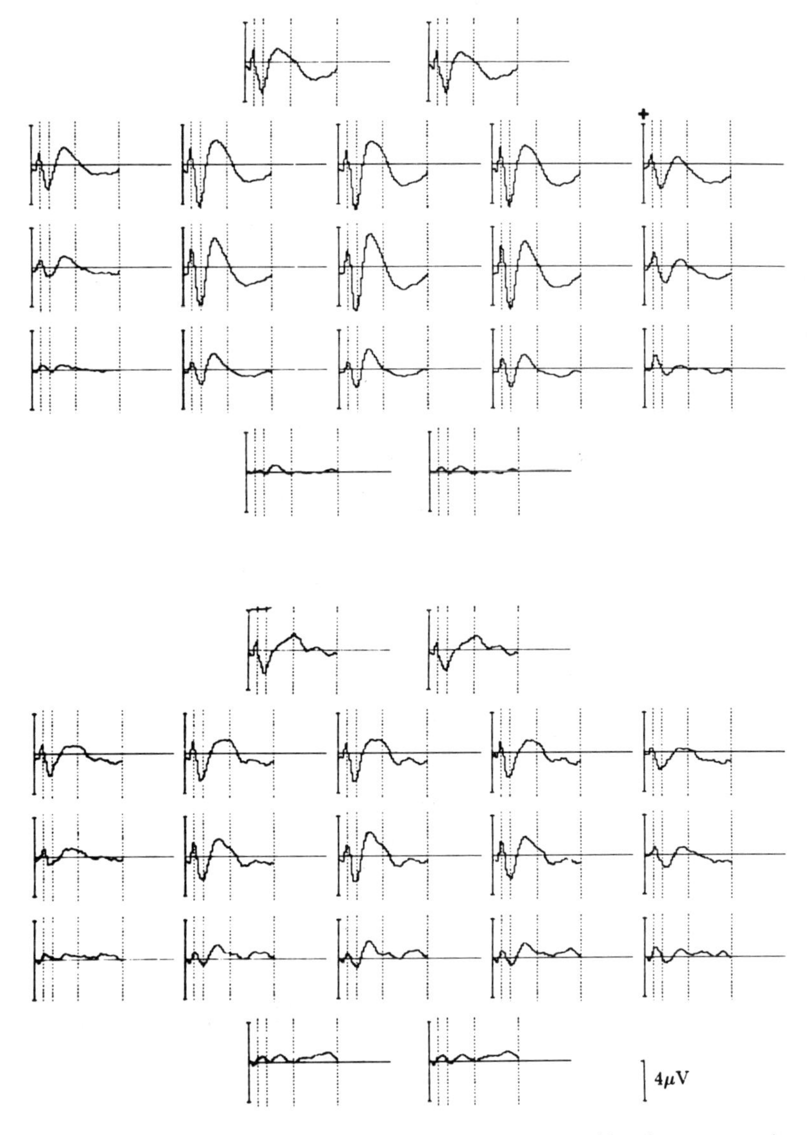

FIGURE 7. *Top*. Group grand average ERPs elicited by 200 artifact-free presentations of the common stimulus (1000 Hz) in a group of 26 patients with primary degenerative dementia. The waveshapes are arrayed topographically as if one were looking down on the 10/20 System with face upward. The *vertical dotted lines* correspond to 50, 100, 250 and 500 milliseconds. *Bottom*. Group grand average ERPs elicited by 100 artifact-free presentations of the rare stimulus (3000 Hz) in the same group of patients with primary degenerative dementia.

Inspection of the group grand average factor Z-score maps for the demented patients, seen in FIGURE 5C, shows that P3 in this group to the rare stimulus was no larger than that found in the normals to the common stimulus in the central and more posterior regions and was, in fact, smaller than that of the normal common ERP in posterior regions of the right hemisphere. In the frontal regions, the late positivity was greater than the response of the normals to "common," more so on the left hemisphere. The third row of TABLE 1 shows this clearly, with a modest percentage (12–23%) of the dementia group displaying significant Factor 3 Z-scores in FP_1, FP_2, F_3, F_7 and F_z, but almost no significant P3 in other regions.

Obsessive-Compulsive Disorder Patients[3]

The OCD group grand average *common* and *rare* ERPs are shown in FIGURE 8, averaged across 14 patients. The response of this group to the common stimuli is much like that of the demented group, except that N1 and P2 are slightly smaller in the frontal regions, P2 is delayed 20–25 ms (seen most clearly in P_3), and the prolonged negativity (N2) seen in all regions of the demented is absent in th OCD patient in all posterior, central and temporal regions. However, it is accentuated and somewhat prolonged in the vertex and frontal regions compared to normal.

The response to the rare stimulus displays a somewhat smaller N100 and P200 than to common in frontal regions. In the central and anterior regions, there appears to be a late positive complex P3 as seen in the normals and dementias, with the first peak at about 255 ms in the centrals, frontals and parietals, delayed to 300 in the posterior temporals and occipitals, and a second peak delayed to 395 ms. Thus, as was seen in the dementias, the P3 complex appears to separate with the first peak showing an increase in latency as one moves posteriorally (sparing the parietals in this group); however, the P395 latency is essentially the same in these regions. This can be seen more clearly in the "rare minus common" difference wave for the OCD patients, shown in FIGURE 2D.

These data indicate that, with respect to attention as reflected in the ERPs, there is relatively little difference in the N1P2 of the frontal regions to rare and common events in the OCD group, similar to the schizophrenics. As in both the normals and the schizophrenics, the common elicits a greater N1P2 than the rare event in the central and more posterior regions. The OCD show little difference in response to rare events compared to the response of normals to common events, with respect to ERPs on the posterior half of the right hemisphere and the vertex, left parietal and occipital regions, as seen in the map presented in FIGURE 5D. Positivities emerge to the rare event in other regions, at both 255–300 and 395 ms and especially on the left hemisphere, which are greater than those elicited by the common event in the normal group.

Thus, the OCD group resembles the schizophrenics and normals with respect to processes related to vigilance. With respect to processes related to the detection of uncertainties, the OCD displays features which resemble those seen in the demented; both are manifested asymmetrically with more abnormality on the right side than the left, suggesting the lack of specificity of this finding.

The last row of TABLE 1 shows this asymmetry clearly, with the percentage of

[3] This population was studied in collaboration with E. Hollander and M. Liebowitz, New York State Psychiatric Institute, New York, NY.

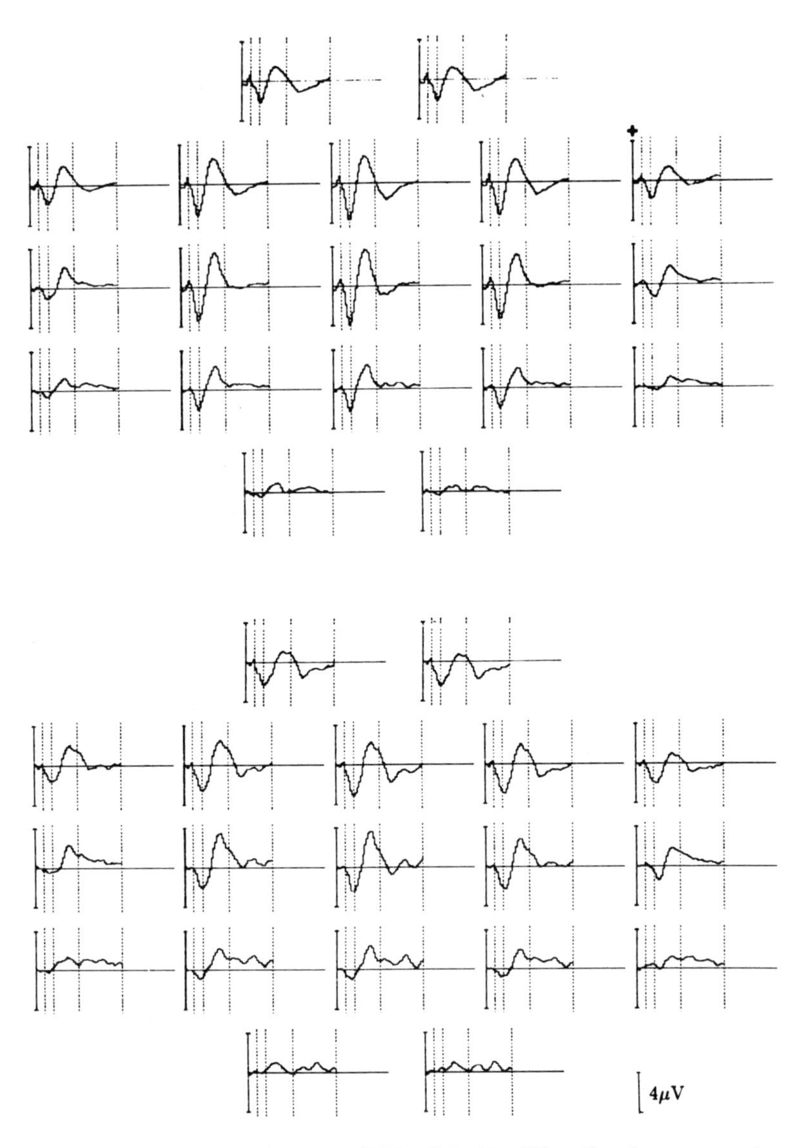

FIGURE 8. *Top*. Group grand average ERPs elicited by 200 artifact-free presentations of the common stimulus (1000 Hz) in a group of 14 patients with obsessive-compulsive disorder. The waveshapes are arrayed topographically as if one were looking down on the 10/20 System with face upward. The *vertical dotted lines* correspond to 50, 100, 250 and 500 milliseconds. *Bottom*. Group grand average ERPs elicited by 100 artifact-free presentations of the rare stimulus (3000 Hz) in the same group of patients with obsessive-compulsive disorder.

patients showing a Factor 3 Z-score which is noteworthy only in the frontal regions (14–50%), with the left side showing 2–4 times more significant detections of P3 than the right.

DISCUSSION

Our original intention in this research was to collect a sufficiently large sample of P3 measurements in psychiatric patients in order to assess their sensitivity and specificity in different disorders and to catalog the salient similarities and dissimilarities of P3 findings among various disorders. Partially because we recorded from the full 10/20 System, which had not been systematically used in many previous P3 studies, and partially because of the differences among patients who are schizophrenic, demented or obsessive-compulsive, we observed phenomena which made this task more difficult than constructing a catalog. A large variance was encountered in normals which was probably due to age effects. For this reason we are currently increasing our normal population, which will allow us to age regress our factor scores. This work is in progress.

Two sets of phenomena emerged from these studies. When the differences between ERPs to common and rare stimuli were examined in different scalp regions in different disorders, it became apparent that frontal and posterior regions behaved differently. Comparing the different categories of patients to normals suggested that these anterior-posterior differences arose from the presence of two functional systems which could be dissociated. A substantial proportion of these psychiatrically/cognitively impaired patients showed preservation of the frontal P3 but absence of the posterior process. Further, the late positive complex elicited by rare more than common events displayed two peaks, "early" at about 250 ms and "late" about 325–350 ms. These early and late components of the late positive complex could also be dissociated. One unexpected finding in this data was the presence of a P3a and P3b in all groups in the context of a passive paradigm. Perhaps due to the lack of instruction it is natural to impose a condition of attention in a paradigm with two different stimuli. It is noted that the waveshapes obtained by Polich (1989) in a passive task also show both early and late P3 components.

When the whole analysis epoch was taken into consideration, the second set of phenomena emerged. Normals show an enormous increase in early negativity, especially in frontal regions, to a *rare* compared to a *common* event. All three groups of patients showed an early negativity in frontal leads to the *common* event comparable to that seen in the normals to the *rare* event, most marked in the dementia patients. Differences between ERPs to common and rare events can clearly be discerned as early as 75–100 ms. Clear *decreases* in the mean amplitude of N1 in central regions appear in responses to the *rare* event in the normals as well as the patients, but are much greater in the three categories of patients. The schizophrenics show the smallest N1 of all four groups to both common and rare, but show a decreased N1 to the rare event. The demented and OCD groups show a central N1 as large as or larger than the normals to the common stimulus, but show a far greater decrement in N1 to the rare events.

In the central regions, the schizophrenics again showed the smallest P2 of the four groups to the common event. Both the normal and the demented groups showed a substantial decrease in P2 to the rare event. With respect to P2, no clear differences in mean amplitude appeared among the four groups in the frontal "common" ERPs.

If one considers N1P2, all four groups showed a substantial reduction of amplitude in the rare compared with the common ERPs of central and more posterior regions. In frontal regions, only the demented group showed such a decrement.

How are we to interpret these findings? There appears to be a brain process which produces a marked *decrement* in the early electrophysiological responses to the *rare* events in this passive P3 paradigm. These changes are so early as to preclude any cognitive process, and fall in the temporal domain of N1, which has been clearly implicated in processes related to vigilance and attention. Since both N1 and N1P2 decrease in response to an *unanticipated* event, the decrement cannot be interpreted to reflect heightened attention to or more active processing of the rare stimulus. The change is more reasonably attributed to the diminution of a process related to the *anticipated* event. This supports the proposal that in the passive P3 paradigm, attention is focused upon the common rather than the rare event. This may be due to the fact that the neural mechanism for detection requires occurrence of the event in order to define the template for recognition of subsequent occurrences of the same event. It may operate as a sort of filter with a moderate time constant, so that the template is established in a few trials and decays after a few trials if not reactivated.

This mechanism appears to be much more strongly reflected in posterior than anterior brain regions. Yet, anterior regions are somehow implicated as shown by the appearance of a frontal decrement of N1 amplitude to rare events in the demented population. In the three groups of psychiatric patients, the posterior process yields a much greater decrement to the rare event than is seen in the normals. This indicates that the filter in these disorders is set so rigidly that the unexpected event is effectively excluded from admission to the brain or at least much diminished in access. This suggestion is intuitively supported by the fact that the exclusion is greatest in the OCDs.

What is the relationship between the N1 suppression and the P3 elicitation by the rare event? What seems quite clear is that the P3 cannot reasonably be attributed to the greater arousal value or heightened attention to the rare event, which ought to result in an increment rather than a decrement of N1 or N1P2. We must conclude that the two processes are independent. The augmented late positive components produced by an uncertain event must arise from the evaluation of the information contained in the afferent input which succeeds in gaining access, rather than the intensity of the input. It seems reasonable to propose that this evaluation necessitates somehow matching the throughput of the template when N1 is decreased with the specification of the template itself, in some feedback circuit, so that a mismatch in this comparator leads to the process reflected as P3. The cognitive disorders examined herein appear to have many features in common with respect to these various processes.

The posterior diminution and delay of the later P3 with more normal early P3 in all three of these patient populations suggests that while P3 may be sensitive to the cognitive dysfunctions in the three different disorders, it may not be specific, as has also been suggested by Pfefferbaum *et al.* (1984b). However, their conclusions were based upon studies confined to ERPs obtained from only 3 channels, F_z, C_z and P_z. The more complete scalp topography surveyed in the present study offers some evidence that differential patterns of the P3 in more lateral locations, together with differences in anterior versus posterior changes in N1 and P2, may permit some useful discrimination among these disorders to be achieved by multivariate techniques. The differential changes in anterior and posterior early versus late peaks of the late positive complex seen in this study are both intriguing and complicated. There appear to be two temporally distinct processes which are independently represented in both anterior and posterior regions of the brain, and which can be differentially affected by psychiatric disorders. The functional differences between these processes and the clinical implications of their differential impairment are presently obscure. We are currently exploring this problem.

ACKNOWLEDGMENTS

The authors wish to acknowledge the help of Dr. Lev Sverdlov with the data analysis and Heather Stein with the preparation of the manuscript.

REFERENCES

Baribeau-Braun, J., T. Picton & J. Gosselin. 1983. Schizophrenia: a neurophysiological evaluation of abnormal information processing. Science **219:** 874–876.

Beech, H., K. Ciesielski & P. Gordon. 1983. Further observations of evoked potential in obsessional patients. Br. J. Psychiatry **142:** 605–609.

Begleiter, H., B. Porjesz, B. Bihari & B. Kissin. 1984. Event-related brain potentials in boys at risk for alcoholism. Science **225:** 1493–1496.

Blackburn, I., D. S. Clair & A. McInnes. 1987. Changes in auditory P3 event-related potential in schizophrenia and depression. Br. J. Psychiatry **150:** 154–160.

Buchsbaum, M., S., Awsare, H. Holcomb, L. DeLisi, E. Hazlett, W. Carpenter, D. Pickar & J. Morihisa. 1986. Topographic differences between normals and schizophrenics: the N120 evoked potential component. Neuropsychobiology **15:** 1–6.

Clair, D. S., I. Blackburn, D. Blackwood & G. Tyrer. 1988. Measuring the course of Alzheimer's disease. A longitudinal study of neuropsychological function and changes in P3 event-related potential. Br. J. Psychiatry **152:** 48–54.

Duncan, C. 1988. Event-related brain potentials: a window on information processing in schizophrenia. Schizophr. Bull. **14**(2): 199–203.

Elmasian, R., H. Neville, D. Woods, M. Schuckit & F. Bloom. 1982. Event-related brain potentials are different in individuals at high and low risk for developing alcoholism. Proc. Natl. Acad. Sci. USA **79:** 7900–7903.

Faux, S., M. Torello, R. McCarley, M. Shenton & F. Duffy. 1988. P300 in schizophrenia: confirmation and statistical validation of temporal region deficit in P300 topography. Biol. Psychiatry **23:** 776–790.

Galderisi, S., M. Maj, A. Mucci, P.Monteleone & D. Kemali. 1988. Lateralization patterns of verbal stimuli processing assessed by reaction time and event-related potentials in schizophrenic patients. Int. J. Psychophysiol. **6:** 167–176.

Goodin, D. & M. Aminoff. 1986. Electrophysiological differences between subtypes of dementia. Brain **109:** 1103–1113.

Goodin, D., K. Squires & A. Starr. 1978. Long latency event-related components of the auditory evoked potential in dementia. Brain **101:** 635–648.

Gordon, E., C. Kraiuhin, A. Harris, R. Meares & A. Howson. 1986. The differential diagnosis of dementia using P300 latency. Biol. Psychiatry **21:** 1123–1132.

John, E., P. Easton, L. Prichep & D. Friedman. 1991a. Standardized Varimax descriptors of event related potentials. II. Evaluation of psychiatric patients. J. Psychiatr. Res. Submitted for publication.

John, E., P. Easton, L. Prichep & J. Friedman. 1991b. Standardized Varimax descriptors of event related potentials. I. Basic considerations. Electroencephalogr. and Clin. Neurophysiol. In press.

John, E., L. Prichep, J. Friedman, & P. Easton. 1988. Neurometrics: computer assisted differential diagnosis of brain dysfunctions. Science **293:** 162–169.

John, E., L. Prichep, J. Friedman & P. Easton. 1989. Neurometric topographic mapping of EEG and evoked potential features: application to clinical diagnosis and cognitive evaluation. *In* Topographic Brain Mapping of EEG and Evoked Potentials. K. Maurer, Ed. 90–111. Springer-Verlag. Berlin & Heidelberg.

JOSIASSEN, R., R. ROEMER, C. SHAGASS & J. STRAUMANIS. 1986. Attention-related effects on somatosensory-evoked potentials in nonpsychotic dysphoric psychiatric patients. *In* Brain Electrical Potentials and Psychopathology. C. Shagass, R. Josiassen & R. Roemer. Eds. 259–277. Elsevier. Amsterdam.

KUTCHER, S., D. BLACKWOOD, D. S. CLAIR, D. GASKELL & W. MUIR. 1987. Auditory P_{300} in borderline personality disorder and schizophrenia. Arch. Gen. Psychiatry **44:** 645–650.

LEVIT, A., S. SUTTON & J. ZUBIN. 1973. Evoked potential correlates of information processing in psychiatric patients. Psychol. Med. **3:** 487–494.

MAURER, K., T. KIERKS, R. IHL & G. LAUX. 1989. Mapping of evoked potentials in normals and patients with psychiatric diseases. *In* Topographic Brain Mapping of EEG and Evoked Potentials. K. Maurer, Ed. 458–473. Springer-Verlag. Berlin & Heidelberg.

MORSTYN, R., F. DUFFY & R. MCCARLEY. 1983. Altered p300 topography in schizophrenia. Arch. Gen. Psychiatry **40:** 729–734.

PATTERSON, J., H. MICHALEWSKI & A. STARR. 1988. Latency variability of the components of auditory event-related potentials to infrequent stimuli in aging, Alzheimer-type dementia, and depression. Electroencephalogr. Clin. Neurophysiol. **71:** 450–460.

PFEFFERBAUM, A., J. FORD, B. WENEGRAT, W. ROTH & B. KOPELL. 1984a. Clinical application of the P3 component of event-related potentials. I. Normal aging. Electroencephalogr. Clin. Neurophysiol. **59:** 85–103.

PFEFFERBAUM, A., B. WENEGRAT, J. FORD, W. ROTH & B. KOPELL. 1984b. Clinical application of the P3 component of event-related potentials. II. Dementia, depression and schizophrenia. Electroencephalogr. Clin. Neurophysiol. **59:** 104–124.

PIERROT-DESEILLIGNY, C., E. TURELL, C. PENET, D. LEGRIGAND, B. PILLON, F. CHAIN & Y. AGID. 1989. Increased wave P300 latency in progressive supranuclear palsy. J. Neurol. Neurosurg. Psychiatry **52:** 656–658.

POLICH, J. 1989. P300 from a passive auditory paradigm. Electroenceph. Clin. Neurophysiol. **74:** 312–320.

POLICH, J. & F. BLOOM. 1986. P300 and alcohol consumption in normals and individuals at risk for alcoholism. A preliminary report. Prog. Neuro-Psychopharmacol. **10:** 201–210.

POLICH, J., C. EHLERS, S. OTIS, A. MANDELL & F. BLOOM. 1986. P300 latency reflects the degree of cognitive decline in dementing illness. Electroencephalogr. Clin. Neurophysiol. **63:** 138–144.

PRITCHARD, W. 1986. Cognitive event-related potential correlates of schizophrenia. Psychol. Bull. **100**(1): 43–66.

RADWAN, M., H. HERMESH & H. MUNITZ. 1991. Event-related potentials in drug-naive schizophrenic patients. Biol. Psychiatry **29:** 265–272.

REISBERG, B., S. FERRIS, M. DE LEON & T. CROOK. 1982. The global deterioration scale for assessment of primary degenerative dementia. Am. J. Psychiatry **139:** 165–173.

ROMANI, A., F. ZERBI, G. MARIOTTI, R. CALLIECO & V. COSI. 1986. Computed tomography and pattern reversal visual evoked potentials in chronic schizophrenic patients. Acta Psychiatr. Scand. **73:** 566–573.

SHAGASS, C. 1983. Evoked potentials in adult psychiatry. *In* EEG and Evoked Potentials in Psychiatry and Behavioral Neurology. J. Hughs & W. Wilson, Eds. 169–210. Butterworth. Boston.

SHAGASS, C., R. JOSIASSEN, R. ROEMER, J. STRAUMANIS & S. SLEPNER. 1983. Failure to replicate evoked potential observations suggesting corpus callosum dysfunction in schizophrenia. Br. J. Psychiatry **142:** 471–476.

SQUIRES, N., K. SQUIRES & S. HILLYARD. 1975. Tow varieties of long-latency positive waves evoked by unpredictable auditory stimuli in man. Electroencephalogr. Clin. Neurophysiol. **38:** 387–401.

STEINHAUER, S. & J. ZUBIN. 1982. Vulnerability to schizophrenia: information processing in the pupil and event-related potential. *In* Biological Markers in Psychiatry and Neurology. I. Hanin & E. Usdin, Eds. 371–385. Pergamon Press.

SURWILLO, W. & V. IYER. 1989. Passively produced P3 components of the average event-related potential in aging and in Alzheimer's type dementia. Neuropsychiatry, Neuropsychol. Behav. Neurol. **1**(3): 177–189.

SUTTON, S., M. BRAREN, E. JOHN & J. ZUBIN. 1965. Evoked potential correlates of stimulus uncertainty. Science **150:** 1187–1188.

TOWEY, J., G. BRUDER, E. HOLLANDER, D. FRIEDMAN, H. ERHAN, M. LIEBOWITZ & S. SUTTON. 1990. Endogenous event-related potentials in obsessive-compulsive disorder. Biol. Psychiatry **28:** 92–98.

VISSER, S., V. V. TILBURG, C. HOOIFER, C. JONKER & W. D. RIJKE. 1985. Visual evoked potentials (VEPs) in senile dementia (Alzheimer type) and in non-organic behavioural disorders in the elderly; comparison with EEG parameters. Electroencephalogr. Clin. Neurophysiol. **60:** 115–121.

Connections: a Search for Bridges between Behavior and the Nervous System

KURT SALZINGER

Department of Psychology
Hofstra University
Hempstead, New York 11550

The historical pendulum never ceases to swing back and forth. Our present era, in which investigation of the nervous system is clearly king of the psychopathology patch, still tends to emphasize one side over the other. Just as in the recent past psychologists were given to the excess of interpreting behavior and spending an equal amount of time speculating about the nervous system—to a degree that resulted in some scientists talking about behavior and the conceptual nervous system, so we now seem to be indulging in the exquisitely detailed investigation of the nervous system and speculating about how it controls behavior (a conceptual behavioral system?). Current advances in both arenas of research, however, provide the opportunity to make connections between data on both sides.

The object of this paper is to review some aspects of this situation and to point to a theory that promotes bridges between body and behavior towards the end of elucidating various forms of psychopathology. As a side note, I would like to say that I feel quite comfortable writing this paper in a volume honoring Samuel Sutton. I had the privilege of discussing research in psychopathology with him on many occasions. Despite the fact that he spent so much of his time working on the physiological aspects of psychophysiology, or perhaps because of it, he always considered the psychological part paramount. Not long before his untimely death, he told me that he disliked being called a neuroscientist and having his area of research referred to in this manner. He is a psychologist, he maintained, and his interest in measuring physiological aspects of an individual in particular psychological situations is to shed light on those psychological aspects and not merely for measuring physiological functioning.

Interaction between psychology and biology (I shall be using these terms as if they were exclusive of one another merely for convenience, not because they really are unrelated) takes a variety of forms:

1) separate but equal, buttressed by mutual benign neglect—each area going along its own way, trying to discern distinct characteristics of various psychiatric diseases in its own way;
2) complete takeover—assuming that all behavior is caused by the biology of the body—as Linus Pauling was reported to have said, every crooked thought has a crooked molecule;

3) deliberate neglect of the effect of biological phenomena on behavior by concentrating on the effect of behavior on biology whether direct or indirect; and
4) attempted full integration of the study of behavior and biology in constant interaction with one another so that each one is always affecting the other.

It is perhaps fair to say that the mutual-benign-neglect condition, like ships that pass in the night, is not rare in research, except for the fact that failing to collect data in the other area does not prevent the investigators from saying a good deal about it. Those working on the biology of schizophrenia, for example, will not refrain from spelling out the behavioral implications of their findings. If the ventricles of schizophrenic patients are larger than those of normal identical twins (Suddath *et al.*, 1990), then the conclusion drawn is that the smaller amount of gray matter is what accounts for the thought disorder and other symptoms of the schizophrenic twin. Although not stated as such, this condition shades over rather quickly into the second form of interaction, with the further implication that, therefore, the behavior need not be studied very carefully except for noting its presence or absence.

Diagnosis in psychopathology is still a problem (Salzinger, 1978; 1986); it is not simply a problem of reliability (or lack of it), it is one of precise specification of behavior as investigators in this field are beginning to learn in specifying the biological variables. It is ironic that as psychiatrists become increasingly more sophisticated in the use of traditionally "scientific" hardware in defining biological variables, they continue to use rating scales which basically employ numbers, not so much to quantify as to cover the underlying phenomenon whose meaning is being investigated with a comforter of respectability.

Stampfer and German (1989) in a recent article on the "neuroscience imperative in psychiatry" state essentially that until now the behavioral and social sciences have made no notable contributions to the treatment of mental illness. In their attempt to equate psychiatry with neuroscience, they eventually equate other approaches with philosophy (a field they believe one must eschew if one is to do real science). Elsewhere in their article, they equate the "other" approaches with the priesthood, counseling or psychotherapy, politics, and "a state of reformist zeal, which, while fluctuating somewhat between frenzy and ecstasy, is characterized more by belief than thought . . ." (p. 153). The same authors misunderstand the behavior-analytic approach to this area. Behaviorists deny neither the existence nor the importance of the brain. What they reject is the usefulness of speculating about behavioral data in terms of physiological and biochemical concepts. Behavior must be explained on its own level.

Although not discussing the behavior-analytic approach, Charlton (1990) also argues against the imperialism of the biological or any other approach above all else in this area. He says (p. 5): "to explain psychiatric illness in terms of biochemistry (for example) is not to describe its underlying cause, but to redescribe it." Speaking about the relationships between different sciences, he maintains that "they are not saying the same things in different ways, but are actually about different things" (p. 5). I would add to this, that to say that the brain adjusts the response rate of a rat working on a fixed interval schedule of reinforcement is to supply no explanation at all. The typical fixed interval behavior found in many different species must be explained by other behavioral variables, such as the other behavior the animal emits during the course of the interval, the number of training sessions, or the degree of deprivation of the reinforcer. The so-called "black box" approach says that only after we have learned to explain the behavioral variables (many of which are found in the environment but some of which are response-produced), can we do the kind of integration suggested by point number 4 above. Behavior unexamined, with its controlling variables unspecified or unknown cannot be fruitfully related to any findings about the brain, no matter how carefully the latter is measured.

Perhaps the most surprising part of Stampfer and German's (1989) article is the authors' use of the statement "we know practically *nothing* about how it (the brain) works" (p. 154, italics the authors') as justification for the belief that a revolution in psychiatry will be forthcoming in this area. Here is a group of investigators who believe, not in benign neglect, but in a takeover approach: knowledge of all psychiatry through an understanding of the brain only. They say (p. 154): "Understanding of the brain, it should be noted, will also shed fresh light on important questions as to *how* it interacts with and is influenced by its social milieu, and this knowledge is likely to revolutionize the social sciences." Finally, let us see what Stampfer and German say about "problems in living:" they are, according to them, the province of priests, counselors and psychotherapists. No place here for science or scientific discoveries, whether produced by studies of the brain or of behavior. But how do we discover that a schizophrenic patient has a problem? By doing a PET scan? When they abdicate problems in living, are they not fleeing their own field?

Zubin (1989) also commented on this article showing that the psychosocial sciences had indeed made some worthwhile discoveries or at least some which the biologically oriented cannot ignore. He points out (p. 159) that "it is the interaction between the wired-in portion of the brain and the environmental forces, including especially the psychosocial, that determines our behavior." Elsewhere, he talks about a neuroplastic portion of the brain that develops as a function of the interaction of the person with his/her environment. Thus, what has to be noted in seeking connections in the area of psychopathology is not only the interaction of the sciences—social, behavioral, and neural—but also the interaction of the organisms that we are all trying to study and understand, the interaction between the insides, the brain, and the outside, their environment, which impinges on it and interacts with it. As the person behaves, his/her brain is surely modified even if not always permanently, and whatever the modifications, they always are in some way related to subsequent behavior. Under those circumstances, how can one ever talk of understanding the person—a goal that is critical in helping someone whose behavior is awry—without taking into account both his/her behavior and neural functioning?

Having made clear that we must make every effort to integrate biology and behavior, it now remains for us to explain how that is to be accomplished. I shall argue that we need not only to measure (quantify) behavior, but also to analyze it by discovering the underlying behavioral mechanism as outlined in FIGURE 1 (Salzinger, 1980). According to this figure, behavior occurs in the presence of certain stimuli and has certain consequences. The variable sets that precede the behavior include the physical state of the organism, a box that the biologically oriented investigators examine with great enthusiasm. Note that the discontinuous lines show feedback circuits, in accordance with which the consequences of behavior are such as to alter the state of the organism. Whether that state of the organism is to be altered by behavior in turn depends on the environment consisting of stimuli both social (for example, the presence of a friend, or a member of the opposite sex, a group of individuals or a threatening letter), and/or physical that promote certain behaviors as a function of their physical effect (for example, the gravitational pull of a slanted walk, the glare of a light in one's eyes, or the presence of delicious food within one's view). The power of these stimuli will depend on their unconditioned eliciting strength as well as on their conditioned significance. They acquired significance by becoming discriminative stimuli, that is, occasions for the occurrence of reinforcement when some behavior is emitted.

Here the importance of the reinforcement history box enters, in that stimuli have different degrees of control over behavior as a function of the person's reinforcement history. If a threatening letter had in the past been followed up by the execution of that threat, then one would expect a different response from the person receiving the letter than if the person had received such letters in jest only.

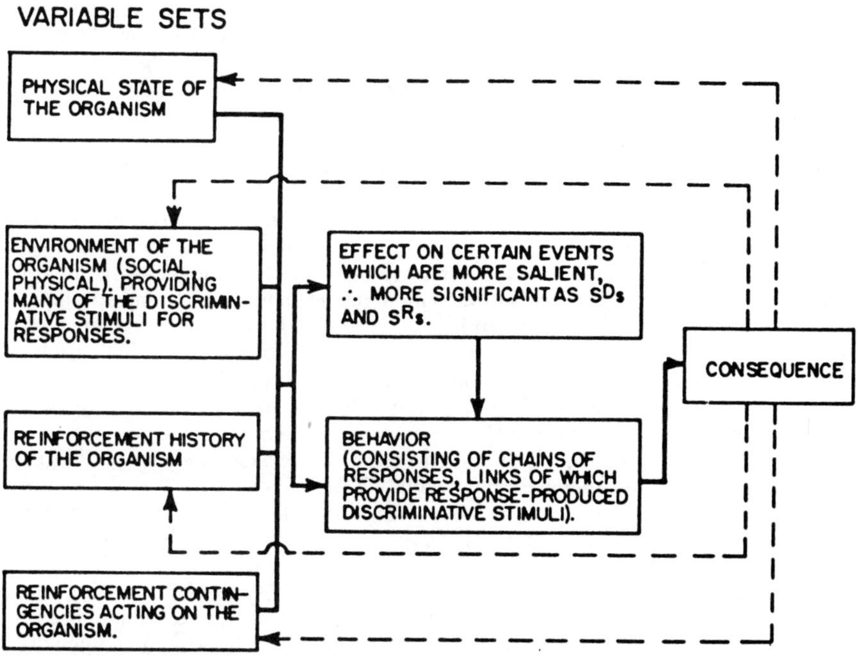

VARIABLE SETS

PHYSICAL STATE OF THE ORGANISM

ENVIRONMENT OF THE ORGANISM (SOCIAL, PHYSICAL), PROVIDING MANY OF THE DISCRIMINATIVE STIMULI FOR RESPONSES.

REINFORCEMENT HISTORY OF THE ORGANISM

REINFORCEMENT CONTINGENCIES ACTING ON THE ORGANISM.

EFFECT ON CERTAIN EVENTS WHICH ARE MORE SALIENT, ∴ MORE SIGNIFICANT AS S^Ds AND S^Rs.

BEHAVIOR (CONSISTING OF CHAINS OF RESPONSES, LINKS OF WHICH PROVIDE RESPONSE-PRODUCED DISCRIMINATIVE STIMULI).

CONSEQUENCE

FIGURE 1. The behavioral mechanism explains the various ways in which behavior is controlled by variables that precede or follow it.

Finally there is the variable set of the current reinforcement contingency. Consequences of one's behavior establish what the current reinforcement contingency is. Does a certain behavior result in positive reinforcement or does it remove a negative reinforcer? All of these boxes on the left hand side of the diagram control the behavior in some way and they also interact with one another, something not indicated in this diagram. For example, the current reinforcement contingency interacts with the person's reinforcement history. Was he/she reinforced in this manner at any time before, or is the current extinction (no reinforcement) of behavior a new and sudden experience? Extinction, we know, produces emotional behavior and thus it can model certain kinds of stressful life events (Salzinger & Chitayat, 1983).

These variable sets are also such as to make certain groups of stimuli more outstanding and, therefore, more likely to function as controlling stimuli than others. The environment may make it difficult for a driver to see a traffic light and, in this way, eliminate some very important control over an individual's driving behavior. This has the consequence that he/she might be injured in an accident, impinging on that person's physical state of being able to drive the next time or of having his/her eyesight injured. The result is that a large set of stimuli no longer controls that person's behavior. The discontinuous lines show how consequences of behavior impinge on the variable sets and thus how the biology of the organism can never be considered as a pristine initiator of a person's actions. Like behavior itself, all behavioral determinants are in dynamic change. The changes that occur in individuals consist of wear and tear, fatigue, development of muscles, practice in mental tasks through learning, and so on. One can also add in this diagram the input of foods, both nutritious and injurious, the intake of drugs, curative, entertaining, or injurious to one's nervous system, and so on.

This diagram shows not only how behavior is controlled but also that one can never consider an organism in any other way but in interaction with the environment. A number of cautions follow. One cannot measure behavior without being able to specify the other conditions of the environment both in terms of the stimuli that precede and the stimuli that follow the behavior of interest. In addition, one cannot measure any physiological functioning without knowing the conditions under which the data were collected. An obvious example is the measurement of blood pressure which will respond to such momentary factors as running up a flight of stairs, being frightened by something, or even by thinking of some incident. This is true even when one looks at permanent changes such as the size of venricles in the brain. Here, one must ask not only whether the ventricles of schizophrenic patients are reliably larger than those of normals, but also whether this reliable difference is due to the disease or the consequences of having the disease, which is, in part, to receive medication that might well produce exactly such a condition.

Let us examine the recent dramatic finding of the ventricle size difference of schizophrenic patients and their normal monozygotic twins. Using magnetic resonance imaging, Suddath and colleagues (1990) found that in 12 of the 15 discordant pairs, the schizophrenic twin had larger ventricles; they found no difference in two of the pairs, and in one, the difference was reversed. They also found significant differences in the size of the hippocampus of the schizophrenic and nonschizophrenic twins. These findings revealed an important nongenetic factor in schizophrenia by using a genetic variable, namely, that of identical genetic make-up found in monozygotic twins. For some time, it has been claimed that schizophrenia is a genetic disease, but the percentage of concordance has varied mightily for identical twins in different studies (Salzinger, 1973). In this study, they simply took advantage of the discordance with respect to schizophrenia to look for other differences between schizophrenic patients and normals.

Such an approach would be equally useful for the study of social and behavioral variables because for monozygotic twins living together, the identity in genetic make-up reduces the variation not only of genetic causation but also the secondary effect of differences in social and behavioral variation. Many of the latter effects are due to such differences as those in physique, physical appearance, and the environmental differences contingent on those differences. Physical characteristics often determine the kinds of environments that people will be exposed to. By way of contrast, similarities of genetic identity and the environmental effects due, for example, to living in the same family, at the same time, and with the same parents, will reduce those secondary differences. But what about the crucial question of getting from biological differences, interesting as they are, to behavior? I believe we must use the behavioral mechanism (FIG. 1) which tells us what underlies the behavioral differences that we find between schizophrenic patients and normal individuals. After we have established the behavioral mechanism in simple enough form, we can search for the biological differences that correspond to it. The indirect information about less brain tissue in some areas is not terribly helpful in suggesting behavioral differences. I shall return to the use of the behavioral mechanism for such purposes later in discussing the Immediacy Theory (Salzinger, 1984).

Two recent papers attracted a great deal of attention (Sherrington *et al.*, 1988; Kennedy *et al.*, 1988). The first showed, in the words of the authors (p. 164) "the first strong evidence for the involvement of a single gene in the causation of schizophrenia." The second article, following immediately upon the first, on a separate sample and using very much the same techniques, again in the words of those authors (p. 167): "found strong evidence against linkage between schizophrenia and the seven loci (on chromosome 5)." Nevertheless, they conclude, not that the first finding is invalid—a conclusion that one might well come to in other areas of science since this appears to qualify as a clear example of a nonreplication—but that (p. 167) "the genetic factors underlying schizophrenia are heterogeneous." What is more, commenting on these two articles in the journal in which they were published, Lander (1988) says in the title of his paper (p. 105): "Two papers in this issue, despite reporting contradictory findings, pave the way for a genetic approach to the diagnosis of schizophrenia."

Both studies are being used to justify the genetic approach to schizophrenia and one investigator (Lander, 1988) even uses the articles as reason to employ genetic measurement as a way of diagnosing schizophrenia, or at least some forms of it. It is of interest to note that not everybody shares this optimism about using "molecular reductionism . . . (which) substitutes biochemistry for behaviour and genetic analysis for psychiatric judgment" (Rose, 1988, p. 512). The latter maintains that biochemical explanations follow clinical interventions, as in the case of the use of the phenothiazines, rather than the other way around. From the point of view of connections, these studies and most of the comments on them are paying attention to behavior only in terms of diagnosis, which is a hotchpotch of interpretations of behavior, and of durations and severity of various symptoms that do not overlap much from patient to patient. What is more, now that psychiatry has adopted so-called operational criteria, these are nevertheless ignored when not enough people are found who fall into a strict schizophrenic diagnosis, with an appeal to the concept of "spectrum disorder." This is not the place to go into a detailed critique of diagnosis (I have discussed those issues elsewhere, Salzinger, 1978; 1986). The point is that no matter how great the value of the isolation of a gene on one chromosome in one group of subjects, none of the authors cited paid sufficient attention to behavior, the area to which they wished to connect their findings.

Let us look next at another way that investigators in this field have tried to connect body and behavior. Holzman and colleagues (1988) have approached the problem of the genetics of schizophrenia by looking at a marker for the disease. Dysfunctions of smooth pursuit eye movements that occur when the eye is tracking a moving target, such as a pendulum, have been nominated for this role. They occur in 51 to 85% of schizophrenic patients, in 45% of their first degree relatives, and in 8% of the normal population. A good deal of data has been collected with this technique and, although not without controversy, it has been increasingly recognized as a marker for schizophrenia. Using a family pedigree approach, Holzman *et al.* (1988) concluded that when one uses eye movement dysfunction together with a diagnosis of schizophrenia as two independent manifestations of a latent trait, then that trait appears to be "genetically transmitted as an autosomal dominant gene with high penetrance" (p. 646). It is to be noted that a marker may not by itself shed any light on the biological relationship to the aberrant behaviors. Its role is merely one of marking those individuals who have a higher probability of having schizophrenia, and of eventually being in a position of locating the critical gene involved in greater susceptibility to schizophrenia.

Finally, let us examine still another approach to studying the relationship between biology and behavior. Weinberger and colleagues (1986) studied dysfunction in the dorsolateral prefrontal cortex in schizophrenia. Although this area of the brain has been implicated by other investigators as the site responsible for the most obvious symptoms of schizophrenia, the current availability of techniques to study regional cerebral blood flow while the subject is engaged in various tasks has made it possible to study schizophrenic patients and control groups under controlled conditions. If this area of the brain is the thinking part of the brain, then the kinds of thinking activity that are taking place ought to significantly affect its activity. As I pointed out elsewhere (Salzinger & Salzinger, 1973), the study of any behavior, including physiological functioning, must take place under conditions that usually control behavior, that is, the stimuli that precede and the stimuli that follow the behavior or physiological function in question. Recognizing this important fact, at least in part, these investigators obtained measures of the prefrontal cortex while the subjects were working on the Wisconsin Sorting Test.

The latter is a test that involves the subject in cognitive behavior. The subject looks at a computer screen which shows characteristic designs and matches those designs to one of a number of switches to indicate whether the match is being made on the basis of color, number, or shape. A green or red light follows each response indicating whether the match is correct or incorrect. After the subject has learned to make a series of ten "correct" responses, the rule is changed so that now the computer signals "correct" in accordance with matching on a different characteristic. This change in contingency is made without warning to the subject.

Although chronic schizophrenic patients and a control group did not differ from one another on the cerebral blood flow measure in the dorsolateral prefrontal cortex during a number-matching task (which was presumably less cognitive than the Wisconsin Sorting Test), the patients showed significantly less cerebral blood flow activity in the pre-frontal cortex when engaged in the sorting task than did the normal control group. The authors concluded that there was a relationship between the abstract reasoning behavior of the sorting test and the physiological functioning in the specific part of the brain. Clearly, this study made an attempt to relate biology to behavior. They also made an attempt to eliminate a number of potential epiphenomena that might have accounted for their results.

Nevertheless, some points need to be made about this experiment. As shown in FIGURE 1, the consequences of behavior are important in controlling behavior. The consequences of behavior used in the sorting test consisted of the onset of two different color lights, signifying whether the response was right or wrong. It seems to me that under those conditions, it cannot be claimed that the two populations were working under similar conditions of motivation. It is quite possible that "being right" might be a more powerful reinforcer in controls than in schizophrenic patients; furthermore, associating a color to "being right" might be stronger in normals than in patients. Thus, one cannot conclude that motivation in terms of the reinforcer contingent on the correct behavior has been excluded as a critical variable determining the results of the study. A point also needs to be made with respect to the attention variable. Weinberger and colleagues discount that variable because they found that schizophrenics did not differ from normal subjects in a continuous performance task. As I shall show below, attention is not merely a matter of attending too much or too little to stimuli, rather it is a matter of which stimuli one attends to. In any case, this study is headed in the right direction in trying to connect body and behavior. What needs to be done in addition is to narrow down the behavior to a behavioral mechanism.

I wish to mention a recent review by Gray and colleagues (1991) of the neuropsychology of schizophrenia which related biology and behavior in a much more precise way, at least on the biology side. Unfortunately, it failed to specify the behavioral side in equal detail. It is recommended that the reader consult the paper for an excellent example of biological specification of what should be examined to correspond to behavior in a schizophrenic patient. My comment (Salzinger, 1991), specifically on that paper, should be read along side it.

Allow me now to discuss what I submit to be an appropriate approach to specifying the behavioral side of the connection. It must first be noted that behavior is multiply determined. FIGURE 1 makes that abundantly clear. What remains to be portrayed is how the basic difference in the patient interacts with all the variables specified in the behavioral mechanism. Let us suppose that we have before us a man who suffers from impaired hearing. Clearly, such a man will be exposed to only a subset of stimuli to which others are subjected. FIGURE 1 shows that stimuli are present in various ways, for example in setting an occasion, as in instructions given or a signal indicating that some form of behavior is now appropriate or inappropriate. It also makes itself felt in having been involved in that way in the past.

Stimuli are also present as consequences so that when this man asks for something and only hears the answer some of the time, he is likely to ask again or to misinterpret the answer. The general consequences of this might be that the man might not do something he should have done or he might ask again and again, annoying the persons around him, with the consequence that he might ultimately be ignored by others, that he might get angry, thus alienating people; alternatively, he might think he hears one thing when, in fact, something else is being said, thus leading to the wrong behavior.

A small problem with one set of stimuli might affect a range of behaviors and might do so in varying ways depending on the environment; for example, if the environment is very quiet then the hearing loss might have minimal effect; when the environment becomes very noisy, however, then the hearing loss would be very much aggravated. If the people are patient, they might well be willing to repeat and to make certain that he did understand what they said as opposed to an environment of impatient people who refused to repeat anything or got angry easily with people who do not hear. This points out how variable the effect of a so-called defect might be as a function not only of how it interacts with the man in question but the way in which that environment responds to him. Even complete deafness is sometimes missed in young children for quite a time with incorrect diagnoses of retardation given to them when nobody thinks to test their hearing. In social situations, people do not go around testing for hearing deficits and thus the dynamic of the behavioral interaction quickly takes over. The general consequences of this is considerable variability of behavioral difficulties that such people display; in some it is the direct consequence of the deficit, but in most it is the result of the interaction of their deficit with its consequences.

Let us turn then to what might be the basic deficit to be found in schizophrenia. Allow me to suggest the immediacy effect (Salzinger, 1984). According to that theory, the behavior of schizophrenic patients is determined by their responding predominantly to stimuli immediate in their environment. When an individual responds primarily to stimuli immediate in his or her environment, only a subset of stimuli controls that person's behavior. In that sense, the person is initially different from others around him or her. It implies that such persons will be conditioned only by a subset of the stimuli by which others around them are conditioned; it means that schizophrenic patients respond to only some discriminative stimuli and their behavior is reinforced by only some stimuli. It is important to note that many of our behaviors are properly controlled by immediate stimuli; it is also true that when we are still young, being controlled by immediate stimuli is expected by people around us; there are also differences in the degree to which immediate stimuli play a role in different occupations and job levels: in some, planning is critical, in others simply responding to what is currently happening is crucial. This is important to note because it tells us that the underlying condition will not be equally noticeable under all conditions and will, therefore, not always get the affected individual into the kind of trouble that will bring him or her to the notice of mental health authorities.

When constructing a theory of schizophrenia, indeed any theory that tries to explain behavior, it is critical that we leave room for determinants other than the biological ones in explaining why the affected individuals behave in the way they are observed to behave. If the biological factors explain entirely why these people behave the way they do, then the theory must be wrong since the interaction of their underlying difference in biology with their environment must play a role as well.

Let us examine how a tendency to respond to immediate stimuli might produce some of the symptoms of schizophrenia. Responding to immediate stimuli implies responding to stimuli out of context—an almost operational definition of a delusion. Even thought broadcasting, a somewhat bizarre symptom reported about schizophrenic patients, can be understood if one assumed that responding to immediate stimuli only might prevent patients from discriminating what they think from what they say. Hallucinations have long ago been shown to be accompanied, if not produced, by subvocal speech or whispering. Again we have the example of immediate stimuli preventing the patients from discriminating their speech from that of others.

Thought disorder, which is often referred to as a loosening of associations or illogical thinking, is better conceptualized as another case of control by immediate stimuli, in this case one of control by response-produced stimuli. When a person speaks in an illogical manner, the problem appears to be one of control of close response-produced stimuli exerting much greater control over the rest of the verbal behavior, rather than one of less control by verbal stimuli over others. In schizophrenic speech and thinking, the words following upon one another can be conceptualized as a short series of closely related words, as in pairs or triplets. As one examines the speech over longer stretches, however, one finds that schizophrenic speech begins to wander; since the connections among words stretch only over a brief series, the speaker drifts off into other meanings of the words than those that rightfully fit into longer stretches of meaningful sentences, paragraphs or stories.

Experimental evidence for thought disorder has been collected over the years in our laboratory and much of it is summarized in my 1984 paper (Salzinger, 1984). The Immediacy Hypothesis, together with the principles of learning and conditioning, presents no difficulty in generating schizophrenic symptoms. This simple principle also explains behavior in different domains of functioning. Schizophrenic patients condition at the same rate as do normal (physically ill) patients, but their behavior extinguishes more rapidly than that of normals. When the critical stimulus, the reinforcer, is no longer there, as is true in extinction, they act as if conditioning had not taken place. They speak in such a way as to produce speech more difficult to predict than the speech of normal individuals. In perceptual constancy experiments, they display greater stimulus constancy than do normals, thus being controlled by retinal stimulation rather than by object constancy, which is the main controlling variable of normal individuals. In semantic generalization experiments, they show greater generalization to sound than to meaning. In sorting experiments, they are more distractible by unrelated stimuli than are normals, again showing that stimuli are responded to, not in terms of their appropriateness, but rather in terms of which stimuli the subject happens to come upon at the time. With respect to memory, schizophrenic patients are as effective as normal in a recognition task where the critical stimulus is immediate, but they do considerably less well when they have to recall—the critical stimulus is now remote. All of these effects and others similar in content can be found in my review paper. Indeed, taking all studies on schizophrenic behavior in one year of the *Journal of Abnormal Psychology*, I found in general that all were interpretable in terms of the Immediacy Theory (Salzinger, 1984).

Another point about the theory is its fit in making connections to biology. There are many ways in which one can characterize schizophrenic behavior. Most of these are picturesque but imprecise, and that includes diagnostic descriptions which unfortunately mix dimensions. Immediacy Theory uses a simple concept in describing the basic fault, the time dimension. This dimension is, of course, one that can easily be applied to the nervous system. The dopamine theory of schizophrenia in fact fits very well with it. A greater concentration of dopamine for whatever reason would allow impulses to pass from neuron to neuron with greater ease making it more likely for the person so beleaguered to respond to the first stimulus producing that impulse rather than waiting for another stimulus or relating it to another stimulus. It also fits well with the deficit in the Wisconsin Sorting Test where the subject has to learn to respond to different stimuli at different times. Since all the stimuli are present simultaneously, that always allows the patient to respond to an inappropriate stimulus, given that his/her eye can easily fall on the wrong one. The fact that schizophrenic patients do only worse at this task, however, suggests that given enough time they can learn to respond to the appropriate stimulus, and suggests that if we devise a therapeutic technique that allows the patient to be reinforced for responding to remote stimuli, that patient might improve with respect to other symptoms as well. There is not space to develop this notion here; it is discussed elsewhere (Salzinger, 1984). The point is that this theory, which allows an easy connection to biology, also allows a connection to therapy.

In conclusion, to make connections between biological findings and behavioral ones, first we must devise theories that lend themselves easily to correspond to biological facts. Second, they must show how the interaction of the basic deficit with the environment, both social and physical, could, by way of the laws of learning and conditioning, produce the symptomatology of the disease in question.

REFERENCES

CHARLTON, B. G. 1990. A critique of biological psychiatry (editorial). Psychol. Med. **20:** 3–6.

GRAY, J. A., J. FELDON, J. N. P. RAWLINS, D. R. HEMSLEY & A. D. SMITH. 1991. The neuropsychology of schizophrenia. Behav. Brain Sci. **14:** 1–20.

HOLZMAN, P. S., E. KRINGLEN, S. MATHYSSE, S. D. FLANAGAN, R. B. LIPTON, G. CRAMER, S. LEVIN, K. LANGE & D. L. LEVY. 1988. A single dominant gene can account for eye tracking dysfunctions and schizophrenia in offspring of discordant twins. Arch. Gen. Psychiatry **45:** 641–647.

KENNEDY, J. L., L. A. GIUFFRA, H. W. MOISES, L. L. CAVALLI-SFORZA, A. PAKSTIS, J. R. KIDD, C. M. CASTIGLIONE, B. SJOGREN, L. WETTERBERG & K. K. KIDD. 1988. Evidence against linkage of schizophrenia to markers on chromosome 5 in a northern Swedish pedigree. Nature **336:** 167–170.

LANDER, E. S. 1988. Splitting schizophrenia. Nature **336:** 105–106.

ROSE, S. 1988. Diagnosis of schizophrenia (letter). Nature 336, 512.

SALZINGER, K. 1973. Schizophrenia: Behavioral Aspects. Wiley. New York.

SALZINGER, K. 1978. A behavioral analysis of diagnosis. *In* Critical Issues in Psychiatric Diagnosis. R. L. Spitzer & D. F. Klein, Eds. 73–84. Raven Press. New York.

SALZINGER, K. 1980. The behavioral mechanism to explain abnormal behavior. Ann. N. Y. Acad. Sci. **340:** 66–87.

SALZINGER, K. 1984. The immediacy hypothesis in a theory of schizophrenia. *In* Theories of Schizophrenia and Psychosis. W. D. Spaulding & J. K. Cole, Eds. 231–282. University of Nebraska Press. Lincoln.

SALZINGER, K. 1986. Diagnosis: distinguishing among behaviors. *In* Contemporary Directions in Psychopathology. T. Millon & G. L. Klerman, Eds. 115–134. Guilford Press. New York.

SALZINGER, K. 1991. What should a theory of schizophrenia be able to do? Behav. Brain Sci. **14:** 44–45.

SALZINGER, K. & D. G. CHITAYAT. 1983. Stressful life events and reduction in reinforcement frequency. Invited address at the Association for Behavior Analysis. Milwaukee, Wisconsin.

SALZINGER, K. & S. SALZINGER. 1973. Behavior theory for the study of psychopathology. *In* Psychopathology. M. Hammer, K. Salzinger, S. Sutton, Eds. 111–125. Wiley. New York.

SHERRINGTON, R., J. BRYNJOLFSSON, H. PETURSSON, M. POTTER, K. DUDLESTON, B. BARRACLOUGH, J. WASMUTH, M. DOBBS & H. GURLING. 1988. Localization of a susceptibility locus for schizophrenia on chromosome 5. Nature **336:** 164–167.

STAMPFER, H. G. & G. A. GERMAN. 1989. The neuroscience imperative in psychiatry. Integr. Psychiatry **6:** 152–164.

SUDDATH, R. L., G. W. CHRISTISON, E. F. TORREY, M. F. CASANOVA & D. R. WEINBERGER. 1990. Anatomical abnormalities in the brains of monozygotic twins discordant for schizophrenia. N. Engl. J. Med. **322:** 789–794.

WEINBERGER, D. R., K. F. BERMAN & R. F. ZEC. 1986. Physiologic dysfunction of dorsolateral prefrontal cortex in schizophrenia. Arch. Gen. Psychiatry **43:** 114–124.

ZUBIN, J. 1989. Commentary on the neuroscience imperative in psychiatry by Stampfer & German. Integr. Psychiatry **6:** 158–161.

Index of Contributors